知天之所为
知人之所为者
至矣

（《庄子·大宗师》）

How To Be Human

A Guided Reading of

Humanities and Social Science Classics

How To Know The Cosmos

A Guided Reading of

Natural Science Classics

武大通识

WHU General Education Center

博雅弘毅

文明以止

成人成才

四通六识

珞 珈 博 雅 文 库

（武大通识教材系列）

A Guided Reading of
Natural Science Classics

自然科学经典导引

（第二版）

主　编　桑建平

副主编　彭　华

WUHAN UNIVERSITY PRESS
武汉大学出版社

图书在版编目(CIP)数据

自然科学经典导引/桑建平主编．—2版.—武汉：武汉大学出版社，2019. 7(2019.9重印)
(珞珈博雅文库.武大通识教材系列)
ISBN 978-7-307-21005-9

Ⅰ.自…　Ⅱ.桑…　Ⅲ. 自然科学—文集　Ⅳ.N53

中国版本图书馆CIP数据核字(2019)第127241号

责任编辑：鲍　玲　李　玚　　　责任校对：李孟潇　　　版式设计：马　佳

出版发行：**武汉大学出版社**　(430072　武昌　珞珈山)
(电子邮箱：cbs22@whu.edu.cn 网址：www.wdp.com.cn)
印刷：湖北恒泰印务有限公司
开本：720×1000　1/16　印张：33　字数：505千字　插页：6
版次：2018年7月第1版　　2019年7月第2版
2019年9月第2版第2次印刷
ISBN 978-7-307-21005-9　　定价：68.00元

主编简介

桑建平，1959年12月生于武汉，山西平顺人，理学博士，武汉大学物理科学与技术学院教授、博士生导师。历任武汉大学物理系副主任、理学院副院长、科技处处长，武汉市科技局副局长，江汉大学副校长，武汉市科协党组书记，教育部高校物理学类专业教学指导委员会委员。

主要从事理论物理科研与教学，曾主持国家自然科学基金、教育部重点基金、教育部骨干教师基金、湖北省科研基金项目，在国内外期刊发表论文八十余篇，主编两本教材，获湖北省教学成果一等奖。

壁画《雅典学院》（*The School of Athens*）：由拉斐尔·圣齐奥（Raffaello Sanzio）于1511年创作完成。
图片来源：https://commons.wikimedia.org/wiki/File:Raphael_School_of_Athens.jpg.

总　序

2015年的“本科教育改革大讨论”确立了武汉大学本科教育的基本理念：人才培养为本，本科教育是根；以成“人”教育统领成“才”教育。在这样一个背景之下，“武大通识3.0”于2016年正式开启。作为一个新的课程体系，“3.0”版本的核心元素是什么？这个核心元素就是《人文社科经典导引》与《自然科学经典导引》(以下简称“两大《导引》”)。因而，两大《导引》的正式出版标志着武汉大学的通识教育进入一个新的历史阶段。

一、两大《导引》的意义

为全校本科生开设两大《导引》课程，这在武汉大学的历史上是开天辟地第一回。武汉大学建校120多年，有过共同的政治课、共同的外语课和共同的体育课，但从未有过共同的通识教育课，而且是以跨学科经典阅读为模式的基础通识课，这将在武汉大学校史上写下浓墨重彩的一笔，必将对武汉大学今后的人才培养产生深远影响，因此毫无疑义具有里程碑意义。

“武大通识3.0”提出了十六字方针，即“博雅弘毅，文明以止，成人成才，四通六识”，并在此基础上构建了“4-2-660”的课程体系，而其中的“2”就是两门基础通识课程(即两大《导引》)。两大《导引》以“人、自然、社会”为关键词，以“如何成人”为问题阈，以“四通六识”为具体目标，集中体现出“3.0”的核心理念，因而构成“武大通识3.0”的核心元素，也是实现十六字方针的重要途径。

两大《导引》作为所有本科生的必修课，已经写进《2018武汉大学本科人才培养方案》。2018级新生品尝的第一道精神大餐就是两大《导引》，这对于他们的博雅弘毅、成人成才意义重大。从中学到大学，是人生最重要的转折期，用哲学家的话说，大一新生必须回答三大终极追问：我是谁？我从哪里来？我到哪里去？因此，两大《导引》首先是对大学生“人生道路”的导引。其次，是对大学生“心灵提升”的导引。大一新生第一次离开父母和家乡，来到陌生的校园，如何与他人、环境和自己相处？如何认知人的天性、理性和悟性？如何养成人的博雅、美感和自由？这些问题，专业教育并不能回答，因此需要通识教育，需要两大《导引》。最后，是对“经典悦读”的导引。从“应试教育”中走出来的中学生，对中外经典的接触大多仅限于教科书。何为经典？为何要读经典？如何阅读经典？阅读经典的意义及乐趣何在？……两大《导引》不仅能提供深邃厚重的理论阐释，而且能提供丰富多彩的课堂实践。

二、学校对两大《导引》高度重视

学校对两大《导引》的建设和实施高度重视。最新版的本科教学培养方案，规定通识教育为 12 个学分，其中两大《导引》课程共 4 个学分，占通识课程总学分的 1/3，可见其地位之重要。学校不仅投入大量经费建设两大《导引》课程，每一门《导引》的课程建设经费达百万元，而且举全校师资之力建设两大《导引》课程。两大《导引》的课程团队约 60 人，来自全校 6 大学部的 20 多个院系，其中有国家“万人计划”教学名师，有“351 人才工程”各个层次的专家，有多次获教学成果奖、业绩奖或竞赛奖的中青年教师。更重要的是，团队全体成员都有共同的人文情怀、博雅理念和奉献精神，已经并将继续为两大《导引》课程的建设尽心尽职，殚精竭虑，努力使武汉大学全体本科生拥有高质量高品位的“共同核心课程”。

三、两大《导引》所面临的挑战

首先是师资的挑战。两大《导引》课程的老师来自不同的专业学院，拥有不同的专业背景和研究方向，如何协调并融通专业课讲授与跨学科经典导引之间的关系，就显得尤为重要。在跨文化、跨语境、跨学科的经典导读之中，需要突破专业教育的束缚及思维惯性，从而在知识结构、学术视野、思维方式和授课方法等不同层面，真正实现由专业教育向博雅教育的创造性转换。

其次是备课的挑战。两大《导引》课程团队的老师均归属于专业院系，各自有着繁重的科研和教学任务。而《导引》的备课任务非常繁重，每门《导引》包括十多部经典，老师要讲解其中的一部经典先须精读多部经典，讲席勒的《审美教育书简》要先读康德，讲《文心雕龙》要先读五经。比如《论法的精神》，孟德斯鸠不仅是在讲法律，而且讲了很多"法律"之外的东西，诸如政治、经济、宗教、社会甚至气候等。因此，在讲授每一部经典时，老师们不仅要对经典本身有着深入的了解，而且要研究经典作者的历史背景、相关的学术前史及成果等，这样才可能抓住每部经典最核心的思想，也才能对学生提供正确的导引。

最后是助教和教室的挑战。两大《导引》将全面实施"大班授课，小班研讨"的教学模式。据初步估算，为满足全校 7200 多名大一学生的修课需求，两大《导引》每学期要开出近 50 个课头，需要近 300 名助教。如何从在读硕、博士研究生中选拔助教，如何对助教进行培训，这是一项艰巨的任务。另外，武汉大学可供研讨的小教室相对偏少，需要本科生院进行规划和协调。

我一直认为，世界上最好的职业就是教师。教师的职责是培养人，而人的健康成长是社会最大的财富。作为教师，我们能够参与武汉大学发展史上具有如此重要意义的事情，是十分荣幸的。从 2018 级本科生开始，我们要将通识课程和通识教育理念一直延续下去。我们有理由相信，在两位首席专家的带领下，在课程团队全体教师和助教的努力下，武汉大学的两大《导引》课程会做得越来越好，会成为海内外一流大学通识教育和通识课程的典范。

周叶中

（作者为武汉大学副校长、武汉大学法学院教授；本文根据作者在武汉大学通识教育两大《导引》课程教师培训开班式上的讲话录音整理而成。）

序 言

《自然科学经典导引》一书是从多部最著名的自然科学经典中分别节选部分内容汇编而成的，是为在高校开设通识教育课程提供的一本参考书。

耶鲁大学1828年推出的《耶鲁报告》被认为是世界通识教育的第一个经典文本。从那时起，有关通识教育的问题被讨论了将近200年，吸引着高等教育界的广泛关注，不同观点林立，实施方法各异，表明无论进行理论探讨，还是实际操作，对通识教育的意义都有共识，但是实施起来都存在困难。

《耶鲁报告》明确提出，通识教育的宗旨不是以既定的职业为目的训练学生，而是着力拓宽学生的视野，培养他们胜任未来职业的能力。1945年，《哈佛通识教育红皮书》强调，通识教育"旨在培养学生成为负责任的人或公民"。芝加哥大学通识教育的特点是重视经典名著的研习和讨论。牛津大学的理念是："既没有纯粹的通识教育，也没有纯粹的专业教育，二者是有机融合的整体。"基于上述理念，牛津大学没有设计专门的通识课程。但是，作为历史悠久的英国著名大学，他们的做法颇能代表英国大学通识教育的特点，在课程目标、课程体系、课程内容与课程实施方面始终贯穿着通识教育精神。

简言之，通识教育应该具有"非职业性、非专业性、非功利性"特征，目的是培养学生独立思考，认知不同学科，并将所学知识融会贯通的能力。事实上，无论什么专业，通识教育与专业教育的主要区别都在教育理念和教学方法上，而不能仅仅根据教学内容作出判断。如果通识课程单纯地变成了知识的传授，那还是通识教育吗？倘若专业课程的教学，除讲授专业知识外，还能把思考和发现知识的过程、研究方法以及同其他学科之间的关联等融入其中，强化能力的培养，那么专业课程也就打上了通识教育的烙印。

“自然科学经典导引”(以下简称“导引”)课程，针对学生一进大学校门就开始一门课、一门课学习，“只见树木、不见森林”的现状而设计。“导引”课程在希腊哲学、物理世界、生命领域和科学方法四方面共选了10本世界经典名著。通过阅读经典，实现完全跨学科的通识教育，重要的不再是知识的传授，而是拓宽学生的视野，让学生学会思考。作为通识课程，“导引”的课程目标既要符合通识教育宗旨，又要体现本课程自身的特点。重点是“课程内容、教学方法和考核评价”能否支撑并证明课程目标的达成。本课程的宗旨是：了解自然科学的起源、方法及发展趋势，提高阅读经典名著的能力；熟悉自然科学的思维方式，初步具有理性判断及批判性思维的能力；提升口头表达能力和写作能力。

什么是自然科学经典？概括地说，就是那些在科学发展中具有里程碑意义的重要著作，或者严密科学体系的构建范式，或者具有永恒价值的理性反思。具体来说，哈钦斯和艾德勒编译的《西方世界伟大著作丛书》中收集的400多部著作，不同时代最伟大科学家的不朽名著，以及科学史中具有重要价值的著作都可以称为经典。

科学起源于西方，而西方科学起源于古希腊。谈科学必言古希腊！古希腊有三位最伟大的哲学家，他们是苏格拉底、柏拉图和亚里士多德。苏格拉底留存于现世的大多是只言片语、断简残篇。柏拉图的《对话录》和亚里士多德的鸿篇巨著已流传百世。想了解柏拉图和亚里士多德，不妨去看一幅世界名画，那就是文艺复兴时期拉斐尔所画的《雅典学院》。拉斐尔以古希腊哲学家柏拉图所建的雅典学院为题，1510年落笔，1511年完成此画，以回忆历史上“黄金时代”的形式，寄托他对美好未来的向往。画中汇集着不同时代、不同地域和不同学派的著名学者。画的中心是两位伟大的学者——柏拉图与亚里士多德，他们似乎边进行着激烈的争论，边向观众方向缓缓走来。在图中，谁是柏拉图？谁是亚里士多德呢？左边的这位年纪较大，手上拿着一本书，另一只手指向天空；右边的这位年龄较轻，手里也拿着一本书，另外一只手掌心向地。亚里士多德有一句名言，他说：“吾爱吾师，吾更爱真理！”可见，年长者是柏拉图，年轻者是亚里士多德。柏拉图手指天空，表明他重视理性思考。亚里士多德掌心向下，表明他偏重现实感受。一个偏重理性，一个偏重经验，在此基础上形成了西方哲学的两大体系。事实上，这也是现实人生中需要思考的问题。由此看来，哲学十分重要。这就是为什么“导引”课程要从柏拉图和亚里士多德开始讲起的原因。朱光潜先生曾说：“《理想国》是西方思想的源泉，也是我向青年推荐的唯一的西方哲学著作。”亚里士多德的《形而上学》被认为是西方哲学的一部奠基性的经典著作，该书全面、深入、详细地探讨了哲学中的各种根本性问题。通常认为，凡是要对西方哲学和西方文化有所了解或进行研究的人，《形而上学》都是一本不可不读的经典名著。

亚里士多德说：“求知是人类的天性。”自人类有文明以来，对物质世界的探索就从来没有停止过。戴维·林德伯格的《西方科学的起源》从古代科学启蒙开始，阐述了柏拉图的理型世界和宇宙论，总结了亚里士多德的自然哲学以及直到中世纪的科学成就，这些成就为新物理以及牛顿物理学的发展奠定了坚实的基础。林德伯格在该书的序言中说：“据我所知，还没有其他著作能在本书的时间跨度和阐述层次上涵盖范围如此之广的材料。”从科学史的视角来看，这本书亦可称之为经典。牛顿的《自然哲学之数学原理》揭示了宇宙中最普遍的规律，被认为是人类科学发展史上最伟大的著作。爱因斯坦说：“至今还没有可能用一个同样无所不包的统一概念，来代替牛顿的关于宇宙的统一概念。而要是没有牛顿的明晰的体系，我们到现在为止所取得的收获就会成为不可能。”爱因斯坦的《狭义与广义相对论浅说》是人类科学史上一部划时代的著作，书中深刻地揭示了时间和空间的本质属性，它是20世纪最伟大的科学理论，改变了人们对宇宙的认识。

在生命科学研究领域，达尔文的《物种起源》以大量的证据阐明了物种不是一成不变的，也不是由上帝所创造的，而是源于共同的祖先，通过自然选择的机制演化而来。这本书问世150多年来，被翻译成30多种文字，印刷量极大，足以说明其传播之广泛、影响之深远。沃森的《DNA：生命的秘密》讲述了DNA结构的发现过程，并最终证实DNA才是生命遗传的密码，至此找到了物种演化的微观解释。亚里士多德认为，理性是人类区别于动物，或者人类之所以成为人类的重要标志。希波克拉底说过：“人类应当懂得，我们的喜怒哀乐不是来自别处，而是来自大脑。”然而，人类对大脑的认识还远远落后于对物质世界和生命领域的探索。弗朗西斯·克里克的《惊人的假说》从神经元的集合出发，对大脑的行为加以解释。到目前为止，这依然是一个十分重要、尚无定论、极其困难的研究领域。有理由预见，其研究成果必然会成为人类知识宝库中的经典。

在科学方法方面，公元前3世纪，欧几里得的《几何原本》就成功地给出了一个严密知识体系的构建范式。牛顿在他的《自然哲学之数学原理》第一版序言里说：“几何学的荣耀在于，他从别处借用很少的原理，就能产生如此众多的成就。”事实上，《自然哲学之数学原理》一书从最基本的定义和公理出发给出运动定律，这种标准的公理化体系构架是《几何原本》的经典物理翻版。从科学研究方法论的角度来看，时至今日，欧几里得的思想方式并未过时。庞加莱的《科学与假设》是科学哲学中一本耀眼的名著，它深刻地阐明了假设在科学研究中的地位和作用。

《导引》所选的这些书对于学生或者教师来说可能有些偏难。不过，英国哲学家、教育家怀特海在《教育的目的》里有一句格言：“教起来不费工夫的书是没有意义的，只配烧掉，因为它根本没有教育价值。”困难一般出现在初始阶段，教师需要不停地调整难度，持续思考所选内容是否合适，程度是否恰当。对于学生来说，在一学期里读这么多书，如何才能读厚？怎样才能读薄？这也很值得思考。《导引》从每一本书中节选部分最精彩的章节，激发学生的兴趣，增强阅读欲望，希望学生会挑选一两本书读完。即使学生未能完整地读完一本书，也没有关系，至少知道有这些经典名著，也读过部分内容，可能某一天会重拾经典，仔细研读。有些书需要长久的思考才会领悟，有些书可能穷尽一生也难以企及，故能否把书读薄也未必重要。人生就应该在读万卷书、行万里路中愉快度过。

关于什么是“原著”的问题需要略加解释。一本书可以有不同版本，不同版本之间通常存在差异；同一本书可以译成不同的文字，即使译成同一种文字，也可以有不同译本。严复先生倡导翻译中应遵循“信、达、雅”原则，但真正做到“达”，则是一件很不容易的事情。优秀的翻译工作是一次再创作的过程，其中融入了译者对“原著”的理解。由于不同文化的差异，有时甚至连对应的词汇都找不到，困难可想而知。《导引》提供中英文对照，目的就是使读者能增加对作者在“原著”中所表达意涵的理解。因此，不同版本、不同文字、不同译文的作品都可看成是“原著”。作为教材，在两种不同文字的著作中进行选择时，不要太在意中英文之间不可避免存在的不完全一一对应，可优先考虑那些比较适合阅读的文本。

在教学方法上，不同时代、不同领域的世界名著如何形成一门课程是需要仔细斟酌的。每一本名著都有自身的主题，以“科学知识、科学思维、科学方法和科学精神”为主线，将这些主题贯穿起来是一种有益的尝试。在这里，科学知识只是诠释科学思维、科学方法和科学精神的载体，知识的广度和深度退居次要地位，够用即可。高校过去未曾开设这种类型的通识教育课程，授课教师自身也需要在学习、思考和实践中成长，建议每位授课教师都要从头到尾坚持完成这门课的教学任务，每一部分内容的教学体验都将大有裨益。对于学生，应该鼓励他们大量阅读经典、积极参加讨论，以期增强面对经典名著时的自信和提高理解复杂文本的能力。

桑建平

目　录

Contents

一、古希腊哲学

二、物理世界

三、生命领域

四、科学方法

一、古希腊哲学

《理想国》导引

什么是科学？在古希腊，科学与哲学都是追根溯源的活动，拥有理性的高贵基因，它的表达形态是思想，是思想的纯粹化、深入化、系统化，是 episteme(希腊语词，意为“知识”、“科学”、“洞见”)。古希腊的“科学”至少包含两层意思，首先是指一切有条理的知识或学问；其次是普遍性的真理，是洞见。

古希腊哲学是科学的母体，孕育了原初的科学思想、科学精神、科学方法。如果我们把科学看成是有生命的整体，种子萌芽成长，枝繁叶茂，开花结果，科学的遗传基因图谱(思想、方法、精神)就镌刻在古希腊哲学中。古希腊有三位最伟大的哲学家，奠定了古希腊哲学的基础，他们是苏格拉底(Socrates，公元前469年—公元前399年)、柏拉图(Plato，公元前427年—公元前347年)和亚里士多德(Aristotle，公元前384年—公元前322年)。苏格拉底没有留下任何文字，柏拉图的“对话录”和亚里士多德的鸿篇巨著已流传百世。现在，就让我们走进柏拉图和亚里士多德的世界，和哲学巨匠一起探究“科学是什么?”

柏拉图是苏格拉底的学生，出身雅典贵族，青年时代投身政治。相传在柏拉图去拜见苏格拉底的前一天夜晚，苏格拉底梦见天鹅落在肩上，翅膀很快长大，展翅高飞，唱着动人的歌。自此柏拉图追随苏格拉底，完全献身于哲学研究。公元前387年，柏拉图在雅典城边一个叫阿加德米的地方建立了世界上第一个学院(Academy)，并在此研究学问，教授知识，这个学院一直传承了900多年，直到公元529年被罗马皇帝下令关闭。

柏拉图著述颇丰，有二十几篇对话录传世，对话多以苏格拉底之名进行，我们选读的《理想国》也是如此。《理想国》讨论理想国家的社会体制和治国原则、正义的本性、理想人格、社会各阶层的教育、哲学家的培养、艺术的本性和社会功用等问题。

柏拉图在《理想国》中提出了著名的"洞喻"理论。"洞喻"构想了一些被束缚在地洞中的囚犯，他们只能看到洞壁上各种手工制品的影子，当有人挣脱了桎梏爬出洞外才看到了真实的事物，看到了照亮世界的太阳。

"那么我想要有一个逐渐适应的过程，他才能看见洞外高处的事物。首先最容易看见的是阴影，其次是那些人和其他事物在水中的倒影，再次是这些事物本身，经过这样一个适应过程，他会继续观察天象和天空本身，他会感到在夜里观察月光和星光比白天观察太阳和阳光要容易些。"①

这样，认识就分为四层次：想象、信念、理智、理性。显然，这里使我们能看见真实世界的关键是要有光，"理性之光"，要想提高认识的层次就是要透过"阴影"或"倒影"去追溯背后的"事物"本身，再进一步追溯照亮事物的"光"和"太阳本身"。

柏拉图区分可见世界与可知世界，"这个囚徒居住的地方就好比可见世界，而洞中的火光就好比太阳的力量。如果你假设从洞穴中上到地面并且看到那里的事物就是灵魂上升到可知世界，那么你没有误解我的解释，因为这正是你想要听的……我在梦境中感到善的型乃是可知世界中最后看到的东西，也是最难看到的东西……"②

如何理解可见世界与可知世界呢？让我们再回到洞中吧。

---

① [古希腊]柏拉图：《柏拉图全集(第二卷)》，王晓朝译，人民出版社2003年版，第512页。

② [古希腊]柏拉图：《柏拉图全集(第二卷)》，王晓朝译，人民出版社2003年版，第514页。

洞中的囚徒被束缚在洞内，只能看见影子，只能听见回声。由于他们从小就生活在那里，因此，将洞中的这些影子和声音当成了真实事物。假如其中一个囚徒得到了自由走出洞穴，这个人会怎样呢？他会十分迷茫，因为他眼前的一切都是他从未看到过的。可是，一旦当他走到湖边，看到自己在水中的倒影时，他会很兴奋，这才是他所熟悉的啊，他很快就会将水中的倒影和自己关联起来。如果湖边有树，树会有倒影，有马，马也会有倒影。最后，他终于看到了过去在洞穴中所看到的影子背后的真实世界。可见，在洞穴寓言中有两个世界，一个是洞穴中所看到的影子的世界；另一个是洞穴外与影子相关的真实世界。那么，柏拉图的这个寓言想告诉我们什么呢？他是想说，其实我们所看到的整个物质世界就好像洞穴中的影子，而我们就好像是洞穴中的囚徒；在这个可见的物质世界之上还应该有一个被称之为“理型”的世界，这个“理型”的世界对应洞穴寓言中洞外的真实世界。

求知的过程是心灵不断从可见世界攀升到可知世界的“上升之旅”，永无止境。最高的知识是用心灵的眼睛看到“善的理型”，是看见“理性之光”，这要付出巨大的辛劳。柏拉图又认为，“善的理型”是彼岸的，现世的人类精神不可能达至，最多只能在现世和理型世界的临界点上，在理智的迷狂状态“直观”到，理智的迷狂状态就如同我们在直视太阳时感到晕眩一样。

毫不夸张，科学的整个历史进程正是不断揭示出可见的感性世界——从微观到宏观、从日月星辰到生命，背后那个看不见的可知世界的过程，并且是一个不断深入的历程。正如怀特海所说，“The safest general characterization of the European philosophical tradition is that it consists of a series of footnotes to Plato（欧洲哲学传统的一般特征是由一系列柏拉图哲学的注脚组成）。”科学传统又何尝不是？

（周祝红）

Republic

Plato

Book 7

1 SOCRATES: Next, then, compare the effect of education and that of the lack of it on our nature to an experience like this. Imagine human beings living in an underground, cavelike dwelling, with an entrance a long way up that is open to the light and as wide as the cave itself. They have been there since childhood, with their necks and legs fettered, so that they are fixed in the same place, able to see only in front of them, because their fetter prevents them from turning their heads around. Light is provided by a fire burning far above and behind them. Between the prisoners and the fire, there is an elevated road stretching. Imagine that along this road a low wall has been built—like the screen in front of people that is provided by puppeteers, and above which they show their puppets.

GLAUCON: I am imagining it.

2 SOCRATES: Also imagine, then, that there are people alongside the wall carrying multifarious artifacts that project above it—statues of people and other animals, made of stone, wood, and every material. And as you would expect, some of the carriers are talking and some are silent.

GLAUCON: It is a strange image you are describing, and strange prisoners.

3 SOCRATES: They are like us. I mean, in the first place, do you think these prisoners have ever seen anything of themselves and one another besides the shadows that the fire casts on the wall of the cave in front of them?

GLAUCON: How could they, if they have to keep their heads motionless throughout life?

本文选自柏拉图著 *Republic*。

理想国*

柏拉图

第七卷

1 我说，接下来让我们把受过教育和缺乏教育的人的本质比作下述情形。①请你想象有这么一个地洞，一条长长的通道通向地面，和洞穴等宽的光线可以照进洞底。一些人从小就住在这个洞里，但他们的脖子和腿脚都捆绑着，不能走动，也不能扭过头来，只能向前看着洞穴的后壁。让我们再想象他们背后远处较高的地方有一些东西在燃烧，发出火光。火光和这些被囚禁的人之间筑有一道矮墙，沿着矮墙还有一条路，就好像演木偶戏的时候，演员在自己和观众之间设有一道屏障，演员们把木偶举到这道屏障上面去表演。

他说，好吧，我全看见了。

2 [我说，]②那么你瞧，有一些人高举着各种东西从矮墙后面走过，这些东西是用木头、石头或其他材料制成的假人和假兽，再假定这些人有些在说话，有些不吭声。

他说，你这个想象倒很新颖，真是一些奇特的囚徒。

3 我说，他们也是和我们一样的人。你先说说看，除了火光投射到他们对面洞壁上的阴影外，他们还能看到自己或同伴吗？

他说，如果他们的脖子一辈子都动不了，那么他们怎么能够看到别的东西呢？

本文选自[古希腊]柏拉图：《柏拉图全集(第二卷)》，王晓朝译，人民出版社 2003 年版，第 510-514 页。

*本文选自《柏拉图全集(第二卷)》国家篇的第七卷，因考虑到中英文标题对照，本《导引》编者将本章章名译为“理想国”。——编者注

① 本卷 514A-521B 是柏拉图著名的“洞喻”，有关图示参阅范明生《柏拉图哲学述评》，第 104 页。

② 为与英文部分保持基本一致及全章基本统一，此处[我说，]及后文的[我说，][他说，]均由本《导引》编者所加。

4 SOCRATES: What about the things carried along the wall? Isn't the same true where they are concerned?

GLAUCON: Of course.

5 SOCRATES: And if they could engage in discussion with one another, don't you think they would assume that the words they used applied to the things they see passing in front of them?

GLAUCON: They would have to.

6 SOCRATES: What if their prison also had an echo from the wall facing them? When one of the carriers passing along the wall spoke, do you think they would believe that anything other than the shadow passing in front of them was speaking?

GLAUCON: I do not, by Zeus.

7 SOCRATES: All in all, then, what the prisoners would take for true reality is nothing other than the shadows of those artifacts.

GLAUCON: That's entirely inevitable.

8 SOCRATES: Consider, then, what being released from their bonds and cured of their foolishness would naturally be like, if something like this should happen to them. When one was freed and suddenly compelled to stand up, turn his neck around, walk, and look up toward the light, he would be pained by doing all these things and be unable to see the things whose shadows he had seen before, because of the flashing lights. What do you think he would say if we told him that what he had seen before was silly nonsense, but that now—because he is a bit closer to what is, and is turned toward things that are more—he sees more correctly? And in particular, if we pointed to each of the things passing by and compelled him to answer what each of them is, don't you think he would be puzzled and believe that

4 [我说,]还有那些在他们后面被人举着过去的东西，除了这些东西的阴影，囚徒们还能看到什么吗?

[他说,]肯定不能。

5 [我说,]那么如果囚徒们能彼此交谈，你难道不认为他们会断定自己所看到的阴影就是真实的物体吗?

[他说,]必然如此。

6 [我说,]如果有一个过路人发出声音，引起囚徒对面洞壁的回声，你难道不认为囚徒们会断定这个声音是在他们对面的洞壁上移动着的阴影发出的吗?

他说：我以宙斯的名义发誓，他们一定会这样想。

7 [我说,]那么这样的囚徒从各方面都会认为实在无非就是这些人造物体的阴影。

他说，必然如此。

8 [我说,]那么请你考虑一下，如果某一天突然有什么事发生，使他们能够解除禁锢，矫正迷误，那会是一种什么样的情景。假定有一个人被松了绑，他挣扎着站了起来，转动着脖子环顾四周，开始走动，而且抬头看到了那堆火。在这样做的时候，他一定很痛苦，并且由于眼花缭乱而无法看清他原来只能看见其阴影的实物。这时候如果有人告诉他，说他过去看到的东西全部都是虚假的，是对他的一种欺骗，而现在他接近了实在，转向比较真实的东西，看到比较真实的东西，那么你认为他听了这话会怎么回答呢?如果再有人把那些从矮墙上经过的东西一样样指给他看，并且逼着他回答这是什么，在这种时候，你难道不认为他会不知所措，并且认为他以前看到的东西比现在指给他看的东西更加真实吗?

他说，对，他会这样想。

the things he saw earlier were more truly real than the ones he was being shown?

GLAUCON: Much more so.

9 SOCRATES: And if he were compelled to look at the light itself, wouldn't his eyes be pained and wouldn't he turn around and flee toward the things he is able to see, and believe that they are really clearer than the ones he is being shown?

GLAUCON: He would.

10 SOCRATES: And if someone dragged him by force away from there, along the rough, steep, upward path, and did not let him go until he had dragged him into the light of the sun, wouldn't he be pained and angry at being treated that way? And when he came into the light, wouldn't he have his eyes filled with sunlight and be unable to see a single one of the things now said to be truly real?

GLAUCON: No, he would not be able to—at least not right away.

11 SOCRATES: He would need time to get adjusted, I suppose, if he is going to see the things in the world above. At first, he would see shadows most easily, then images of men and other things in water, then the things themselves. From these, it would be easier for him to go on to look at the things in the sky and the sky itself at night, gazing at the light of the stars and the moon, than during the day, gazing at the sun and the light of the sun.

GLAUCON: Of course.

12 SOCRATES: Finally, I suppose, he would be able to see the sun—not reflections of it in water or some alien place, but the sun just by itself in its own place—and be able to look at it and see what it is like.

GLAUCON: He would have to.

9 [我说，]如果强迫他看那火光，那么他的眼睛会感到疼痛，他会转身逃走，回到他能看得清的事物中去，并且认为这些事物确实比指给他看的那些事物更加清晰、更加精确，难道不会吗？

他说，他会这样做。

10 我说，再要是有人硬拉着他走上那条陡峭崎岖的坡道，直到把他拉出洞穴，见到了外面的阳光，你难道不认为他会很恼火地觉得这样被迫行走很痛苦，等他来到阳光下，他会觉得两眼直冒金星，根本无法看见任何一个现在被我们称作真实事物的东西？

他说，是的，他不可能马上就看见。

11 [我说，]那么我想要有一个逐渐适应的过程，他才能看见洞外高处的事物。首先最容易看见的是阴影，其次是那些人和其他事物在水中的倒影，再次是这些事物本身，经过这样一个适应过程，他会继续观察天象和天空本身，他会感到在夜里观察月光和星光比白天观察太阳和阳光要容易些。

[他说，]那当然了。

12 [我说，]经过这样一番适应，我认为他最后终于能观察太阳本身，看到太阳的真相了，不是通过水中的倒影或影像来看，也不借助于其他媒介，而是直接观察处在原位的太阳本身。

他说，必定如此。

13 SOCRATES: After that, he would already be able to conclude about it that it provides the seasons and the years, governs everything in the visible world, and is in some way the cause of all the things that he and his fellows used to see.

GLAUCON: That would clearly be his next step.

14 SOCRATES: What about when he reminds himself of his first dwelling place, what passed for wisdom there, and his fellow prisoners? Don't you think he would count himself happy for the change and pity the others?

GLAUCON: Certainly.

15 SOCRATES: And if there had been honors, praises, or prizes among them for the one who was sharpest at identifying the shadows as they passed by; and was best able to remember which usually came earlier, which later, and which simultaneously; and who was thus best able to prophesize the future, do you think that our man would desire these rewards or envy those among the prisoners who were honored and held power? Or do you think he would feel with Homer that he would much prefer to "work the earth as a serf for another man, a man without possessions of his own," and go through any sufferings, rather than share their beliefs and live as they do?

GLAUCON: Yes, I think he would rather suffer anything than live like that.

16 SOCRATES: Consider this too, then. If this man went back down into the cave and sat down in his same seat, wouldn't his eyes be filled with darkness, coming suddenly out of the sun like that?

GLAUCON: Certainly.

13 [我说,]这时候他会做出推论，认为正是太阳造成了四季交替和年岁周期，并主宰着可见世界的所有事物，太阳也是他们过去曾经看到过的一切事物的原因。

他说，这很明显，他接下去就会做出这样的推论。

14 [我说,]如果在这种时候他回想起自己原先居住的洞穴，想起那时候的智力水平和一同遭到禁锢的同伴，那么他会为自己的变化感到庆幸，也会对自己的同伴感到遗憾，你难道不这样认为吗?

[他说,]他确实会这样想。

15 [我说,]如果洞穴中的囚徒之间也有某种荣誉和表扬，那些敏于识别影像，能记住影像出现的通常次序，而且最能准确预言后续影像的人会受到奖励，那么你认为这个已经逃离洞穴的人还会再热衷于取得这种奖励吗?他还会妒忌那些受到囚徒们的尊重并成为领袖的人，与他们争夺那里的权力和地位吗?或者说，他会像荷马所说的那样，宁愿活在世上做一个穷人的奴隶，一个没有家园的人，受苦受难，也不愿再和囚徒们有共同的看法，过他们那样的生活，是吗?

他说，是的，我想他会宁愿吃苦也不愿再过囚徒的生活。

16 我说，再请你考虑一下这种情况，如果他又下到洞中，再坐回他原来的位置，由于突然离开阳光而进入洞穴，他的眼睛难道不会因为黑暗而什么也看不见吗?

[他说,]他一定会这样。

17 SOCRATES: Now, if he had to compete once again with the perpetual prisoners in recognizing the shadows, while his sight was still dim and before his eyes had recovered, and if the time required for readjustment was not short, wouldn't he provoke ridicule? Wouldn't it be said of him that he had returned from his upward journey with his eyes ruined, and that it is not worthwhile even to try to travel upward? And as for anyone who tried to free the prisoners and lead them upward, if they could somehow get their hands on him, wouldn't they kill him?

GLAUCON: They certainly would.

18 SOCRATES: This image, my dear Glaucon, must be fitted together as a whole with what we said before. The realm revealed through sight should be likened to the prison dwelling, and the light of the fire inside it to the sun's power. And if you think of the upward journey and the seeing of things above as the upward journey of the soul to the intelligible realm, you won't mistake my intention—since it is what you wanted to hear about. Only the god knows whether it is true. But this is how these phenomena seem to me: in the knowable realm, the last thing to be seen is the form of the good, and it is seen only with toil and trouble. Once one has seen it, however, one must infer that it is the cause of all that is correct and beautiful in anything, that in the visible realm it produces both light and its source, and that in the intelligible realm it controls and provides truth and understanding; and that anyone who is to act sensibly in private or public must see it.

GLAUCON: I agree, so far as I am able.

19 SOCRATES: Come on, then, and join me in this further thought: you should not be surprised that the ones who get to this point are not willing to occupy themselves with human affairs, but that, on the contrary, their souls are always eager to spend their time above. I mean, that is surely what we

17 [我说，]如果这个时候那些终生监禁的囚徒要和他一道"评价"洞中的阴影，而这个时候他的视力还很模糊，还来不及适应黑暗，因为重新习惯黑暗也需要一段不短的时间，那么他难道不会招来讥笑吗？那些囚徒难道不会说他上去走了一趟以后就把眼睛弄坏了，因此连产生上去的念头都是不值得的吗？要是那些囚徒有可能抓住这个想要解救他们，把他们带出洞穴的人，他们难道不会杀了他吗？

他说，他们一定会这样做。

18 [我说，]亲爱的格老孔，我们必须把这番想象整个地用到前面讲过的事情上去，这个囚徒居住的地方就好比可见世界，而洞中的火光就好比太阳的力量。如果你假设从洞穴中上到地面并且看到那里的事物就是灵魂上升到可知世界，那么你没有误解我的解释，因为这正是你想要听的。至于这个解释本身对不对，那只有神知道。但不管怎么说，我在梦境中感到善的型乃是可知世界中最后看到的东西，也是最难看到的东西，一旦善的型被我们看见了，它一定会向我们指出下述结论：它确实就是一切正义的、美好的事物的原因，它在可见世界中产生了光，是光的创造者，而它本身在可知世界里就是真理和理性的真正源泉，凡是能在私人生活或公共生活中合乎理性地行事的人，一定看见过善的型。

他说，就我能理解的范围来说，我同意你的看法。

19 我说，那么来吧，和我一起进一步思考，而且你看到下面这种情况也别感到惊奇！那些已经达到这一高度的人不愿意做那些凡人的琐事，他们的灵魂一直有一种向上飞升的冲动，渴望在高处飞翔。如果我们可以作此想象，那么这样说我认为是适宜的。

[他说，]没错，可以这么说。

would expect, if indeed the image I described before is also accurate here.

GLAUCON: It is what we would expect.

20 SOCRATES: What about when someone, coming from looking at divine things, looks to the evils of human life? Do you think it is surprising that he behaves awkwardly and appears completely ridiculous, if—while his sight is still dim and he has not yet become accustomed to the darkness around him—he is compelled, either in the courts or elsewhere, to compete about the shadows of justice, or about the statues of which they are the shadows; and to dispute the way these things are understood by people who have never seen justice itself?

GLAUCON: It is not surprising at all.

20 我说，再说，如果有人从这种神圣的凝视转回到苦难的人间，以猥琐可笑的面貌出现，当他两眼昏花，还不习惯黑暗环境时，就被迫在法庭或在别的什么地方与人争论正义的影子或产生影子的偶像，而他的对手却从未见过正义本身，那么你会感到这一切都很奇怪吗？

他说，不，一点也不奇怪。

《形而上学》导引

虽然很舍不得，我们也不得不离开柏拉图，走进他最为卓越的学生，无与伦比的百科全书式哲学家——亚里士多德。

亚里士多德公元前 384 年生于与马其顿毗邻的色雷斯，他的父亲是马其顿国王腓力二世的御医。亚里士多德很早就失去双亲，17 岁来到雅典柏拉图学院，追随柏拉图 20 年直到柏拉图去世。

柏拉图称赞亚里士多德为“学院之灵”，可是柏拉图并未把自己的学院传给亚里士多德。通常的推断是“如果柏拉图的学派的继续，是企望能在其中把柏拉图自己所主张的哲学更确切地维持下去，那么，柏拉图当然不能任命亚里士多德为其继承人。”①可是黑格尔却认为，其实亚里士多德的思想是沿着柏拉图奠定的方向继续往前走，推动柏拉图哲学更加纯粹化和系统化，较之柏拉图更深刻更完善。② 对此我们不多讨论，只特别提及亚里士多德的名言“吾爱吾师，吾更爱真理”，体现了西方哲学独有的精神特质：理性批判。

① 参见[德]黑格尔：《哲学史讲演录(第二卷)》，贺麟等译，商务印书馆 1997 年版，第 272 页。

② 参见[德]黑格尔：《哲学史讲演录(第二卷)》，贺麟等译，商务印书馆 1997 年版，第 272 页。

亚里士多德也有一个著名的学生——亚历山大。亚历山大15岁时，父亲马其顿王腓力给亚里士多德写了一封著名的邀师信，流传至今。

“我有一个儿子，我感谢神灵赐我此子，还不若我感谢他们让他生于你的时代。我希望你的关怀和智慧将使他配得上我，并无负于他未来的王国。”① 有谁能拒绝这样的邀请呢？

亚里士多德之所以能成为百科全书式的思想家与这位著名学生亚历山大也有莫大关系。亚历山大在征服世界的征战中还牵挂着科学和艺术，牵挂着自己的老师，满世界收集动植物交给老师做研究，“凡在亚细亚发现了什么有关新的动物和植物的材料，便必须把原物或该物的绘图或详细的描述寄送给亚里士多德。亚历山大的这种关怀使得亚里士多德有了一个很好的条件来收集他对自然研究的宝贵资料。普里尼记述说：‘亚历山大命令近一千个以打猎、捕鱼、捕鸟为生的人，波斯帝国境内动物园、禽鸟园、鱼塘的监督者，经常供给亚里士多德以每个地方值得注意的东西。这样，亚历山大在亚洲的征战对于亚里士多德有了进一步的作用，使得他能够成为博物学的始祖，而且据普里尼说，他著了五十部博物学的书。’”②

亚里士多德给我们留下了“一大堆”著作。据公元2世纪著名传记作家记载，亚里士多德的著作有164种400余卷，我们现在能够见到的有47种，内容涉及很多领域，可以说大部分哲学和科学的划分、产生源于亚里士多德的工作。

选读的文本是亚里士多德《形而上学》第一卷的前三章，内容主要涉及“四因说”和有关科学本性的论述。我们追随亚里士多德探究的问题是：什么是“形而上学”？什么是“科学”？什么是“四因说”？

---

① ［德］黑格尔：《哲学史讲演录（第二卷）》，贺麟等译，商务印书馆1997年版，第273页。

② ［德］黑格尔：《哲学史讲演录（第二卷）》，贺麟等译，商务印书馆1997年版，第277页。

什么是“形而上学”？

西方哲学中自然哲学传统最为古老，可以说西方哲学由自然哲学开端。当泰勒斯说出“水是万物的本原”，便开启了哲学的历史。泰勒斯的“水”已不是单纯的感性事物，而是万物存在的根据，万物产生于它又复归于它，是变化中那不变的普遍性，这已经是哲学性的“水”了，因为哲学正是对最高普遍性的叩问。

追根溯源，叩问最高普遍性，也是亚里士多德关于“作为存在的存在”的学问，意思有三层：其一，万物产生变化的最普遍法则；其二，万物创生的力量；其三，对整体性、统一性的理解。

关于“作为存在的存在”的学问就叫做形而上学，这也规定了西方哲学的任务和特质。在亚里士多德，形而上学是哲学中回答第一原因、第一原理的部分，是“第一哲学”。

什么是“科学”？

亚里士多德《形而上学》开篇第一句：求知是人的天性，凸显了求知的重要地位。他接着把知识分为经验、技艺和智慧：经验，由感觉和记忆积累，是关于个别事物的知识，知其然不知其所以然；经验炉火纯青可造就技艺，技艺是知其所以然的普遍知识；智慧(wisdom)是“关于原理和原因的知识”，最高的智慧是关于第一原理、第一原因的知识。科学属于智慧，是关于原理和原因的知识，也是追求智慧的活动。

在亚里士多德看来，值得我们去寻求的是智慧，是科学，是关于“原因和原理的知识”。追求智慧只是为了满足好奇心，只是为了理性自身的完满，而不是为了实用或利益。亚里士多德一再强调科学的非功利性，“因为人们是由于惊奇，才从现在开始并且也从最初开始了哲学思考……因为他们是为了免除无知而进行哲学思考，显然他们是为了认识而追求科学，而不是为了任何实用的目的。”“由于其自身的缘故以及为了认识它而加以追求的科学，比之于为了它的结果而加以追求的科学，更具有智慧的本性”，“为认识而追求科学，而不是为了任何实用的目的”。①

亚里士多德是通过总结古希腊早期哲学各学派关于“本原”、“第一原因”的思想提出“四因说”的。认为第一原因可以在四种意义上表达：形式因(本质)；质料因(基质)；始动因(变化的来源)；目的因(生成与变化的目的)。

“显然我们必须寻求原初的原因的知识，因为只有在我们认为我们认识了事物的第一原因时，我们才说我们认识了该事物。原因在四种意义上被述说，其一是指实体，亦即本质(因为这个“为什么”最终可还原为定义，而终极的“为什么”即是一个原因和原理)；另一个意义是指质料或基质；第三个意义是变化的来源；第四个意义是与此相对立的原因，即目的与善(因为这是所有生成和变化的目的)。”②

“四因说”多用来说明生命、宇宙整体和人类活动的目的性行为。

例如，那个著名的古希腊雕塑《掷铁饼者》：形式因——运动家的形态，符合数的和谐，对立的和谐 ；质料因——石材；始动因——雕塑家；目的因——展现美。

重读亚里士多德，重要的不是为了学习具体的知识，而是重新思考科学的本性，理解科学的思维方式，特别是科学精神。

科学精神是批判的。科学从不自诩为真理，而是首先大无畏地承认自己可能有错，但找到了清除错误逼近真理的原则和方法，能够在错误和迷途中探索通往真理的道路。科学总是行走在迷途之中，但那是伟大的迷途。从事科学就必须具备问题意识和理性批判精神，科学从来都认为追求知识是困难的，需要不断经受经验和思想的批判。

科学精神也是求真的。科学源于理解自然和生命本身的好奇心，科学追求知识，追求真理，一旦失去求真精神，科学就会沦为实现其他目的的手段，也就失去了科学之成为科学的理据，科学自身便荡然无存。

① [古希腊]亚里士多德：《形而上学》，李真译，上海人民出版社 2005 年版，第 18-19 页。
② [古希腊]亚里士多德：《形而上学》，李真译，上海人民出版社 2005 年版，第 21 页。

科学是求真的学问，科学是理性批判，科学真正的原创力是好奇心，是我们理解这个世界和我们自己的万丈雄心。

（周祝红）

Metaphysics

Aristotle

I

1 All men by nature desire to know. An indication of this is the delight we take in our senses; for even apart from their usefulness they are loved for themselves; and above all others the sense of sight. For not only with a view to action, but even when we are not going to do anything, we prefer sight to almost everything else. The reason is that this, most of all the senses, makes us know and brings to light many differences between things.

2 By nature animals are born with the faculty of sensation, and from sensation memory is produced in some of them, though not in others. And therefore the former are more intelligent and apt at learning than those which cannot remember; those which are incapable of hearing sounds are intelligent though they cannot be taught, e.g. the bee, and any other race of animals that may be like it; and those which besides memory have this sense of hearing, can be taught.

3 The animals other than man live by appearances and memories, and have but little of connected experience; but the human race lives also by art and reasonings. And from memory experience is produced in men; for many memories of the same thing produce finally the capacity for a single experience. Experience seems to be very similar to science and art, but really science and art come to men *through* experience; for "experience made art", as Polus says, "but inexperience luck." And art arises, when from many notions gained by experience one universal judgement about similar

---

本文选自 Walter Kaufmann 著 *Philosophic Classic*: *from Plato to Nietzsche*。

形而上学

亚里士多德

第一章

1 所有人在本性上都愿求知。其标志就是我们对感觉的爱好；因为除了它们的用处之外，它们本身就被喜爱；在诸感觉中，尤其喜爱视觉。因为不仅着眼于行动，即使我们不打算进行任何活动，比之于任何事情，我们也更喜欢观看。其理由是，在所有感觉中，视觉最能帮助我们认识事物并揭示事物之间的差别。

2 动物由于本性都生而具有感觉能力，而且它们之中有一些就从感觉产生记忆，而另一些则没有记忆。因而与不能记忆的比较，前者更为聪明而易于学习；那些不能听到声音的尽管是聪明的，但它们不能被教导，例如蜜蜂以及任何可能与之相似的其他族类的动物；但是，那些在记忆之外还具有这个听觉的动物则可以被教导。

3 除了人之外，动物都凭表象与记忆而生活，只有很少有联系的经验；但是，人类的生活还凭技术与推理。人们从记忆产生出经验；因为对同样事物的多次记忆，最后产生出关于某一单个经验的能力。经验似乎极其像科学与技艺，其实，人们是通过经验而获得科学与技艺的。因为正如波鲁斯(Πῶλος/Polus)所说："经验造就技艺，而无经验则造就机遇。"当其由经验获得的许多概念得出一个关于一类对象的一般判断时，技艺就出现了，因为我们有一个判断断定：卡里亚(Καλλία/Callias)患这种病时，这个东西对他有益，并且在苏格拉底以及许多其他个别的人患这

本文选自[古希腊]亚里士多德：《形而上学》，李真译，上海人民出版社2005年版，第15-24页。

objects is produced. For to have a judgement that when Callias was ill of this disease this did him good, and similarly in the case of Socrates and in many individual cases, is a matter of experience; but to judge that it has done good to all persons of a certain constitution, marked off in one class, when they were ill of this disease, e. g. to phlegmatic or bilious people when burning with fever—this is a matter of art.

4 With a view to action experience seems in no respect inferior to art, and we even see men of experience succeeding more than those who have theory without experience. The reason is that experience is knowledge of individuals, art of universals, and actions and productions are all concerned with the individual; for the physician does not cure a man, except in an incidental way, but Callias or Socrates or some other called by some such individual name, who happens to be a man. If, then, a man has theory without experience, and knows the universal but does not know the individual included in this, he will often fail to cure; for it is the individual that is to be cured. But yet we think that *knowledge* and *understanding* belong to art rather than to experience, and we suppose artists to be wiser than men of experience (which implies that wisdom depends in all cases rather on knowledge); and this because the former know the cause, but the latter do not. For men of experience know that the thing is so, but do not know why, while the others know the "why" and the cause. Hence we think that the master-workers in each craft are more honourable and know in a truer sense and are wiser than the manual workers, because they know the causes of the things that are done (we think the manual workers are like certain lifeless things which act indeed, but act without knowing what they do, as fire burns,—but while the lifeless things perform each of their functions by a natural tendency, the labourers perform them through habit); thus we view them as being wiser not in virtue of being able to act, but of

种病时也是如此，这是经验；但是，断定对于所有某种类型的人(标志出一个类)当他们患这种病时，例如黏液质的人或胆汁质的人发烧时，这个东西对于他们都有益处，这就是技艺了。

4 就从事活动来说，经验似乎并不亚于技艺，而且有经验的人比起那些有理论而无经验的人更能获得成功。(理由在于经验是个别的知识，技艺是普遍的知识，而行动和生产都是涉及个别事物的；因为医生并不是给人看病，而是给卡里亚或苏格拉底或者某个别的具有这类个别名称的人治病，而他们恰好都是人。这样，如果一个人有理论而无经验，认识普遍而不知道包含于其中的个别，他就会经常在治病时失败；因为他治疗的恰恰是个别的人。)然而，我们认为，知识与理解属于技艺更甚于属于经验，并且我们设想技艺家比有经验的人更有智慧(这意味着智慧在所有情况下都依赖于知识)。这是由于前者知道原因，但是后者却不知道。因为有经验的人认识事情是这样的，但并不知道为什么，而另外的人知道这个“为什么”及其原因。由此，我们也认为在每一项手艺中，匠师比之一般工匠更为可敬，知道得更多，而且更有智慧，因为他们知道他们从事的工作的原因。(我们认为工匠像某种无生命的东西，的确，他们在活动，但是他们活动而不知他们在做什么，正如火在燃烧，但是，无生命的东西以一种自然趋势演示它们的每一种功能，而工匠是依据习惯来活动的。)①这样，我们把他们看作更有智慧，不是由于能够行动，而是由于他们具有理论和知道原因。一般说来，一个人知道或不知道[原因]的一个标志，是前者能够教别人，因而我们认为技艺比之经验是更加真的知识，因为技艺家能教别人，而仅仅有经验的人则不能。

① 罗斯认为括号内这一段话可能是后来的增添。

having the theory for themselves and knowing the causes. And in general it is a sign of the man who knows, that he can teach, and therefore we think art more truly knowledge than experience is; for artists can teach, and men of mere experience cannot.

5 Again, we do not regard any of the senses as wisdom; yet surely these give the most authoritative knowledge of particulars. But they do not tell us the "why" of anything—e.g. why fire is hot; they only say that it is hot.

6 At first he who invented any art that went beyond the common perceptions of man was naturally admired by men, not only because there was something useful in the inventions, but because he was thought wise and superior to the rest. But as more arts were invented, and some were directed to the necessities of life, others to its recreation, the inventors of the latter were always regarded as wiser than the inventors of the former, because their branches of knowledge did not aim at utility. Hence when all such inventions were already established, the sciences which do not aim at giving pleasure or at the necessities of life were discovered, and first in the places where men first began to have leisure. This is why the mathematical arts were founded in Egypt; for there the priestly caste was allowed to be at leisure.

7 We have said in the *Ethics* what the difference is between art and science and the other kindred faculties; but the point of our present discussion is this, that all men suppose what is called wisdom to deal with the first causes and the principles of things; This is why, as has been said before, the man of experience is thought to be wiser than the possessors of any perception whatever, the artist wiser than the men of experience, the master-worker than the mechanic, and the theoretical kinds of knowledge to be more of the nature of wisdom than the productive. Clearly then wisdom is knowledge about certain causes and principles.

5 再有，我们并不把任何感觉看作智慧，然而肯定地说这些感觉提供了特殊东西的最权威的知识。但是它们不能告诉我们任何事物的“为什么”（τὸδιὰτί/Why），例如，火为什么是热的；它们仅仅说，它是热的。

6 那个最先发明了无论何种技艺，超过了人的普通知觉的人，自然地受到人们的崇敬，不仅由于其发明中有某种有用的东西，而且由于他被认为是有智慧的并优于其他的人。但是当更多的技艺被发明时，其中有些是直接指向生活的必需的，另一些是为了娱乐的，后者的发明者自然地总是被看作比前者的发明者更加聪明，因为他们的知识的分支并不是指向实用的。因此，当所有那样的发明都已建立起来时，那种并不为了提供愉悦或为了生活必需的科学就被发现了，并且是在那些人们最先有了闲暇的地方，这就是为什么数学技艺首先在埃及被发现；因为那里的祭司等级被允许享有闲暇。

7 我们在《伦理学》中说过①，在技艺与科学以及其他类似能力之间的差别是什么；但是，我们现在讨论它的理由是，所有人都假定那被称为智慧的东西是处理事物的第一原因和原理；所以，如前面所说，一个有经验的人被认为比拥有无论什么感觉知觉的人更有智慧，技艺家比有经验的人更有智慧，匠师比机械地工作的人更有智慧，而理论性的知识比起生产知识来，具有更多的智慧本性。那么显然，智慧是关于某种原理和原因的知识。

① 参看《尼各马可伦理学》V1.3—7，$1139^{b}14$—$1141^{b}8$。

Ⅱ

8 Since we are seeking this knowledge, we must inquire of what kind are the causes and the principles, the knowledge of which is wisdom. If we were to take the notions we have about the wise man, this might perhaps make the answer more evident. We suppose first, then, that the wise man knows all things, as far as possible, although he has not knowledge of each of them individually; secondly, that he who can learn things that are difficult, and not easy for man to know, is wise (sense-perception is common to all, and therefore easy and no mark of wisdom); again, he who is more exact and more capable of teaching the causes is wiser, in every branch of knowledge; and of the sciences, also, that which is desirable on its own account and for the sake of knowing it is more of the nature of wisdom than that which is desirable on account of its results, and the superior science is more of the nature of wisdom than the ancillary; for the wise man must not be ordered but must order, and he must not obey another, but the less wise must obey him.

9 Such and so many are the notions, then, which we have about wisdom and the wise. Now of these characteristics that of knowing all things must belong to him who has in the highest degree universal knowledge; for he knows in a sense all the subordinate objects. And these things, the most universal, are on the whole the hardest for men to know; for they are farthest from the senses. And the most exact of the sciences are those which deal most with first principles; for those which involve fewer principles are more exact than those which involve additional principles, e.g. arithmetic than geometry. But the science which investigates causes is also more capable of reaching, for the people who teach are those who tell the causes of each thing. And understanding and knowledge pursued for their own sake are found most in

第二章

8 因为我们正在寻求这种知识，我们必须研究什么种类的原因和原理的知识是智慧(σοφία/Wisdom)。如果我们采用我们关于智慧的人的概念，这也许会使得答案更为明白。这样，我们首先假定智慧的人知道所有的事物，尽可能地广泛，尽管他没有关于每一事物的细节的知识；其次，那些能够学习困难的、一般人不容易懂得的事物的人，是智慧的(感觉—知觉对于所有人都是共同的，因此是容易的而非智慧的标志)；再次，在知识的每一个分支中，那些能够更确切、更有能力教导原因的人是更为智慧的；而且在各门科学中，由于其本身的缘故以及为了认识它而加以追求的科学，比之于为了它的结果而加以追求的科学，更具有智慧的本性，而高级的科学比之于辅助的科学，更具有智慧的本性；因为智慧的人不应接受命令而应发出命令，而且他不应服从别人，相反，较少智慧的人应当服从他。

9 这就是我们所有的关于智慧以及智慧的人的这样一些概念。现在，在这样一些特征中，知道所有的事物必定属于那有着最高程度的普遍知识的人；因为在一种意义上他知道归属于普遍的所有事例。并且这些事物，这些最普遍事物，总的来说，都是人们最难认识的；因为它们都是离感觉最遥远的。并且那些最严格的科学几乎都是处理第一原理的；因为那包含较少原理的知识比之于那些包含附加原理的是更为严格的，例如算术较几何学更严格。而且毫无疑问，研究原因的知识也是更加有能力教导的，因为那教导的人都是述说每一事物的原因的人。并且由于其自身缘故而加以追求的理解和知识大多在最可知的知识之中(因为为了认识的缘故而选择认识的人将会最确定地选择最真实的知识，而那就是最可知的知识)，而第一原理和原因就是最可知的知识，因为通过它们以及

the knowledge of that which is most knowable; for he who chooses to know for the sake of knowing will choose most readily that which is most truly knowledge, and such is the knowledge of that which is most knowable; and the first principles and the causes are most knowable; for by reason of these, and from these, all other things are known, but these are not known by means of the things subordinate to them. And the science which knows to what end each thing must be done is the most authoritative of the sciences, and more authoritative than any ancillary science; and this end is the good in each class, and in general the supreme good in the whole of nature. Judged by all the tests we have mentioned, then, the name in question falls to the same science; this must be a science that investigates the first principles and causes; for the good, i.e. that for the sake of which, is one of the causes.

10 That it is not a science of production is clear even from the history of the earliest philosophers. For it is owing to their wonder that men both now begin and at first began to philosophize; they wondered originally at the obvious difficulties, then advanced little by little and stated difficulties about the greater matters, e.g. about the phenomena of the moon and those of the sun and the stars, and about the genesis of the universe. And a man who is puzzled and wonders thinks himself ignorant (whence even the lover of myth is in a sense a lover of wisdom, for myth is composed of wonders); therefore since they philosophized in order to escape from ignorance, evidently they were pursuing science in order to know, and not for any utilitarian end. And this is confirmed by the facts; for it was when almost all the necessities of life and the things that make for comfort and recreation were present, that such knowledge began to be sought. Evidently then we do not seek it for the sake of any other advantage; but as the man is free, we say, who exists for himself and not for another, so we pursue this as the only free science, for it alone exists for itself.

从它们出发，所有其他事物都得以认识，而不是借助于从属于它们的事物而认识它们。那知道每一事物应当达到的目的的知识是知识中最有权威性的，比任何辅助性的知识更具有权威性；这个目的就是每一事物的善(τἀγαθὸν εκάσον)，整个说来就是在整个自然中的最高的善(τὸνἄριστον)。于是，从我们所说的所有的考虑来判断，我们所研究的项目就属于同一门知识；它必定是一门研究第一原理和原因的知识；因为善，亦即为了它缘故的那个东西，乃是诸原因中的一个。

它不是一门生产的科学，即使从最早的哲学家的历史来看也是很清楚的。因为人们是由于惊奇(διὰ γὰρ τὸ θαυμάζειν/Oweing to their wonder)，才从现在开始并且也从最初开始了哲学思考(φιλοσοφεῖν/Philosophize)。他们最初惊奇于明显的困难，然后一点一点逐步进展并陈述关于较重大问题的困难，例如关于月亮、太阳和星辰的现象，以及关于宇宙的生成的问题。而且一个人在困惑和惊奇的时候，认为他自己是无知的(因而即使是神话的爱好者，在一种意义上也是智慧的爱好者，因为神话是由惊奇组成的)；因此，他们是为了免除无知而进行哲学思考，显然他们是为了认识而追求科学，而不是为了任何实用的目的。并且这一点是由事实加以确证的：因为那是在几乎所有的生活必需品以及提供舒适和娱乐的事物都已得到保障时，才开始寻求那样的知识的。那么，很明显，我们不是为了任何其他利益的缘故而寻求它；而是当人们自由的时候，人们是为了自己的缘故而不是为了别的人而存在时，所以我们追求这门作为唯一自由的科学，因为它只是为了它自身的缘故而存在的。

11 Hence the possession of it might be justly regarded as beyond human power; for in many ways human nature is in bondage, so that according to Simonides "God alone can have this privilege," and it is unfitting that man should not be content to seek the knowledge that is suited to him. If there is something in what the poets say, and jealousy is natural to the divine power, it would probably occur in this case above all, and all who excelled in this knowledge would be unfortunate. But the divine power cannot be jealous (indeed, according to the proverb, "bards tell many a lie"), nor should any science be thought more honourable than one of this sort. For the most divine science is also most honourable; and this science alone is, in two ways, most divine. For the science which it would be most meet for God to have is a divine science, and so is any science that deals with divine objects; and this science alone has both these qualities; for God is thought to be among the causes of all things and to be a first principle, and such a science either God alone can have, or God above all others. All the sciences, indeed, are more necessary than this, but none is better.

12 Yet the acquisition of it must in a sense end in something which is the opposite of our original inquiries. For all men begin, as we said, by wondering that the matter is so (as in the case of automatic marionettes or the solstices or the incommensurability of the diagonal of a square with the side; for it seems wonderful to all men who have not yet perceived the explanation that there is a thing which cannot be measured even by the smallest unit). But we must end in the contrary and, according to the proverb, the better state, as is the case in these instances when men learn the cause; for there is nothing which would surprise a geometer so much as if the diagonal turned out to be commensurable.

11 也由于这个缘故，拥有它可以公正地被认为是超出人类的能力的；因为在许多方面人的本性是受束缚的，所以正如西蒙尼德(Σιμωνίδης/Simonides)所说，“只有神能有此特权”①，并且人们应当不满足于寻求适合于他们的知识，则是不恰当的。如果诗人们的确说出了某些[道理]，而对于神的权能来说，妒嫉是自然的话，那么它也许首先会发生在这个情况下了，而所有在这种知识中超越了的人将会是不幸的，但是神的权能不可能是妒嫉的(的确，如谚语②所说，“行吟诗人说了许多谎”)，任何其他科学也必定不会被认为比这样一门科学更为荣誉。因为最神圣的科学也是最荣耀的，而唯有这门科学在两种方式中是最神圣的。因为最适合于神具有的科学是一门神圣的科学，因而任何处理神圣对象的科学也是如此；而只有这门科学具有这两种性质；因为(1)神被认为是在所有事物的原因中间，并且是一个第一原理，而且(2)这样一门科学或者只有神能具有，或者是在所有事物中神首先具有。的确，所有科学都比这门科学更必需，但是没有任何科学比它更好。

12 然而对它的掌握必定在某一种意义上终止于某种与我们原来的探索相对立的东西。因为正如我们所说，所有人以惊奇于事物是它们那样的而开始，如像他们惊奇于自动的牵线木偶，或关于冬至、夏至的至点，或者关于四边形的对角线与其一边的长度的不可通约性的惊奇。因为对于所有还没有看到原因的人来说，它显得是令人惊奇的，竟然有一种东西，即使用最小的单位也不能度量。但是，我们必须终止于相对立的状态，正如谚语所说，较好的状况就是(在这些事例中也是如此)当人们学习到原因的时候；因为没有什么事情比对角线如果变得可以通约会更加使几何学家吃惊的了。

① 残篇 3(Hiller 版本)。

② 参看梭伦，残篇 26(Hiller 版本)。

13 We have stated, then, what is the nature of the science we are searching for, and what is the mark which our search and our whole investigation must reach.

Ⅲ

14 Evidently we have to acquire knowledge of the original causes (for we say we know each thing only when we think we recognize its first cause), and causes are spoken of in four senses. In one of these we mean the substance, i. e. the essence (for the "why" is referred finally to the formula, and the ultimate "why" is a cause and principle); in another the matter or substratum, in a third the source of the change, and in a fourth the cause opposed to this, that for the sake of which and the good (for this is the end of all generation and change). We have studied these causes sufficiently in our work on nature, but yet let us call to our aid those who have attacked the investigation of being and philosophized about reality before us. For obviously they too speak of certain principles and causes; to go over their views, then, will be of profit to the present inquiry, for we shall either find another kind of cause, or be more convinced of the correctness of those which we now maintain.

15 Of the first philosophers, most thought the principles which were of the nature of matter were the only principles of all things; that of which all things that are consist, and from which they first come to be, and into which they are finally resolved (the substance remaining, but changing in its modifications), this they say is the element and the principle of things, and therefore they think nothing is either generated or destroyed, since this sort of entity is always conserved, as we say Socrates neither comes to be absolutely when he comes to be beautiful or musical, nor ceases to be when he loses

13 这样我们就陈述了我们寻求的科学的本性是什么，而且我们的寻求以及我们整个的研究必须达到的目标是什么。

第三章

14 显然我们必须寻求原初的原因(ἀρχ ήα' ίτια/Original Cause)的知识，因为只有在我们认为我们认识了事物的第一原因(τήυ πρώ την αἰ τί ταν/First cause)时，我们才说我们认识了该事物。原因在四种意义上被述说，其一是指实体，亦即本质(εἰναι τήν οὐσίαν καί τὸτ ἰῆν εἶ ναι)(因为这个“为什么”最终可还原为定义，而终极的“为什么”即是一个原因和原理)；另一个意义是指质料或基质(τὴ ν ὕλην καὶ τὸ ὑποκεί μενον)；第三个意义是变化的来源(ἡἀρχὴ της κενέσεως)，第四个意义是与此相对立的原因，即目的与善(τὸ οὗ ἕνεκα καὶ τἀγαθόν)(因为这是所有生成和变化的目的[τέλος])。在我们的论自然的著作中①，我们已经充分地研究了这些原因，不过，还是让我们利用那些在我们之前研究过实在并对真理进行过哲学思考的人们的看法作为证据吧。因为他们显然也谈到某些原理和原因；那么，重温他们的看法，对于目前的研究将会是有益的，因为我们或者会发现另一类原因，或者对于我们现在主张的那些原因的正确性会更加确信。

15 在第一批哲学家中间，大多数认为质料性的本原是所有事物的唯一本原。所有事物由之构成，它们最先从它产生，最后它们又消融于它(实体保持着，但是在它的变形中变化着)，他们说，这就是元素(στοιχεῖ ον/Element)，这就是事物的本原(ἀρχήν)。因此，他们认为既没有什么东西被生成，也没有什么东西被摧毁，因为这类自然物永远保存着。这正像我们说，当苏格拉底变得漂亮和有教养时，我们不说他绝对地生成出来，而当他失去这些特性时，我们也不说他停止存在了，因为其基质，即苏格拉底自身保持着。正因为如此，他们说没有什么东西生成出

① 参看《物理学》Ⅱ·3，7。

these characteristics, because the substratum, Socrates himself, remains. So they say nothing else comes to be or ceases to be; for there must be some entity—either one or more than one—from which all other things come to be, it being conserved.

16 Yet they do not all agree as to the number and the nature of these principles. Thales, the founder of this school of philosophy, says the principle is water (for which reason he declared that the earth rests on water), getting the notion perhaps from seeing that the nutriment of all things is moist, and that heat itself is generated from the moist and kept alive by it (and that from which they come to be is a principle of all things). He got his notion from this fact, and from the fact that the seeds of all things have a moist nature, and that water is the origin of the nature of moist things.

17 Some think that the ancients who lived long before the present generation, and first framed accounts of the gods, had a similar view of nature; for they made Ocean and Tethys the parents of creation, and described the oath of the gods as being by water, which they themselves call Styx; for what is oldest is most honourable, and the most honourable thing is that by which one swears. It may perhaps be uncertain whether this opinion about nature is primitive and ancient, but Thales at any rate is said to have declared himself thus about the first cause. Hippo no one would think fit to include among these thinkers, because of the paltriness of his thought.

18 Anaximenes and Diogenes make air prior to water, and the most primary of the simple bodies, while Hippasus of Metapontium and Heraclitus of Ephesus say this of fire, and Empedocles says it of the four elements, adding a fourth—earth—to those which have been named; for these, he says, always remain and do not come to be, except that they come to be more or fewer, being aggregated into one and segregated out of one.

来或不复存在；因为有某种自然物（φύσις）——或一种或多于一种——永远持续存在，所有其他事物都从它产生出来。

16 然而，关于这些本原的数目和性质，他们并不完全一致。泰勒斯（Θαλῆs/Thales）是这种类型的哲学的创始人。他说，本原是水（由于这个理由，他宣称大地浮在水上）。他得到这个概念也许是由于看到所有事物的营养物都是潮湿的，而且热本身是从湿气中产生出来，并且生命也依靠它而得以保持（而那个它由之产生的东西就是所有事物的本原）。他从这个事实并且从所有事物的种子都具有潮湿的性质这个事实，以及水是潮湿的事物的性质的来源，得出他的概念。

17 有人①认为甚至那些生活于距现代很久以前的古人，以及编订诸神的传闻的古代人，也曾具有关于自然的相似的观点。因为他们使奥克安诺斯（'Ωκειανὸs/Ocean）和德蒂斯（Τηθ ύν/Tethys）成为创世的双亲②，而且描写众神皆以水起誓③，并且称之为"斯蒂克斯"（Στύξ/Styx）④；因为最古老的就是最受尊敬的，而最受尊敬的东西就是一个人以之起誓的东西。这个关于自然的看法究竟是否原始的和古老的，也许是不确定的，但是无论如何，据说泰勒斯宣称他本人主张这样的第一原因，没有人认为希波（'Ἵππων/Hippo）宜于列入这些思想家中间，因为他的思想微不足道。

18 阿那克西美尼和第欧根尼使气先于水，认为气是简单物体中最基本的，而墨达蓬梯的希巴索和爱菲斯的赫拉克利特（'Ηράκλειτοs ὁ'Εφέσιοs）说它是火，恩培多克勒（'Εμπεδοκλῆ）则说它乃四种元素，即在已经说过的那些之外加上第四种［即土］。因为他说，这些元素永远保持而不是被产生出来，只是它们变得多些或少些，即聚集为一和从一分离出来。

① 也许是指柏拉图（参看《克拉底鲁篇》，402B；《泰阿泰德篇》152E，162D，180C）。

② 奥克安诺斯和德蒂斯为希腊神话中的海洋之神和海洋女神，参看荷马：《伊里亚特》XIV·201，204。

③ 参看荷马：《伊里亚特》Ⅱ·755，XIV·271，XV·37。

④ 据希腊神话，斯蒂克斯为冥界的一条河，Στύξ意为"可恨"，因此意为恨河。据神话，亡灵饮此水则前事尽忘。

19 Anaxagoras of Clazomenae, who, though older than Empedocles, was later in his philosophical activity, says the principles are infinite in number; for he says almost all the things that are homogeneous are generated and destroyed (as water or fire is) only by aggregation and segregation, and are not in any other sense generated or destroyed, but remain eternally.

20 From these facts one might think that the only cause is the so-called material cause; but as men thus advanced, the very facts showed them the way and joined in forcing them to investigate the subject. However true it may be that all generation and destruction proceed from some one or more elements, why does this happen and what is the cause? For at least the substratum itself does not make itself change; e. g. neither the wood nor the bronze causes the change of either of them, nor does the wood manufacture a bed and the bronze a statue, but something else is the cause of the change. And to seek this is to seek the second cause, as we should say—that from which comes the beginning of movement. Now those who at the very beginning set themselves to this kind of inquiry, and said the substratum was one, were not at all dissatisfied with themselves; but some at least of those who maintain it to be one—as though defeated by this search for the second cause—say the one and nature as a whole is unchangeable not only in respect of generation and destruction (for this is an ancient belief, and all agreed in it), but also of all other change; and this view is peculiar to them. Of those who said the universe was one, none succeeded in discovering a cause of this sort, except perhaps Parmenides, and he only insomuch that he supposes that there is not only one but in some sense two causes. But for those who make more elements it is more possible to state the second cause, e. g. for those who make hot and cold, or fire and earth, the elements; for they treat fire as having a nature which fits it to move things, and water and earth and such things they treat in the contrary way.

19 克拉佐美尼的阿那克萨哥拉('Αναξαγὸρας ὁ Κλαζομένιος)虽然年长于恩培多克勒，但其哲学活动则晚于后者。他说，本原在数目上是无限的；因为他说几乎所有事物都是由与其自身相同的微粒(τὰ ὁμοιομερῆ/Homoeomerous)构成的，就像水或火一样，并且仅仅在这种方式中生成和消灭，即通过结合和分散，而并非在任何其他意义上生成和消灭，而是永恒地保持着。

20 从这些事实，人们可能认为唯一的原因就是这类所谓的质料因；但是当人们这样向前进展时，许多事实为他们开辟了道路，并一起迫使他们进一步研究。所有的生成和消灭从某一个或者(对于那种质料说)从更多的元素开始进行，无论这可以是怎样的真实，那么，为什么这会发生而且它的原因是什么呢？因为至少基质自身并不造成自身的变化。我的意思是，既不是木头也不是铜引起它们各自的变化，木头也不会造出一张床，铜也不会造成一座雕像，而是别的某种东西是这个变化的原因，而寻求这个原因就是寻求另一种原因，即我们应当问：运动的源泉从何而来。现在那些最早从事这类研究，并且主张基质是一个的①，对于这个问题没有什么疑虑；但是至少他们中的有些人，主张其[基质]是一②——似乎被这个关于第二个原因的研究所困扰了——他们说这一[个基质]和自然作为一个整体是不变的，不仅就生成和毁灭来说(因为这是一个原始的信念，而且所有人都同意)，而且也是就所有其他变化说的。这个观点是他们所特有的。那些说宇宙是一的人当中，没有人继续探寻这类原因，也许巴门尼德是个例外，因为只有他假设不仅有一种[原因]，而且在某种意义上有两种原因。但是对于那些主张有多种元素的人③来说，更有可能陈述第二种原因，例如，那些把热和冷，或火和土作为元素的人就是这样。因为他们把火当作具有适合于推动事物的性质，而水和土之类的事物，则是相反的。

① 指泰勒斯、阿那克西美尼和赫拉克利特。

② 指埃利亚派。

③ 大约是指恩培多克勒。

21 When these men and the principles of this kind had had their day, as the latter were found inadequate to generate the nature of things, men were again forced by the truth itself, as we said, to inquire into the next kind of cause. For surely it is not likely either that fire or earth or any such element should be the reason why things manifest goodness and beauty both in their being and in their coming to be, or that those thinkers should have supposed it was; nor again could it be right to ascribe so great a matter to spontaneity and luck. When one man said, then, that reason was present—as in animals, so throughout nature—as the cause of the world and of all its order, he seemed like a sober man in contrast with the random talk of his predecessors. We know that Anaxagoras certainly adopted these views, but Hermotimus of Clazomenae is credited with expressing them earlier. Those who thought thus stated that there is a principle of things which is at the same time the cause of beauty, and that sort of cause from which things acquire movement.

21 在这些人以及这一类的本原之后，因为它们不能充分地揭示事物本性的生成，人们被真理所迫(如我们已经说过的)，重新研究另一类的原理。因为火或者土或者任何这样的元素应当是事物在其存在和在其生成方面均表现为善和美的理由；或者这些思想家应该假定它是这样的；这似乎都不妥当。再有，把这样重大的问题委之于自发性和机遇，也是不对的。于是，当有一个人说理性(νοῦs/Reason)表现为——正如在动物中一样，它充斥于整个自然界——秩序和所有安排的原因时，他与那些随意谈论他的先辈比较，似乎是一位严肃的人。我们知道阿那克萨哥拉肯定地采取了这个观点，但是克拉佐美尼的赫尔摩提谟('ΕρμὸτιμοsόΚλαζόμενιοs)被认为表述这个观点更为早些。那些像这样思考的人陈述说：有一个诸事物的本原，它同时就是美的原因，而且从这一类原因中，事物因之而获得运动。

二、物理世界

《西方科学的起源》导引

《西方科学的起源》的作者戴维·林德伯格(David C. Lindberg)(1935—2015)拥有美国西北大学物理学学位和印第安纳大学历史和科学哲学博士学位，曾任威斯康辛—麦迪逊大学历史科学研究所的希尔德代尔教授和人文科学研究所所长。他是著名的科学史学家，主要从事中世纪和现代科学早期历史的研究。林德伯格教授笔耕不辍，是很多书籍的作者和编辑，比如他是《剑桥科学史(八卷)》的总编辑，也是《中世纪科学》杂志的米迦勒编辑。他获得了包括约翰·西蒙古根海姆纪念基金会、美国国家科学基金会、美国国家人文基金会、普林斯顿大学高级研究所等众多组织的奖励和资助，并于1999年获得世界科学史界终身学术成就最高奖项——萨顿奖章。

《西方科学的起源》这本书的第一版基于林德伯格教授"为大学本科生讲授古代和中世纪科学史的20年经验写成"，获得过约翰·坦普莱顿基金会神学和自然科学杰出著作奖。它有一个很长的副标题——"公元前六百年至公元一千四百五十年宗教、哲学和社会建制大背景下的欧洲科学传统"，基本体现了作者在书中探讨的历史和空间范围。我们阅读的版本是这本书的修订版，距离第一版的出版时间又过了20年。在这后面的20年中，林德伯格教授又有了新的授课经验，并钻研了更新的学术著作，对第一版书的“每一页几乎都有修改”。这种对待科学著作的认真而专注的态度，是否对正在读书的你有所触动？

相较于柏拉图的《理想国》和亚里士多德的《形而上学》，这本书具有更强的可读性。不仅因为它是大学的教材，更是由于作者注重用浅显易懂的语言“引领我们走过两千多年的西方科学发展的历史长廊”，“告诉我们应如何正确认知和理解历史”：我们不能“按照与现代科学的相似性来研究过去的做法和信念”，“我们必须尊重前人研究自然的方式，承认它虽然可能不同于现代方式，但仍然是有趣的，因为它是我们思想来源的一部分”。

也许你们会问：我们为什么要读这本书？或许两千年前的古罗马哲学家西塞罗的话是一个参考答案。他说，“一个人不了解生下来以前的事，那他始终只是一个孩子。”这本书的作者也说：“倘若我们希望理解科学事业的本质、科学与周围更广大文化背景的关系、人类对科学所涉内容的认知程度，那么历史研究，包括对早期科学的研究，就是必不可少的。”尽管这本书讲述的事情发生在很久以前遥远的西方，但这些事情所带来的科学对我们的国家和我们的生活产生了深刻的影响，科学的发展也已经成为我们关注的核心事件，书中涉及的“批判性评价”，是西方科学极为重要的传统。所以，回首西方科学的源头，对我们来说，也许正当时！

林德伯格教授在这本书中描绘了一幅科学发展史诗般的宏阔图景：一方面，时间跨越千年，从古希腊时期科学的起源描绘至中世纪的科学遗产，以翔实的史料证据回答了若干我们看似拥有“唯一的正确答案”，但事实上可能并非如此的问题，例如“中世纪是不是一个在科学上长期无知和迷信的时期”？另一方面，地域上横跨欧亚，迄今为止也是空前。

比如第一章描写史前人类以及古埃及社会和两河流域文明在促进人类科学进步中所作的贡献。后面几章具体阐述以上成就对希腊自然哲学的渗透和影响。在第二、三章中大家可以读到散文式的关于柏拉图与亚里士多德所提出的哲学问题的小结与评价。亚里士多德哲学与科学体系如此具有说服力，在希腊人与罗马人的数学、医学和中世纪早期科学中都可以看到它们的身影。

不同时期的社会环境对科学的影响同样巨大，作者通过回答疑问来说明自己的观点。这些疑问分别是：为什么近代科学产生在欧洲？欧洲是如何吸收并发展古希腊文明的？古希腊之后为何没有接着出现近现代科学，反而经历了漫长的中世纪？中世纪与早期近代科学之间究竟是连续的还是断裂的？除此之外，《西方科学的起源》一书提供了百余幅插图和众多参考文献，不仅可以让阅读变得直观、令人遐想，更极大地方便了读者的深入理解与探究。这些问题是否引起了你的好奇？让我们现在就开始阅读，去书中寻找上述问题的答案吧！

编者在这本不可多得的优秀科学史著作中为大家选择了两个部分作为阅读文本，所占篇幅不多，但尤为重要。在第一部分，林德伯格教授阐述了他对“科学”的理解与定义。关于“科学”的定义，被作者放在《西方科学的起源》的第一章“希腊人之前的科学”的卷首，说明他认为科学的起源可以追溯到古希腊，认可中世纪对早期近代科学的贡献。到底什么是科学？这是一个非常好的问题。直到今天，这个问题仍然可以引发我们的思考，激起大家的讨论。

文本的第二部分描述的是亚里士多德的“宇宙论”。在前面的文本中，大家学习了亚里士多德“研究和理解世界的方法和原则”，这里要介绍的是他关于“自然现象的理论，上至天空，下至大地及其居住者”。这一部分既可以承接古希腊文明，又可以开启鸿篇巨著《自然哲学之数学原理》的序幕。

凡是对科学感兴趣的人，也一定会对这本书产生极大的兴趣！

（祁宁）

The Beginnings of Western Science

David C. Lindberg

Chapter 1 Science before the Greeks

What is science?

1 The opinion that there was *no science* in the two thousand years covered by this book continues to be stated with considerable regularity and dogmatic fervor. If the claim is true, I have written a book about a nonexistent subject—no mean feat, but not my goal. This book proclaims in its title that it will portray the beginnings of Western science over the approximately three millennia ending about the year A.D. 1450. Was there truly such a thing as science in those times? And if the answer is affirmative, was there enough of it to merit book-length coverage?

Before we can answer these questions, we need a definition of "science"—something that turns out to be surprisingly difficult to come by. There is, of course, the dictionary definition, according to which "science" is organized, systematic knowledge of the material world. But this proves to be so general as to be of little help. For example, do craft traditions and technology count for science, or are science and technology to be distinguished from one another—the former dedicated to theoretical knowledge, the latter to its application? If only theoretical knowledge counts as genuine science, we then need to decide which theories (or which kinds of theory) pass the test. Do astrology and parapsychology, both of which are chock full of theories, count as sciences?

本文选自 David C. Lindberg 著 *The Beginnings of Western Science*，芝加哥大学出版社 2007 年版。

西方科学的起源

戴维·林德伯格

第一章　希腊人之前的科学

什么是科学

1　时至今日，仍然有人会教条地认为，在本书所涵盖的2000年里没有科学。倘若这种断言是正确的，那么本书讨论的便是一个莫须有的主题——这虽然绝非易事，但并非我的目标。本书标题即已言明，它将论述西方科学在公元1450年之前大约3000年的时间里的起源。在那些时代真有科学这样一种东西吗？即使答案是肯定的，它是否值得用一本书来讨论？

在回答这些问题之前，我们需要对“科学”作出定义——事实证明，这种定义出奇地难下。当然字典上有定义，它说，“科学”是关于物质世界的有组织的系统知识。但这种说法过于笼统，无甚帮助。例如，手艺传统和技术算科学吗？抑或科学和技术是有区别的——科学致力于理论知识，而技术致力于科学的应用？即使真的只有理论知识算真正的科学，我们也需要确定哪些理论（或哪种理论）是够格的。占星术和超心理学中都充斥着理论，它们算科学吗？

本文选自[美]戴维·林德伯格：《西方科学的起源》，张卜天译，湖南科学技术出版社2016年版，第1-5页，第54-60页，第70-76页，第83-89页。

2 Perceiving that the "theoretical knowledge" criterion is heading toward a dead end, some participants in the debate argue that true science can be recognized by its methodology—specifically, the experimental method, according to which a theory, if it is to be truly scientific, must be built on and tested against the results of observation and experiment. (In the minds of many of its advocates, a series of rigorously defined steps must be employed.) Theories that meet this test are often credited with superior epistemological status or warrant and thus are representative of a privileged way of knowing. Finally, for many people—scientists and general public alike—true science is defined simply by its content—the current teachings of physics, chemistry, biology, geology, anthropology, psychology, and so forth.

3 This brief foray into lexicography ought to remind us that many words, especially the most interesting ones, have multiple meanings that shift with the contexts of usage or the practices of specific linguistic communities. Every meaning of the term "science" discussed above is a convention accepted by a sizable group of people, who are unlikely to relinquish their favored usage without a fight. From which it follows that we have no choice but to accept a diverse set of meanings as legitimate and do our best to determine from the context of usage what the term "science" means on any specific occasion.

4 But where does that leave us? Was there anything in Europe or the Near East in the twenty centuries covered by this book that merits the name "science"? No doubt! Many of the ingredients of what we now regard as science were certainly present. I have in mind languages for describing nature, methods for exploring or investigating it (including the performance of experiments), factual and theoretical claims (stated mathematically wherever possible) that

2　由于察觉到“理论知识”的标准正在走向死胡同，一些人认为，真正的科学可以根据其方法来辨别，尤其是实验方法。它主张，一种理论如果是真正科学的，就必须建立在观察和实验结果的基础上并接受其检验。(许多持这种看法的人认为，必须采取一系列严格规定的步骤)能够通过这种检验的理论常被认为具有卓越的知识论地位或保证，从而代表一种优越的认识方式。最后，在许多人——无论是科学家还是广大公众看来，真正的科学纯粹是通过其内容来定义的，即物理学、化学、生物学、地质学、人类学、心理学等目前所讲授的东西。

3　关于词语含义的这一简短讨论应当提醒我们注意，许多词(尤其是最有趣的词)有多重含义，因使用语境或特定语言共同体的实践而异。“科学”一词的上述每一种含义都是相当多的人所接受的一种约定，不经历一番斗争，他们不大可能放弃自己所偏爱的用法。因此我们只能认为各种含义都是合法的，并试图从使用语境中确定“科学”一词在某一特定场合的含义。

4　那么我们该怎么做？在本书所涵盖的2000年里，欧洲或近东是否有某种东西值得被称为“科学”？毫无疑问！我们现在所谓的科学中肯定有许多内容在当时是存在的。我指的是描述自然的语言，探索或研究自然的方法(包括做实验)，由这些研究作出的事实断言和理论断言(尽可能作数学表述)，以及用什么标准来判别这些断言是正确的或有效的。不仅如此，古代和中世纪由此获得的某些知识与现在公认的真正科学就其实际目的而言完全相同。行星天文学、几何光学、博物学和某些医学分支便是很好的例子。

emerged from such explorations, and criteria for judging the truth or validity of the claims thus made. Moreover, it is clear that pieces of the resulting ancient and medieval knowledge were, for all practical purposes, identical to what all parties would now judge to be genuine science. Planetary astronomy, geometrical optics, field biology or natural history, and certain branches of medicine are excellent examples.

5 This is not to deny significant differences—in motivation, instrumentation, institutional support, methodological preferences, mechanisms for the dissemination of theoretical results, and social function. Despite these differences, I believe that we can comfortably employ the expression "science" or "natural science" in the context of antiquity and the Middle Ages. In so doing, we declare that the ancient and medieval activities that we are investigating are the ancestors of modern scientific disciplines and therefore an integral part of their history. It is like my relationship to my paternal grandfather. The differences between us may outweigh the similarities; but I am his descendant, bearing to some extent both his genetic and his cultural stamp. And both of us may honorably claim the family name.

6 There is a danger that must be avoided. If historians of science were to investigate past practices and beliefs only insofar as those practices and beliefs resemble modern science, the result would be serious distortion. We would not be responding to the past as it existed, but examining it through a modern grid. If we wish to do justice to the historical enterprise, we must take the past for what it was. And that means that we must resist the temptation to scour the past for examples or precursors of modern science. We must respect the way earlier generations approached nature, acknowledging that although it may differ from the modern way, it is

5 这并不是要否认它们在动机、仪器、体制支持、方法偏好、理论成果的传播机制以及社会功能等方面存在着显著差异。尽管如此，我认为仍然可以在古代和中世纪的背景下安心使用“科学”或“自然科学”这一表述。我们由此宣布这些古代和中世纪活动是现代科学学科的前身，因而是其历史不可或缺的一部分。这就像我与我祖父的关系。我们之间的差异可能大于相似之处，但我是他的后代，在一定程度上带有他的遗传印记和文化特质。我可以光明正大地要求与他冠以同一个家庭姓氏。

6 我们必须避免一种危险。如果科学史家仅仅按照与现代科学的相似性来研究过去的做法和信念，将会导致严重歪曲。那样一来，我们不是对过去的实际情况作出反应，而是透过一个现代框架来考察它。要想公正地对待历史，就必须如实地对待过去。这意味着我们必须抵制诱惑，避免到历史中搜寻现代科学的例子或前身。我们必须尊重前人研究自然的方式，承认它虽然可能不同于现代方式，但仍然是有趣的，因为它是我们思想来源的一部分。这是了解我们如何变成现在这样的唯一恰当的方式。于是，历史学家需要一种非常宽泛的“科学”定义，它将允许我们研究其背后的各种做法和信念，并且有助于理解现代科学事业。我们需要

nonetheless of interest because it is part of our intellectual ancestry. This is the only suitable way of understanding how we became what we are. The historian, then, requires a very broad definition of "science"—one that will permit investigation of the vast range of practices and beliefs that lie behind, and help us to understand, the modern scientific enterprise. We need to be broad and inclusive, rather than narrow and exclusive; and we should expect that the farther back we go, the broader we will need to be.

7 I will do my best to heed my own advice, adopting a definition of "science" as broad as that of the historical actors whose intellectual efforts we are attempting to understand. This does not mean, of course, that all distinctions are forbidden. I will distinguish between the craft and theoretical sides of science—a distinction that many ancient and medieval scholars would themselves have insisted upon—and I will focus my attention on the latter. The exclusion of technology and the crafts from this narrative is not meant as a commentary on their importance, but rather as an acknowledgment of the magnitude of the problems confronting the history of technology and its status as a distinct historical specialty having its own skilled practitioners. My concern will be with the beginnings of scientific theories, the methods by which they were formulated, and the uses to which they were put; and that will prove a sufficient challenge.

8 A final word about terminology. Until now, I have consistently employed the word "science" to denote the object of our historical study. The time has come, however, to introduce the alternative expressions "natural philosophy" and "philosophy of nature," which will also appear frequently in this book. These are expressions that ancient and medieval scholars themselves applied to investigations of the natural world that concentrated on questions of material causation, as opposed to mathematical analysis. For the

广泛和包容，而不是狭隘和排他；可以预见，我们往回追溯得越远，就越需要开阔的视野。①

7 我将尽可能地采取一种宽泛的“科学”定义，使之符合我们试图理解的历史人物的思想倾向。当然，这并不意味着抹去一切差别。我将对科学的技艺方面和理论方面加以区分(许多古代和中世纪学者也坚持这一区分)，并且集中在理论方面。② 从叙事中排除技术和技艺并不意味着我对它们的重要性作出了相应评价，而是承认在面对技术史及其地位时还存在许多问题，因为技术史作为一门清楚的历史专业自有其行家里手。我将关注科学理论的起源、表述科学理论的方法以及对科学理论的应用。事实证明，这是一项极大的挑战。

8 关于术语再说一句。到目前为止，我一直用“科学”一词来指我们历史研究的对象，然而现在应该引入替代性的术语“自然哲学”，它在本书中也将频繁出现。古代和中世纪学者在研究自然界时，如果关注的是物质性的因果关系而不是数学分析，就会使用“自然哲学”这个术语，对于数学分析则可用“数学”一词。最后，用来指天文学、光学、气象学、冶金

① David Pingree，“Hellenophilva versus the history of science”很好地指出了这一点。

② 关于古代和中世纪对待技术的态度，见 Elspeth whiteny，*Paradise Restored*。

latter, the term "mathematics" did service. And finally, a vocabulary developed for identifying subdisciplines such as astronomy, optics, meteorology, metallurgy, the science of motion, the science of weights, geography, natural history (including both plants and animals), and medicine. Close attention by the reader to context should make the meaning clear in every case.

PLATO'S World of Forms

9 The death of Socrates in 399 B.C., coming as it did around the turn of the century (not on their calendar, of course, but on ours), has made it a convenient point of demarcation in the history of Greek philosophy. Thus Socrates' predecessors of the sixth and fifth centuries (the philosophers who have occupied us until now in this chapter) are commonly called the "pre-Socratic philosophers." But Socrates' prominence is more than an accident of the calendar, for Socrates represents a shift in emphasis within Greek philosophy, away from the cosmological concerns of the sixth and fifth centuries toward political and ethical matters. Nonetheless, the shift was not so dramatic as to preclude continuing attention to the major problems of pre-Socratic philosophy. We find both the new and the old in the work of Socrates' younger friend and disciple, Plato (fig. 2.4).

10 Plato (427-348/347) was born into a distinguished Athenian family, active in affairs of state; he was undoubtedly a close observer of the political events that led up to Socrates' execution. After Socrates' death, Plato left Athens and visited Italy and Sicily, where he seems to have come into contact with Pythagorean philosophers. In 388 Plato returned to Athens and founded a school of his own, the Academy, where young men could pursue advanced studies. Plato's literary output appears to have consisted almost entirely of dialogues, the majority of which have survived. We will find it necessary to

学、运动科学、重量科学、地理学、博物学(包括植物和动物)和医学等学科分支的一套词汇被发展出来。读者只要细心注意语境，就能明确每一种情况下术语的含义。

柏拉图的理式世界

9 苏格拉底(Socrates)死于公元前399年，正好在世纪之交(当然是按照我们的而不是他们的日历)，他的死成为希腊哲学史上一个方便的分界点。于是，在苏格拉底之前的公元前5、6世纪的前辈们(本章前面讨论的那些哲学家)一般被称为“前苏格拉底哲学家”。但苏格拉底具有突出地位并非只是时间上的偶然，因为苏格拉底代表着希腊哲学的重点转移，从公元前5、6世纪的宇宙论关切转向了政治伦理议题。尽管如此，这种转移并未剧烈到使前苏格拉底的那些重要问题不再受到关注。在苏格拉底的年轻朋友和学生柏拉图(图2.4)的著作中，我们可以发现新旧两种问题。

10 柏拉图(前427—前348/347)生于雅典的一个名门望族。他积极参与城邦事务，无疑亲眼目睹了导致苏格拉底被处死的那些政治事件。苏格拉底死后，柏拉图离开雅典，访问了意大利和西西里，在那里似乎接触到了毕达哥拉斯学派的哲学家。公元前388年，柏拉图回到雅典创建了他自己的学校——学园(Academy)，年轻人可以在那里从事高等研究。柏拉图的作品几乎全是对话录，其中大部分保留了下来。我们将会看到，对柏拉图哲学的考察必须有高度的选择性，我们先从他对基本实在的探究开始。①

① 关于柏拉图的学术研究浩如烟海。我深深地得益于 Vlastos，*Plato's Universe* 以及 Francis M. Cornford 对各种柏拉图对话的翻译加评注。新近的简要介绍见 R. M. Hare，*Plato*；David J. Melling，*Understanding Plato*。

be highly selective in our examination of Plato's philosophy; let us begin with his quest for the underlying reality.

11 In a passage in one of his dialogues, *Republic*, Plato reflected on the relationship between the actual tables constructed by a carpenter and the idea or definition of a table in the carpenter's mind. The carpenter replicates the mental idea as closely as possible in each table he makes, but always imperfectly. No two manufactured tables are alike down to the smallest detail, and limitations in the material (a knot here, a warped board there) ensure that none will fully measure up to the ideal.

12 Now, Plato argued, there is a divine craftsman who bears the same relationship to the cosmos as the carpenter bears to his tables. The divine craftsman (the Demiurge) constructed the cosmos according to an idea or plan, so that the cosmos and everything in it are replicas of eternal ideas or forms—but always imperfect replicas because of limitations inherent in the materials available to the Demiurge. In short, there are two realms: a realm of forms or ideas, containing the perfect form of everything; and the material realm in which these forms or ideas are imperfectly replicated.

13 Plato's notion of two distinct realms will seem strange to many people, and we must therefore stress several points of importance. The forms are incorporeal, intangible, and insensible; they have always existed, sharing the property of eternality with the Demiurge; and they are absolutely changeless. They include the form, the perfect idea, of everything in the material world. One does not speak of their location, since they are incorporeal and therefore not spatial. Although incorporeal and imperceptible by the senses, they objectively exist; indeed, true reality (reality in its fullness) is located only in the world of forms. The sensible, corporeal

图 2.4　柏拉图(公元 1 世纪的复制品)，藏于梵蒂冈博物馆

11　柏拉图在他的对话录《理想国》(*Republic*)中反思了木匠实际制造的桌子与他心灵中关于桌子的理式(idea)或定义之间的关系。木匠在其制作的每一张桌子中都尽可能精确地复制心灵中的理式，但这种复制总是不完美的。任何两张桌子都不可能在最小的细节上一模一样，材料限制(比如节疤、翘曲的木板)使任何一张桌子都不可能完全符合理想的情况。

12　柏拉图指出，有一位神匠，他与宇宙的关系就如同木匠与桌子的关系。这位神匠——巨匠造物主(the Demiurge)——根据一种理式或方案构建了宇宙，因此宇宙万物都是永恒理式或形式的复制品——但是由于神匠所获得材料的内在局限性，它们总是不完美的复制品。简而言之，存在着两个世界：一个是形式或理式的世界，包含着一切事物的完美形式；另一个则是物质的世界，这些形式或理式在其中得到了不完美的复制。

13　许多人可能会觉得柏拉图关于两个迥异世界的观念很奇怪，因此我们必须强调几点。形式是无形的、不可触的、不可感的；它们总是存在着，与巨匠造物主都具有永恒性；它们是绝对不变的。“形式”包括世间万物的形式或完美理式。我们无法谈及它们的位置，因为它们是无形的，从而不在空间之中。虽然无形和不可感，但它们客观存在着；事实上，真

world, by contrast, is imperfect and transitory. It is less real in the sense that the corporeal object is a replica of, and therefore dependent for its existence upon, the form. The form has primary existence, its corporeal replica secondary existence.

14 Plato illustrated this conception of reality in his famous "allegory of the cave," found in book VII of *Republic*. Men are imprisoned within a deep cave, chained so as to be incapable of moving their heads. Behind them is a wall, and beyond that a fire. People walk back and forth behind the wall, holding above it various objects, including statues of humans and animals; the objects cast shadows on the wall that is visible to the prisoners. The prisoners see only the shadows cast by these objects; and, having lived in the cave from childhood, they no longer recall any other reality. They do not suspect that these shadows are but imperfect images of objects that they cannot see; and consequently they mistake the shadows for the real.

15 So it is with all of us, says Plato. We are souls imprisoned in bodies. The shadows of the allegory represent the world of sense experience. The soul, peering out from its prison, is able to perceive only these flickering shadows, and the ignorant claim that this is all there is to reality. However, there do exist the statues and other objects of which the shadows are feeble representations and also the humans and animals of which the statues are imperfect replicas. To gain access to these higher realities, we must escape the bondage of sense experience and climb out of the cave, until we find ourselves able, finally, to gaze on the eternal realities, thereby entering the realm of true knowledge.

16 What are the implications of these views for the concerns of the pre-Socratic philosophers? First, Plato equated his forms with the underlying reality,

正的实在(完全的实在)只存在于形式世界中。相反，可感的有形世界是不完美的和短暂的。物体是形式的复制品，因此它的存在依赖于理式，在这个意义上物体具有更少的实在性。形式的存在是首要的，其有形复制品的存在则是派生的。

14 柏拉图在《理想国》第七卷著名的“洞喻”说中阐明了这种实在观。一些人被囚禁在一个深深的的洞穴中，受锁链所缚，无法转动头部。他们身后有一面墙，墙后面有一团火。有人在墙后走来走去，把各种东西举过墙头，包括人和动物的雕像；这些东西在囚徒们可见的墙壁上投下影子；囚徒们只能看到这些东西的投影，而且他们从小就生活在洞穴中，记不起任何其他东西。他们不会想到这些影子仅仅是他们看不到的物体的不完美影像，因此误把影子当成了真实的东西。

15 柏拉图说，我们所有人也是这样。我们是囚禁在肉体之中的灵魂。“洞喻”中的影子代表感觉经验世界。从其牢笼向外凝视的灵魂只能感知到这些闪烁不定的影子，无知者却宣称这些影子就是实在。然而，确实存在着雕像和其他物体，这些影子只是其微弱的表现；也确实存在着人和动物，那些雕像乃是其不完美的复制品。为了把握这些更高的实在，我们必须摆脱感觉经验的束缚，爬出这个洞穴，直到最终能够看到永恒的实在，从而进入真知的世界。①

16 这些看法对于前苏格拉底哲学家的关切有何含义呢？首先，柏拉图将他所说的形式等同于基本实在，从而将可感的有形世界指为派生的或第二性的存在。其次，通过给变与不变分别指定不同的实在性层次，柏拉图给变与不变都留下了空间：有形世界展现的是不完美和变化，而形式世

① Plato, *Republic*, bk. Ⅶ. 514a-521b.

while assigning derivative or secondary existence to the corporeal world of sensible things. Second, Plato has made room for both change and stability by assigning each to a different level of reality: the corporeal realm is the scene of imperfection and change, while the realm of forms is characterized by eternal, changeless perfection. Both change and stability are therefore genuine; each characterizes something; but changelessness belongs to the forms and thus shares their fuller reality.

17 Third, as we have seen, Plato addressed epistemological questions, placing observation and true knowledge (or understanding) in opposition. Far from leading upward to knowledge or understanding, the senses are chains that tie us down; the route to knowledge is through philosophical reflection. This is explicit in the Phaedo, where Plato maintains the uselessness of the senses for the acquisition of truth and points out that when the soul attempts to employ them it is inevitably deceived.

18 Now the short account of Plato's epistemology frequently ends here; but there are important qualifications that it would be a serious mistake to omit. Plato did not, in fact, dismiss the senses altogether, as Parmenides had done and as the passage from the Phaedo might suggest Plato did. Sense experience, in Plato's view, served various useful functions. First, sense experience may provide wholesome recreation. Second, observation of certain sensible objects (especially those with geometrical properties) may serve to direct the soul toward nobler objects in the realm of forms. Plato used this argument as justification for the pursuit of astronomy. Third, Plato argued (in his theory of reminiscence) that sense experience may actually stir the memory and remind the soul of forms that it knew in a prior existence, thus stimulating a process of recollection that will lead to actual knowledge of the forms.

界则以永恒不变的完美性为特征。因此，变与不变都是真实的，都是事物的特征；但不变属于形式世界，从而有更大的实在性。

17 第三，正如我们所看到的，柏拉图提出了认识论问题，把观察与真知（或理解）对立了起来。感官远没有导向知识或理解，而是限制我们的锁链；通过哲学反思才能获得知识。这种观点显见于《斐多篇》（*Phaedo*）中，柏拉图在其中强调感官对于获得真理没有用处，并指出灵魂在试图运用感官时不可避免会受骗。

18 对柏拉图认识论的简短说明常常在这里结束；但有一些重要的限定，如果遗漏将是一个严重的错误。事实上，柏拉图并不像巴门尼德所做的和《斐多篇》中可能暗示的那样完全摒弃感官。在柏拉图看来，感觉经验有各种有用的功能。首先，感觉经验可以提供有益健康的消遣。其次，对某些可感物体（尤其是那些具有几何属性的物体）的观察可以将灵魂引向形式世界中更高贵的对象；柏拉图用这个论证来为天文学研究辩护。第三，柏拉图（在其回忆说中）主张，感觉经验可以实际唤起回忆，使灵魂回想起它在之前存在时认识的形式，从而激起一种回忆过程，导向对形式的真正认识。

19 Finally, although Plato firmly believed that knowledge of the eternal forms (the highest, and perhaps the only true, form of knowledge) is obtainable only through the exercise of reason, the changeable realm of matter is also an acceptable object of study. Such studies serve the purpose of supplying examples of the operation of reason in the cosmos. If this is what interests us (as it sometimes did Plato), the best method of exploring it is surely to observe it. The legitimacy and utility of sense experience are clearly implied in *Republic*, where Plato acknowledged that a prisoner emerging from the cave first employs his sense of sight to apprehend living creatures, the stars, and finally the most noble of visible (material) things, the sun. But if he aspires to apprehend "the essential reality," he must proceed "through the discourse of reason unaided by any of the senses." Both reason and sense are thus instruments worth having; which one we employ on a particular occasion will depend on the object of study.

20 There is another way of expressing all of this, which may shed light on Plato's achievement. When Plato assigned reality to the forms, he was, in fact, identifying reality with the properties that classes of things have in common. The bearer of true reality is not (for example) this dog with the droopy left ear or that one with the menacing bark, but the idealized form of a dog shared (imperfectly, to be sure) by every individual dog—those characteristics by virtue of which we are able to classify all of them as dogs. Therefore, to gain true knowledge, we must set aside all characteristics peculiar to things as individuals and seek the shared characteristics that define them into classes. Now stated in this modest fashion, Plato's view has a distinctly modern ring. Idealization is a prominent feature of a great deal of modern science; we develop models or laws that overlook the incidental in favor of the essential. However, Plato went beyond this, maintaining not merely that true reality is to be found in the common properties of classes of

19 最后，虽然柏拉图坚信关于永恒形式的知识(最高的也许是唯一真实的知识)只有通过运用理性才能获得，但可变的物质世界也是一种可接受的研究对象。这些研究是为了提供理性在宇宙中运作的范例。如果这是使我们感兴趣的东西(正如它有时使柏拉图感兴趣那样)，那么研究它的最好方法肯定是观察。《理想国》显然蕴含着感觉经验的正当性和用处，柏拉图在其中承认，从洞穴中走出来的囚徒先是用视觉来把握生物、星辰，最后理解最高贵的(物质性的)可见物——太阳，但如果渴望理解"本质性的实在"，他就必须"凭借不受任何感官帮助的理性"。因此，理性和感觉都是值得拥有的工具，在特定场合运用哪一种工具将取决于研究对象。①

20 表达所有这些还有另一种方式，也许有助于说明柏拉图的成就。当柏拉图为形式指定实在性时，他实际上是把实在性等同于一类事物所共有的属性。真正实在性的承载者既不是(例如)这只左耳下垂的狗，也不是那只狂吠的狗，而是每只个别的狗(当然是不完美地)所共有的一只狗的理想化形式——正是凭借那些特征，我们才能把它们全都称为"狗"。因此，要想获得真正的知识，我们必须抛开个体事物所特有的特征，而去寻求使它们成为一类的那些共有的特征。以这种温和的方式来表述，柏拉图的观点便有了明显的现代意味：理想化是许多现代科学的一个突出特征；我们发展出了注重本质而忽略偶性的模型或定律。然而，柏拉图比这走得更远，他不仅认为真正的实在可见于一类事物的共同属性中，而且强调这些共同属性(理式或形式)有客观、独立和在先的存在性。

① Lloyd, *Early Greek Science*, pp. 68-72; Plato, *Phaedo*, 65b; Plato, *Republic*, bk. Ⅶ. 532, trans. Francis M. Cornford, p. 252.

things, but also that this common property (the idea or form) has objective, independent, and indeed prior existence.

Chapter 3 Aristotle's Philosophy of Nature

Life and Works

21 Aristotle (fig. 3.1) was born in 384 B.C. in the northern Greek town of Stagira, into a privileged family. His father was personal physician to the Macedonian king, Amyntas II (grandfather of Alexander the Great). Aristotle had the advantage of an exceptional education: at age seventeen, he was sent to Athens to study with Plato. He remained in Athens as a member of Plato's Academy for twenty years, until Plato's death about 347. Aristotle then spent several years in travel and study, crossing the Aegean Sea to Asia Minor and its coastal islands. During this period he undertook biological studies, and he encountered Theophrastus (from the island of Lesbos), who was to become his pupil and lifetime colleague. He returned to Macedonia in 342 to become the tutor of the young Alexander (later "the Great"). In 335, when Athens fell under Macedonian rule, Aristotle returned to the city and began to teach in the Lyceum, a public garden frequented by teachers. He remained there, establishing an informal school, until shortly before his death in 322.

第三章 亚里士多德的自然哲学

生平和著作

21 公元前384年，亚里士多德(图3.1)出生在希腊北部斯塔吉拉(Stagira)镇的一个特权家族。他的父亲是马其顿国王阿敏塔斯二世(Amyntas II，亚历山大大帝的祖父)的御医。亚里士多德有机会得到特殊教育：17岁时，他被送往雅典，师从柏拉图。作为柏拉图学园的成员，他在雅典待了20年，直到公元前347年左右柏拉图逝世。此后数年，亚里士多德四处游历和研究，穿过爱琴海来到小亚细亚及其沿岸诸岛。在此期间，他开

图3.1　亚里士多德，藏于罗马国家博物馆

展了生物学研究，并遇到了后来成为其学生和终生共事的(来自莱斯博斯[Lesbos]岛的)特奥弗拉斯特。公元前342年，亚里士多德回到马其顿，成为年轻的亚历山大(后来的“亚历山大大帝”)的私人教师。公元前335年，雅典陷于马其顿统治，亚里士多德回到雅典，开始在吕克昂(Lyceum)——老师们经常光顾的一个公共花园教学。直到公元前322年逝世前不久，亚里士多德一直生活在那里，并且建立了一个非正式的学派。①

① 关于吕克昂的更多内容，见[美]戴维·林德伯格著，张卜天译，湖南科学技术出版社2016年出版的《西方科学的起源》的第四章。关于亚里士多德与亚历山大大帝的关系，见 Peter Green, *Alexander of Macedon*, pp. 53-62。

22 In the course of his long career as student and teacher, Aristotle systematically and comprehensively addressed the major philosophical issues of his day. He is credited with more than 150 treatises, approximately 30 of which have come down to us. The surviving works appear to consist mainly of lecture notes or unfinished treatises not intended for wide circulation; whatever their exact origin, they were obviously directed to other philosophers, including advanced students. In modern translation, they occupy well over a foot of bookshelf, and they present a philosophical system overwhelming in power and scope. It is out of the question for us to survey the whole of Aristotle's philosophy, and we must be content with examining the fundamentals of his philosophy of nature—beginning with his response to positions taken by the pre-Socratics and Plato.

Metaphysics and Epistemology

23 Through his long association with Plato, Aristotle had, of course, become thoroughly versed in Plato's theory of forms. Plato had drastically diminished (without totally rejecting) the reality of the material world observed by the senses. Reality in its perfect fullness, Plato argued, is found only in the eternal forms, which are dependent on nothing else for their existence. The objects that make up the sensible world, by contrast, derive their characteristics and their very being from the forms; it follows that sensible objects exist only derivatively or dependently.

24 Aristotle refused to accept this diminished, dependent status that Plato assigned to sensible objects. They must exist fully and independently, for in Aristotle's view they were what make up the real world. Moreover, the traits that give an individual object its character do not, Aristotle argued, have a prior and separate existence in a world of forms, but belong to the object itself. There is no perfect form of a dog, for example, existing independently

22 在长期的学习和教学生涯中，亚里士多德系统而全面地提出了他那个时代重要的哲学问题。据说他写了150多部论著，其中大约有30部流传至今。留存下来的著作似乎主要是一些讲课笔记或未打算广为流传的未完成的论著；无论确切来源如何，这些著作显然是面向包括高级学生在内的其他哲学家的。其现代译本可以摆满一层书架，代表着一个极为广泛和强大的哲学体系。这里我们不可能考察亚里士多德的全部哲学，而只能考察其自然哲学的基本原理——从他对前苏格拉底哲学家和柏拉图立场的反应开始。①

形而上学和认识论

23 与柏拉图长期交往的亚里士多德当然非常精通柏拉图的形式理论。柏拉图强烈贬低(但没有完全拒斥)感官所觉察到的物质世界的实在性。柏拉图主张，完满的实在只见于不依赖其他任何东西而存在的永恒形式。而构成可感世界的物体则由形式获得了它们的特征甚至是存在本身；因此，可感物体的存在仅仅是派生的或从属的。

24 亚里士多德拒绝接受柏拉图为可感物体指定的这种从属地位。可感物体必定可以独立存在，因为在亚里士多德看来，它们就是构成实际世界的东西。此外，亚里士多德指出，使个体物体具有自身特点的那些特性并非在形式世界有在先的分离存在，而是属于物体本身。例如，并没有狗的完美形式独立存在于形式世界，并且被不完美地复制于个体的狗从而

① 关于亚里士多德有大量优秀的介绍性文献，特别参见 G. E. R. Lloyd，*Aristotle：The Growth and Structure of His Thought*；Jonathan Barnes，*Aristotle*；Abraham Edel，*Aristotle and His Philosophy*。

in the world of forms and replicated imperfectly in individual dogs, imparting to them their attributes. For Aristotle, there were just individual dogs. These dogs certainly shared a set of attributes—for otherwise we would not be entitled to call them "dogs"—but these attributes exist in, and belong to, individual dogs.

25 Perhaps this way of viewing the world has a familiar ring. Making individual sensible objects the primary realities ("substances", Aristotle called them) will seem like good common sense to most readers of this book, and probably struck Aristotle's contemporaries the same way. But if it makes good common sense, can it also be good philosophy? That is, can it deal successfully, or at least plausibly, with the difficult philosophical issues raised by the pre-Socratics and Plato—the nature of the fundamental reality, epistemological concerns, and the problem of change and stability? Let us take up these problems one by one.

26 The decision to locate reality in sensible, corporeal objects does not yet tell us very much about reality—only that we should look for it in the sensible world. Already in Aristotle's day, any philosopher would demand to know more: one thing he would demand to know was whether the corporeal materials of daily experience (wood, water, air, stone, metal, flesh, etc.) are themselves the fundamental, irreducible constituents of things, or whether they are composites of still more fundamental stuff. Aristotle addressed this question by drawing a distinction between properties and their subjects. He maintained (as most of us would) that a property has to be the property of something; we call that something its "subject." To be a property is to belong to a subject; properties cannot exist independently.

赋予其属性。对亚里士多德来说，存在的仅仅是个体的狗。这些狗肯定有一些共性，否则我们便无权称之为“狗”，但这些属性存在于个体的狗之中并属于个体的狗。

25 对于这种看待世界的方式，我们也许并不陌生，在本书的大多数读者看来，认为可感的个别物体是第一性的实在（亚里士多德所谓的“实体”）似乎是常识，亚里士多德的同时代人或许也会产生同样的印象。但如果它是常识，它是否也是好的哲学？也就是说，它能否成功地或至少是看似合理地解决前苏格拉底哲学家和柏拉图所提出的那些困难的哲学问题——基本实在的本性、认识论关切以及变与不变的问题？让我们逐一讨论这些问题。①

26 决定把实在定位于可感的有形物体，并没有就实在说出很多东西——它仅仅告诉我们应当在可感世界中寻找实在。在亚里士多德的时代，哲学家已经要求知道得更多：哲学家会要求知道，日常经验中的有形物质（木头、水、气、石头、金属、肉等）本身是事物不可还原的基本组分，还是由更基本的东西复合而成。亚里士多德通过区分属性及其基体（subject）解决了这个问题。他（就像我们大多数人会做的那样）坚持认为，属性必定是某种东西的属性；我们把那种东西称为属性的“基体”。身为一种属性就要属于一个基体；属性不能独立存在。

① Barnes, *Aristotle*, pp. 32-51; Edel, *Aristotle*, chaps. 3-4; Lloyd, *Aristotle*, chap. 3。

27 Individual corporeal objects, then, have both properties (color, weight, texture, and the like) and something other than properties to serve as their subject. These two roles are played by "form" and "matter," respectively. Corporeal objects are "composites" of form and matter—form consisting of the properties that make the thing what it is, matter serving as the subject or substratum for the form. A white rock, for example, is white, hard, heavy, and so forth, by virtue of its form; but matter must also be present, to serve as subject for the form, and this matter brings no properties of its own to its union with form. (Aristotle's doctrine will be further discussed in chap. 12, below, in connection with medieval attempts to clarify and extend it.)

28 We can never, in actuality, separate form and matter; they are presented to us only as a unitary composite. If they were separable, we should be able to put the properties (no longer the properties of anything) in one pile, the matter (absolutely propertyless) in another—an obvious impossibility. But if form and matter can never be separated, is it not meaningless to speak of them as the real constituents of things? Isn't this a purely logical distinction, existing in our minds, but not in the external world? Surely not for Aristotle, and perhaps not for us; most of us would think twice before denying the real existence of cold or red, although we can never collect a bucket of either one. In short, Aristotle once again surprises us by using commonsense notions to build a persuasive philosophical edifice.

29 Aristotle's claim that the primary realities are concrete individuals surely has epistemological implications, since true knowledge must be knowledge of truly real things. By this criterion, Plato's attention was naturally directed toward the eternal forms, knowable through reason or philosophical reflection. Aristotle's metaphysics of concrete individuals, by contrast, directed his quest for knowledge toward the material world of individuals, of nature, and of change—a world encountered through the senses.

27 于是，个体的有形物体既有属性(颜色、重量、质地等)，也有某种东西来充当属性的基体。这两种角色分别由“形式”和“质料”来承担。有形物体是形式与质料的“复合物”——形式是由使物体是其所是的那些属性构成的，而质料则充当着形式的基体或基底。例如，一块白色的岩石因其形式而是白的、硬的、重的等；但质料也必须存在，以充当形式的基体，而且在与形式的结合中，质料并未带入自身的属性。①（我们将在第十二章②联系中世纪学者的澄清和拓展工作对亚里士多德学说作进一步讨论。）

28 在现实中，我们永远也无法把形式与质料分离，它们只能呈现为一个统一的复合体。倘若可以分离，我们就可以把属性(不再是某种东西的属性)放在一堆，而把(完全没有属性的)质料放在另一堆——这显然是不可能的。但如果形式与质料永远不可分离，那么说它们是事物的实际组成部分不就毫无意义了吗？这难道不是存在于我们心灵之中而非外部世界之中的一种纯逻辑区分吗？对亚里士多德来说肯定并非如此，对我们来说可能也不是；在否认冷或红的实际存在性之前，我们大多数人会三思而行，尽管我们永远也收集不到一桶冷或红。简而言之，亚里士多德根据常识观念构建了一幢哲学大厦，其说服力着实令我们惊讶。

29 亚里士多德断言第一性的实在是具体的个别事物，这肯定有认识论的意涵，因为真正的知识一定是关于真实存在的事物的知识。根据这种标准，柏拉图的注意力自然转向了永恒的形式，它们可以通过理性或哲学反思而被认识。而亚里士多德关于具体个体的形而上学则把他的知识探求引向了个体、自然和变化的物质世界——一个由感官感知的世界。

① 亚里士多德这一学说的专业名称是“形式质料说”(hylomorphism)——来自“质料”(*hyle*)和“形式”(*morphe*)。

② 是[美]戴维·林德伯格著，张卜天译，湖南科学技术出版社 2016 年出版的《西方科学的起源》的第十二章，非本导引第十二章。

30 Aristotle's epistemology is complex and sophisticated. It must suffice here to indicate that the process of acquiring knowledge begins with sense experience. From repeated sense experience follows memory; and from memory, by a process of "intuition" or insight, the experienced investigator is able to discern the universal features of things. By the repeated observation of dogs, for example, an experienced dog breeder comes to know what a dog really is; that is, he comes to understand the form or definition of a dog, the crucial traits without which an animal cannot be a dog. Note that Aristotle, no less than Plato, was determined to grasp the universal traits or properties of things; but, unlike his teacher, Aristotle argued that one must start with the individual material thing. Once we grasp the universal properties or definition, we can put it to use as the premise of deductive demonstrations.

31 Knowledge is thus gained by a process that begins with experience (a term broad enough, in some contexts, to include common opinion or the reports of distant observers). In that sense knowledge is empirical; nothing can be known apart from such experience. But what we learn by this "inductive" process does not acquire the status of true knowledge until put into deductive form; the end product is a deductive demonstration (nicely illustrated in a Euclidean proof) beginning from universal definitions as premises. Although Aristotle discussed both the inductive and deductive phases (the latter far more than the former) in the acquisition of knowledge, he stopped considerably short of later methodologists, especially in the analysis of induction.

32 This is the theory of knowledge outlined by Aristotle in the abstract. Is it also the method actually employed in Aristotle's own scientific investigations? Probably not—with perhaps an occasional exception. Like modern scientists, Aristotle did not proceed by following a methodological recipe book, but

30 亚里士多德的认识论复杂而缜密。这里只需指出，获得知识的过程始于感觉经验。由重复的感觉经验产生了记忆；通过一种“直觉”或洞察过程，有经验的研究者可以由记忆识别出事物的普遍特征。例如，通过对狗的反复观察，有经验的养狗人逐渐认识到狗到底是什么，即他逐渐理解了狗的形式或定义，也就是将一种动物称其为“狗”的那些关键特性。请注意，亚里士多德和柏拉图一样决心把握事物的普遍特性或属性；但与柏拉图不同，亚里士多德认为必须从个体的物质事物开始。一旦我们把握了普遍的属性或定义，我们就可以把它用作演绎证明的前提。①

31 因此，获得知识的过程始于经验(在某些语境下，这个词宽泛到足以包括常识或传闻)。在这种意义上，知识是经验的，离开这些经验我们什么也无法知道。但我们通过这一“归纳”过程所了解到的东西只有以演绎形式表达出来，才能成为真正的知识；最终结果是以普遍定义为前提进行的演绎证明(它在欧几里得几何学的证明中得到了很好的展示)。亚里士多德虽然讨论了知识获取过程中的归纳阶段和演绎阶段(后者远多于前者)，但尤其在对归纳的分析方面还远远不及后来的方法论家。

32 抽象地说，这就是亚里士多德所概述的知识理论。它是否也是亚里士多德本人在科学研究中实际运用的方法呢？可能不是——不过也许会有例外。和现代科学家一样，亚里士多德并未遵循一部方法论的指导手册行事，而是依照粗糙的现成方法，完成那些业已被实践证明的常用程序。有人曾把科学定义为“尽你所能，百无禁忌”；当亚里士多德进行(比如说)广泛的生物学研究时，他无疑正是这样做的。在思考知识的本性和

① 关于亚里士多德的认识论，见 Edel，*Aristotle*，chap. 12-15；Lloyd，*Aristotle*，chap. 6；Jonathan Lear，*Aristotle*：*The Desire to Understand*，chap. 4；Marjorie Grene，*A Portrait of Aristotle*，chap. 3。

rather by rough and ready methods, familiar procedures that had proved themselves in practice. Somebody has defined science as "doing your damnedest, no holds barred"; when it came (for example) to his extensive biological researches, this is exactly what Aristotle did. It is not a surprise, and certainly no character defect, that Aristotle should, in the course of thinking about the nature and the foundations of knowledge, formulate a theoretical scheme (an epistemology) not perfectly consistent with his own scientific practice.

Cosmology

33 Aristotle not only devised methods and principles by which to investigate and understand the world: form and matter, nature, potentiality and actuality, and the four causes. In the process, he also developed detailed and influential theories regarding an enormous range of natural phenomena, from the heavens above to the earth and its inhabitants below.

34 Let us start with the question of origins. Aristotle adamantly denied the possibility of a beginning, insisting that the universe must be eternal. The alternative—that the universe came into being at some point in time—he regarded as unthinkable, violating (among other things) Parmenidean strictures about something coming from nothing. Aristotle's position on this question would prove troublesome for medieval Christian Aristotelians.

35 Aristotle considered this eternal universe to be a great sphere, divided into an upper and a lower region by the spherical shell in which the moon is situated. Above the moon is the celestial region; below is the terrestrial region; the moon, spatially intermediate, is also of intermediate nature. The terrestrial or sublunar region is characterized by birth, death, and transient change of all kinds; the celestial or supralunar region, by contrast, is a region of eternally

基础时，亚里士多德提出了一种与自己的科学实践并不完全一致的理论体系(一种认识论)，这并不让人意外，也肯定不是性格缺陷所致。①

宇宙论

33 亚里士多德不仅设计了研究和理解世界的方法和原则：形式与质料、本性、潜能与现实、四因，而且在此过程中还发展出了一些详细而有影响的关于大量自然现象的理论，上至天空，下至大地及其居住者。②

34 让我们从起源问题开始。亚里士多德坚决否认开端的可能性，坚称宇宙是永恒的。他认为另一种看法是不可思议的，即宇宙在某个时间点产生，比如这违反了巴门尼德对无中生有的限制。事实证明，亚里士多德对这个问题的看法将使中世纪的基督教评注者们感到非常棘手。

35 亚里士多德认为这个永恒的宇宙是一个大球，月球所处的球壳将其分成上下两个区域。月亮以上是天界，以下是地界；在空间上居间的月球也具有居间的本性。地界或月下区的特征是生、死和各种短暂变化，而天界或月上区则是永恒不变的循环区域。这种图示似乎显然源于观察；在《论天》(*On the Heavens*)中，亚里士多德指出："在整个历史上，就我们的记录所及，无论是整个最外层天还是它的任何一个固有部分都没有

---

① 关于这一主题，见 Jonathan Barnes，"Aristotle's Theory of Demonstration"；G. E. R. Lloyd，*Magic*，*Reason*，*and Experience*，pp. 200-220。

② 特别参见 Friedrich Solmsen，*Aristotle's System of the Physical World*；Lloyd，*Aristotle*，chaps，7-8。

unchanging cycles. That this scheme had its origin in observation would seem clear enough; in his *On the Heavens*, Aristotle noted that "in the whole range of time past, so far as our inherited records reach, no change appears to have taken place either in the whole scheme of the outermost heaven or in any of its proper parts." If in the heavens we observe eternally unvarying circular motion, he continued, we can infer that the heavens are not made of the terrestrial elements, the nature of which (observation reveals) is to rise or fall in transient rectilinear motions. The heavens must consist of an incorruptible fifth element (there are four terrestrial elements): the quintessence (literally, the fifth essence) or aether. The celestial region is completely filled with this quintessence (no void space) and divided, as we shall see, into concentric spherical shells bearing the planets. It had, for Aristotle, a superior, quasi-divine status.

36 The sublunar region is the scene of generation, corruption, and impermanence. Aristotle, like his predecessors, inquired into the basic element or elements to which the multitude of substances found in the terrestrial region can be reduced. He accepted the four elements originally proposed by Empedocles and subsequently adopted by Plato—earth, water, air, and fire. He agreed with Plato that these elements are in fact reducible to something even more fundamental; but he did not share Plato's mathematical inclination and therefore refused to accept Plato's regular solids and their constituent triangles. Instead, he expressed his own commitment to the reality of the world of sense experience by choosing sensible qualities as the ultimate building blocks. Two pairs of qualities are crucial: hot-cold and wet-dry. These combine in four pairs, each of which yields one of the elements (see fig. 3.2). Notice the use made once again of contraries. There is nothing to forbid any of the four qualities being replaced by its contrary, as the result of outside influence. If water is heated, so that the cold of water yields to hot,

发生过任何变化。”①他又说，如果我们观察到天上有永恒不变的圆周运动，我们就能推断出，天不是由地界元素构成的，(观察表明)地界元素的本性是短暂地直线上升和下降。天必定是由不朽的第五元素(地界元素有四种)或以太构成的。天界完全被这种第五元素所充满(没有虚空)，且被分成携带行星的若干同心球壳，就像我们后面将要看到的那样。在亚里士多德看来，天具有一种更高的、类似于神的地位。②

36 月下区则是生灭无常的舞台。和前人一样，亚里士多德也在探究构成月下区众多实体的基本元素。他接受了最初由恩培多克勒提出，后被柏拉图采纳的四元素：土、水、火、气。他和柏拉图一样认为这些元素实际上可以被还原成某种更为基本的东西，但他没有柏拉图那种数学倾向，因此拒绝接受柏拉图的正多面体及其三角形组分。相反，他选择可感性质作为最终的建筑材料，从而表明自己相信感觉世界的实在性。有两对性质是至关重要的：热—冷和干—湿。它们可以结合成四对，每一对都产生出一种元素(见图 3.2)。请注意，亚里士多德又一次使用了对立面。

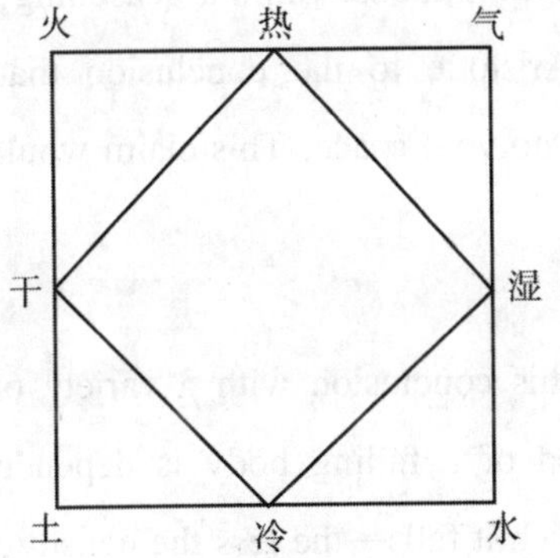

冷和干=土
冷和湿=水
热和湿=气
热和干=火

图 3.2　亚里士多德关于元素与性质对立面的正方形示意图
(这幅图的一个中世纪〈公元 9 世纪〉版本见
John E. Murdoch, *Album of Science: Antiquity and Middle Ages*, p. 352)

① *On the Heavens*, I. 4. 270b13-16, quoted from *The complete works of Aristotle*, ed. Barnes, 1: 451.

② Lloyd, *Aristotle*, chap. 7.

the water is transformed into air. Such a process easily explains changes of state (from solid to liquid to vapor, and conversely), but also more general transmutation of one substance into another. On such a theory as this, alchemists could easily build.

37 The various substances that make up the cosmos totally fill it, leaving no empty space. To appreciate Aristotle's view, we must lay aside our almost automatic inclination to think atomistically; we must conceive material things not as aggregates of tiny particles but as continuous wholes. If it is obvious that, say, a loaf of bread is composed of crumbs separated by small spaces, there is no reason not to suppose that those spaces are filled by some finer substance, such as air or water. And there is certainly no simple way of demonstrating, nor indeed any obvious reason for believing, that water and air are anything but continuous. Similar reasoning, applied to the whole of the universe, led Aristotle to the conclusion that the universe is full, *a plenum*, containing no void space. This claim would be attacked by medieval scholars.

38 Aristotle defended this conclusion with a variety of arguments, such as the following. The speed of a falling body is dependent on the density of the medium through which it falls—the less the density, the swifter the motion of the falling body. It follows that in avoid space (density zero), there is nothing to slow the descent of the body, from which we would be forced to conclude that the body would fall with infinite speed—a nonsensical notion, since it implies that the body could be at two places at the same time. Critics have frequently noted that this argument can just as well be taken to prove that the absence of resistance does not entail infinite speed as to prove that void does not exist. The point is, of course, well taken. However, we need to understand that Aristotle's denial of the void did not rest on this single

没有任何东西能够阻止这四种性质中的任何一种因外界影响而被其对立面所取代。如果水被加热，即水的冷屈服于热，则水就变成了气。这一过程很容易解释物态变化(从固体到液体到气体，以及反之)，以及从一种东西到另一种东西的更一般转变。炼金术士很容易以这样一种理论为基础。①

37 宇宙万物完全充满了宇宙，没有留下任何空的空间。要想理解亚里士多德的观点，就必须抛开那种近乎无意识的原子论倾向；我们绝不能把物质性的东西设想成微粒的聚集，而应想象成连续的整体。如果说组成面包的显然是由细小空间隔开的面包屑，那么就没有理由不推测这些空间被某种更精细的东西如气和水所充满。肯定没有简单的证明方式，事实上也没有任何显然的理由使我们相信水和气绝不是连续的。亚里士多德将类似推理应用于整个宇宙，得出结论说：宇宙是充满的，是一种实满(*plenum*)，不包含任何虚空。这种断言将会遭到中世纪学者的抨击。

38 亚里士多德用各种论证来捍卫这一结论，比如下面这个论证：落体速度依赖于它所通过介质的密度——介质密度越小，落体的运动就越快。因此在虚空中(密度为零)，没有任何东西可以减慢物体的下落，由此我们不得不得出结论，物体将以无限大的速度下落——这是荒谬的，因为它意味着物体可以在同一时间处于两个位置。批评者常常指出，这一论证既可以证明虚空不存在，也可以证明无阻力并不必然带来无穷大的速度。这当然是有道理的。但我们要知道，亚里士多德对虚空的否认并不依赖于这一条推理。事实上，这只是与原子论者所作的长期斗争的一小部分，在这场斗争中，亚里士多德用各种论证反驳了虚空(或空的空间)

① Ibid，chap. 8。关于炼金术，见[美]戴维·林德伯格著，张卜天译，湖南科学技术出版社 2016 年出版的《西方科学的起源》的第十二章。

piece of reasoning. In fact, this was but one small part of a lengthy campaign against the atomists, in which Aristotle battled the notion of void space (or void place) with a variety of arguments, some more and some less persuasive.

39 In addition to being hot or cold and wet or dry, each of the elements is also heavy or light. Earth and water are heavy, but earth is the heavier of the two. Air and fire are light, fire being the lighter of the two. In assigning levity to two of the elements, Aristotle did not mean (as we might, if we were making the claim) simply that they are less heavy, but that they are light in an absolute sense; levity is not a weaker version of gravity, but its contrary. Because earth and water are heavy, it is their nature to descend toward the center of the universe; because air and fire are light, it is their nature to ascend toward the periphery (that is, the periphery of the terrestrial region, the spherical shell that contains the moon). If there were no hindrances, therefore, earth and water would collect at the center; because of its greater heaviness, earth would achieve a lower position, forming a sphere at the very center of the universe; water would collect in a concentric spherical shell just outside it. Air and fire naturally ascend, but fire, owing to its greater levity, occupies the outermost region, with air as a concentric sphere just inside it. In the ideal case (in which there are no mixed bodies and nothing prevents the natures of the four elements from fulfilling themselves), the elements would thus form a set of concentric spheres: fire on the outside, followed by air and water, and finally earth at the center (see fig. 3.3). But in reality, the world is composed largely of mixed bodies, one always interfering with another, and the ideal is never attained. Nonetheless, the ideal arrangement defines the natural place of each of the elements; the natural place of earth is at the center of the universe, of fire just inside the sphere of the moon, and so forth.

概念，说服力各有不同。①

39 除了热冷、湿干，每一种元素还是轻的或重的。土和水重，但土又比水更重。气和火轻，但火比气更轻。亚里士多德把轻归于火和气时，并不只是说它们不够重，而是说它们在一种绝对意义上是轻的；轻并非较弱的重，而是重的对立面。因为土和水重，所以向宇宙中心下落是其本性；气和火轻，所以向外围(即地界的外围，包含月球的球壳)上升是其本性。因此，如果没有阻碍，土和水将在宇宙中心聚集；土由于更重，将会到达更低的位置，在宇宙中心形成一个球体；水则聚集成土球外部的同心球壳。气和火自然上升，但火因为更轻，将会占据最外层区域，气则聚集成火内部的同心球壳。理想情况下(元素没有混合，也没有任何东西阻碍四种元素的本性实现自己)，这些元素将形成一组同心球：火在最外面，然后是气和水，土则位于中心(图 3.3)。但在现实中，世界主要由混合物组成，元素之间相互干扰，理想状况从未实现。不过，理想排列决定了每一种元素的自然位置；土的自然位置在宇宙中心，火的自然位置恰在月亮天球之内，等等。②

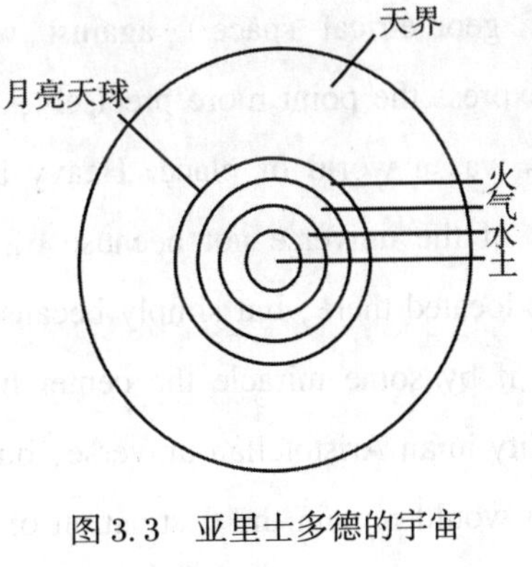

图 3.3　亚里士多德的宇宙

① 关于亚里士多德对虚空的论述，见 Solmsen，*Aristotle's system of the Physical World*，pp. 135-143；David Furley，*Cosmic Problems*，pp. 77-90.

② Furley，*Cosmic Problems*. chaps. 12-13.

40 It must be emphasized that the arrangement of the elements is spherical. Earth collects at the center to form the earth, and it too is spherical. Aristotle defended this belief with a variety of arguments. Arguing from his natural philosophy, he pointed out that since the natural tendency of earth is to move toward the center of the universe, it must arrange itself symmetrically about that point. But he also called attention to observational evidence, including the circular shadow cast by the earth during a lunar eclipse and the fact that north-south motion by an observer on the surface of the earth alters the apparent position of the stars. Aristotle even reported an estimate by mathematicians of the earth's circumference (400,000 stades = about 45,000 miles, roughly 1.8 times the modern value). The sphericity of the earth, thus defended by Aristotle, would never be forgotten or seriously questioned. The widespread myth that medieval people believed in a flat earth is of modern origin.

41 Finally, we must note one of the implications of this cosmology, namely that space, instead of being a neutral, homogeneous backdrop (analogous to our modern notion of geometrical space) against which events occur, has properties. Or to express the point more precisely, ours is a world of space, whereas Aristotle's was a world of place. Heavy bodies move toward their place at the center of the universe not because of a tendency to unite with other heavy bodies located there, but simply because it is their nature to seek that central place; if by some miracle the center happened to be vacant (a physical impossibility in an Aristotelian universe, but an interesting imaginary state of affairs), it would remain the destination of every heavy body.

40 必须强调的是，元素的排列是球形的。土聚集在中心形成地球，地球也是球形的。亚里士多德用各种论证来捍卫这种信念。他根据其自然哲学指出，既然土的自然倾向是移向宇宙的中心，那么它必定会围绕宇宙中心对称排列。他还注意到观察证据，比如月食期间地球所投的圆形阴影，以及观察者在大地表面南北运动会改变星体的视位置。亚里士多德甚至记述了数学家们估计的地球周长(400 000 斯塔德[stades]=大约45 000英里，约为现代值的 1.8 倍)。亚里士多德为大地球形的主张所作的辩护从未被遗忘，也从未遭到严重质疑。那个广为流传的神话，即中世纪的人认为地球是平的，是现代人杜撰的。①

41 最后，我们必须注意这种宇宙论的一个意涵，即空间并非事件发生的一个中性的、同质的背景(类似于我们现代的几何空间概念)，而是具有属性。或者更确切地说，我们的世界是一个空间世界，而亚里士多德的世界则是一个位置世界。重物之所以会移向它在宇宙中心的位置，不是因为它倾向于与位于那里的其他重物相结合，而仅仅是因为寻求那个中心位置是其本性；即使由于某种奇迹，那个中心碰巧未被占用(在亚里士多德的宇宙中这在物理上是不可能的，而仅仅是一种有趣的想象事态)，它仍将是每一个重物的目的地。②

① 亚里士多德在《论天》Ⅱ.13 中讨论了地球的形状。另见 D. R. Dicks, *Early Greek Astronomy to Aristotle*, pp. 196-198。《西方科学的起源》第七章中讨论了那个神话，即认为古代和中世纪的人相信地球是平的。

② Waterlow, *Nature*, *Change*, *and Agency in Aristotle's Physics*, pp. 103-104.

《自然哲学之数学原理》导引

艾萨克·牛顿爵士(Sir Isaac Newton)是我们熟知的英国数学家、科学家和哲学家。1643 年 1 月 4 日，他出生于英格兰林肯郡乡下的一个小村落的伍尔索普(Woolsthorpe)庄园。在牛顿出生之时，英格兰并没有采用教皇的最新历法，因此他的生日被记载为 1642 年的圣诞节。牛顿和苹果之间的故事广为流传。他的万有引力定律在人类历史上第一次把“天界”的运动和“地界”的运动统一起来，改变了人们一千多年来在宗教统治下对物质世界的看法。法国文学家、哲学家伏尔泰(Voltaire)说牛顿是最伟大的人，因为“他用真理的力量统治我们的头脑，而不是用武力奴役我们”。不仅如此，牛顿和莱布尼茨几乎同时独立发展出了微积分(牛顿称之为“流数术”)，为物理学的飞速发展扫清了数学障碍；他提出了二项式定理；奠定了变分法基础；对光色散的认知，使光谱学的建立成为可能。莱布尼茨说：“从世界的开始直到牛顿生活的时代为止，对数学发展的贡献绝大部分是牛顿作出的”。牛顿的一生著作颇丰，如《流数术》(1671 年)、《物体在轨道中之运动》(*De Motu Corporum in Gyrum*)(1684 年)、《自然哲学之数学原理》(1687 年)、《光学》(*Opticks*)(1704 年)、《广义算术》

(*Arithmetica Universalis*)(1707 年)和《两处著名圣经讹误的历史变迁》(*An Historical Account of Two Notable Corruptions of Scripture*)(1754 年),等等。

因为牛顿对人类的贡献如此之大,故而当他 85 岁高龄离去时,和很多著名的英国人一样被葬于伦敦威斯敏斯特教堂。英国诗人亚历山大·波普(Alexander Pope)为他写的墓志铭是这样的:Nature, and Nature's Laws lay hid in night: God said, let Newton be! And all was light。我们译为:天地初开,万物处于混沌;上帝说:降生吧,牛顿!于是,世界初见光明。但牛顿自己似乎很谦虚,他说:If I have seen further, it is by standing on the shoulders of giants …(如果我看得比别人更远些,那是因为我站在巨人的肩膀上)。我很好奇,"巨人"是谁?是一个人吗?抑或一群人?正在读书的你,脑海中是否浮现出和我一样的问题?

在牛顿之前,人们力求了解世界的体系、物体运动以及运动表象后面蕴含的原理。柏拉图和亚里士多德认为地球是宇宙的中心,各大行星均绕地球作圆周运动;托勒密利用本轮-均轮的思想解释行星运行的奥秘;到了 17 世纪中期,在一大批学者如哥白尼、第谷、开普勒、伽利略的努力下,"日心说"得以确立,行星的运行规律被总结为开普勒三定律,它们分别是:椭圆定律(所有行星绕太阳的轨道都是椭圆,太阳在椭圆的一个焦点上)、面积定律(行星和太阳的连线在相等的时间间隔内扫过的面积相等)与和谐定律(所有行星绕太阳一周的恒星时间 T 的平方与它们轨道长半轴 a 的立方成比例)。英国科学家哈雷(Edmond Halley)和胡克(Robert Hooke)都猜想,使行星保持在轨道上运动的原因是它们受到的力"与距离的平方成反比",但他们都无法由这个猜测推导出天球运动的规律(开普勒三定律)。

这个问题在 1687 年 7 月问世的《自然哲学之数学原理》(以下简称《原理》)一书中得到了圆满的解决。按照当时的传统,该书由拉丁文写成(书名为 *Philosophiæ Naturalis Principia Mathematica*),1713 年出第 2 版,1725 年出第

3版，1729年译为英文付印，中译本最早出版于1931年。它是人类掌握的第一个完整的科学的宇宙论和科学理论体系，其影响遍及经典自然科学的众多领域。法国科学院院士、著名数学家拉普拉斯(Pierre-Simon Laplace)说：“牛顿的《原理》解释了宇宙的最伟大定律，它将永远成为深邃智慧的纪念碑”。爱因斯坦(Albert Einstein)说：“自然在牛顿面前好像是一本内容浩瀚的书本，他毫不费力地遨游其中。他的伟大之处在于，他集艺术家、试验者、机械师和理论家于一身。牛顿的各种发现已进入公认的知识宝库，成为‘伟大的人类之光’。”

为何《原理》会获得如此高的评价？是因为这本书给出了科学理论体系的经典研究范式：提出定义，给出公理，得出相关定律，最后将之运用于实践？或是打破宗教的束缚，给出三大运动定律，推导出万有引力定律，并用它解释太阳系各个天球运动，得出描述“天地合一”的普适规律？还是因为它不仅可以完美解释当时对宇宙与现实生活提出的众多问题，更能准确地预测后续发生的事情，如哈雷彗星的回归？

让我们打开书，看看牛顿自己怎么说吧。在序言中牛顿说：“本书的宗旨在于从各种运动现象探究自然力，再用这些力说明各种自然现象。”全书共分五部分。第一部分是对书中用到的一些概念，如物质的量、运动的量、惯性、力等给出了定义和必要的说明。第二部分是“运动的公理或定律”，这部分内容是我们非常熟悉的牛顿三大定律。第三部分为该书的第一编“物体的运动”，讨论了物体在无阻力的自由空间中的运动，尤其是在向心力作用下的运动。这些为后续对宇宙体系的讨论做出理论上的准备。第四部分为该书的第二编，讨论不同“物体(在阻滞介质中)的运动”等问题。通过这一部分内容的展开，牛顿得出结论：“行星们不是由物质涡旋带动的”，从而否定了当时欧洲占据着主导地位的笛卡儿(René Descartes)的“以太涡旋”学说。第五部分为该书的第三编，使用数学的语言探索“宇宙体系”的结构。他把开普勒的和谐定律作为观测到的“现象”，利用第一编中得出的力学定理，推导出万有引力定律，并根据该定律证明了太阳系所有成员都必然围绕着整个系统的公共重心运转，而这个重心必然处于质量最大的太阳附近。这是人类第一次从物理学上完成了对日心说的理论证明！后面牛顿接着利用它分析了大量观测到的自然事实（如月球运动的偏差、海洋潮汐的大小变化、岁差的长短不一等），来检验其理论的正确性。

在编者为大家选择的文本中，包含该书第一和第二部分的"定义"和三大运动定律。这些中学生都熟悉的内容为什么会入选？那是因为我们希望大家试着遗忘掉脑海里所有的知识，试着把自己看作牛顿，当我们亲自去总结万物运行的规律时，会提出哪些最基础最本质的概念，会如何用纯粹的语言来描述物体运动背后的根本原因。请试着这样去实践吧！也许思索的过程会很艰难。这都没有关系。只要思索了，我们就会有进步；只要思索了，我们就更能体会出牛顿在书中用词遣句之精确，处理问题之巧妙，看待问题之透彻！

如果说前面谈到的"定义"和三条基本运动定律是《原理》全书的基本出发点，那么，教材文本中选择的"附注"则阐明了牛顿对时间、空间的基本认识，现代科学称之为牛顿的绝对时空观。牛顿认为时间是"绝对、真实和数学的"，它在均匀地流逝。至于空间也是绝对的，是"具有自己的特性而无需外在参照物，永远保持各向同性以及不可移动性"。简单地说，牛顿把空间看成一个空旷的舞台或者一个框架，宇宙的一切都在这个舞台中/框架内发生；舞台/框架是永恒不变的，演出不会影响舞台，同样舞台也不能影响演出。他说，时间和空间是互不相干的。通过这样的描述，牛顿得以理解我们能看到的几乎所有的运动形式，再根据原理，几乎可以预测一切，小到苹果落地，大至天上星星的轨迹。《原理》给出的规律至今仍然指导着我们的生活，无论是建构大厦还是卫星发射。

文本中牛顿给出了四条"哲学的推理规则"，摘录第二条如下："因此对于相同的自然现象，必须尽可能地寻求相同的原因。例如人与野兽的呼吸；欧洲与美洲的石头下落；炊事用火的光亮与阳光，地球反光与行星反光。”读到这里，大家是否茅塞顿开？原来，牛顿深信宇宙万物是按简单、和谐与统一的原则构成的；原来，今天我们所遵循的科学研究中的基本准则出自这里！

开卷有益，更何况是阅读牛顿的《自然哲学之数学原理》！

（祁宁）

The Principia: Mathematical Principles of Natural Philosophy

Isaac Newton

Definition

1 Definition 1: Quantity of matter is a measure of matter that arises from its density and volume jointly.

2 If the density of air is doubled in a space that is also doubled, there is four times as much air, and there is six times as much if the space is tripled. The case is the same for snow and powders condensed by compression or liquefaction, and also for all bodies that are condensed in various ways by any causes whatsoever. For the present, I am not taking into account any medium, if there should be any, freely pervading the interstices between the parts of bodies. Furthermore, I mean this quantity whenever I use the term "body" or "mass" in the following pages. It can always be known from a body's weight, for—by making very accurate experiments with pendulums—I have found it to be proportional to the weight, as will be shown below.

3 Definition 2: Quantity of motion is a measure of motion that arises from the velocity and the quantity of matter jointly.

4 The motion of a whole is the sum of the motions of the individual parts, and thus if a body is twice as large as another and has equal velocity there is twice as much motion, and if it has twice the velocity there is four times as much motion.

5 Definition 3: Inherent force of matter is the power of resisting by which every body, so far as it is able, perseveres in its state either of resting or of moving uniformly straight forward.

本文选自艾萨克·牛顿著 *The Principia: Mathematical Principles of Natural Philosophy*。

自然哲学之数学原理

艾萨克·牛顿

定义

1 定义1：物质的量是物质的度量，可由其密度和体积共同求出。

2 所以空气的密度加倍，体积加倍，它的量就增加到四倍；体积加到三倍，它的量就增加到六倍。因挤紧或液化而压缩起来的雪、微尘或粉末，以及由任何原因而无论怎样不同地压缩起来的所有物体，也都可以作同样的理解。我在此没有考虑可以自由穿透物体各部分间隙的介质，如果有这种物质的话。此后我不论在何处提到“物体”或“质量”这一名称，指的就是这个量。从每一物体的重量可推知这个量，因为它正比于重量，正如我在很精确的单摆实验中所发现的那样，后面我将加以详述。

3 定义2：运动的量是运动的度量，可由速度和物质的量共同求出。

4 整体的运动是所有部分运动的总和。因此，速度相等而物质量加倍的物体，其运动量加倍；若其速度也加倍，则运动量加到四倍。

5 定义3：Vis insita，或物质固有的力，是一种起抵抗作用的力，它存在于每一物体当中，大小与该物体相当，并使之保持其现有的状态，或是静止，或是匀速直线运动。

本文选自[英]艾萨克·牛顿：《自然哲学之数学原理》(科学素养文库·科学元典丛书)，王克迪译，北京大学出版社 2006 年版，第 1-9 页，第 256-257 页。

6 This force is always proportional to the body and does not differ in any way from the inertia of the mass except in the manner in which it is conceived. Because of the inertia of matter, every body is only with difficulty put out of its state either of resting or of moving. Consequently, inherent force may also be called by the very significant name of force of inertia. Moreover, a body exerts this force only during a change of its state, caused by another force impressed upon it, and this exercise of force is, depending on the viewpoint, both resistance and impetus: resistance insofar as the body, in order to maintain its state, strives against the impressed force, and impetus insofar as the same body, yielding only with difficulty to the force of a resisting obstacle, endeavors to change the state of that obstacle. Resistance is commonly attributed to resting bodies and impetus to moving bodies; but motion and rest, in the popular sense of the terms, are distinguished from each other only by point of view, and bodies commonly regarded as being at rest are not always truly at rest.

7 Definition 4: Impressed force is the action exerted on a body to change its state either of resting or of moving uniformly straight forward.

8 This force consists solely in the action and does not remain in a body after the action has ceased. For a body perseveres in any new state solely by the force of inertia. Moreover, there are various sources of impressed force, such as percussion, pressure, or centripetal force.

9 Definition 5: Centripetal force is the force by which bodies are drawn from all sides, are impelled, or in any way tend, toward some point as to a center.

6 这个力总是正比于物体，它来自于物体的惯性，与之没有什么区别，在此按我们的想法来研究它。一个物体，由于物体的惯性，要改变其静止或运动的状态不是没有困难的。由此看来，这个固有的力可以用最恰当不过的名称——惯性或惯性力来称呼它。但是，物体只有当有其他力作用于它，或者要改变它的状态时，才会产生这种力。这种力的作用既可以看做是抵抗力，也可以看做是推斥力。当物体维持现有状态，反抗外来力的时候，即表现为抵抗力；当物体不易于向外来力屈服，并要改变外来力的状态时，即表现为推斥力。抵抗力通常属于静止物体，而推斥力通常属于运动物体。不过正如通常所说的那样，运动与静止只能作相对的区分，一般认为是静止的物体，并不总是真的静止。

7 定义 4：外力是一种对物体的推动作用，使其改变静止的或匀速直线运动的状态。

8 这种力只存在于作用之时，作用消失后并不存留于物体中，因为物体只靠其惯性维持它所获得的状态。不过外力有多种来源，如来自撞击、来自挤压、来自向心力。

9 定义 5：向心力使物体受到指向一个中心点的吸引、或推斥或任何倾向于该点的作用。

10 One force of this kind is gravity, by which bodies tend toward the center of the earth; another is magnetic force, by which iron seeks a lodestone; and yet another is that force, whatever it may be, by which the planets are continually drawn back from rectilinear motions and compelled to revolve in curved lines. A stone whirled in a sling endeavors to leave the hand that is whirling it, and by its endeavor it stretches the sling, doing so the more strongly the more swiftly it revolves; and as soon as it is released, it flies away. The force opposed to that endeavor, that is, the force by which the sling continually draws the stone back toward the hand and keeps it in an orbit, I call centripetal, since it is directed toward the hand as toward the center of an orbit. And the same applies to all bodies that are made to move in orbits. They all endeavor to recede from the centers of their orbits, and unless some force opposed to that endeavor is present, restraining them and keeping them in orbits and hence called by me centripetal, they will go off in straight lines with uniform motion. If a projectile were deprived of the force of gravity, it would not be deflected toward the earth but would go off in a straight line into the heavens and do so with uniform motion, provided that the resistance of the air were removed. The projectile, by its gravity, is drawn back from a rectilinear course and continually deflected toward the earth, and this is so to a greater or lesser degree in proportion to its gravity and its velocity of motion. The less its gravity in proportion to its quantity of matter, or the greater the velocity with which it is projected, the less it will deviate from a rectilinear course and the farther it will go. If a lead ball were projected with a given velocity along a horizontal line from the top of some mountain by the force of gunpowder and went in a curved line for a distance of two miles before falling to the earth, then the same ball projected with twice the velocity would go about twice as far and with ten times the velocity about ten times as far, provided that the resistance of the air were removed.

10 属于这种力的有重力，它使物体倾向于落向地球中心；磁力，它使铁趋向于磁石；以及那种使得行星不断偏离直线运动，否则它们将沿直线运动，进入沿曲线轨道环形运动的力，不论它是什么力。系于投石器上旋转的石块，企图飞离使之旋转的手，这企图张紧投石器，旋转越快，张紧的力越大，一旦将石块放开，它就飞离而去。那种反抗这种企图的力，使投石器不断地把石块拉向人手，把石块维持在其环行轨道上，由于它指向轨道的中心——人手，我称为向心力。所有环行于任何轨道上的物体都可作相同的理解，它们都企图离开其轨道中心；如果没有一个与之对抗的力来遏制其企图，把它们约束在轨道上，它们将沿直线以匀速飞去，所以我称这种力为向心力。一个抛射物体，如果没有引力牵制，将不会回落到地球上，而是沿直线向天空飞去，如果没有空气阻力，飞离速度是匀速的。正是引力使其不断偏离直线轨道，向地球偏转，偏转的强弱，取决于引力和抛射物的运动速度。引力越小，或其物质的量越小，或它被抛出的速度越大，它对直线轨道的偏离越小，它就飞得越远。如果用火药力从山顶上发射铅弹，给定其速度，方向与地平面平行，铅弹将沿曲线在落地前飞行 2 英里；同样，如果没有空气阻力，发射速度加倍或加到十倍，则铅弹飞行距离也加倍或加到十倍。

11 And by increasing the velocity, the distance to which it would be projected could be increased at will and the curvature of the line that it would describe could be decreased, in such a way that it would finally fall at a distance of 10 or 30 or 90 degrees or even go around the whole earth or, lastly, go off into the heavens and continue indefinitely in this motion. And in the same way that a projectile could, by the force of gravity, be deflected into an orbit and go around the whole earth, so too the moon, whether by the force of gravity—if it has gravity—or by any other force by which it may be urged toward the earth, can always be drawn back toward the earth from a rectilinear course and deflected into its orbit; and without such a force the moon cannot be kept in its orbit. If this force were too small, it would not deflect the moon sufficiently from a rectilinear course; if it were too great, it would deflect the moon excessively and draw it down from its orbit toward the earth. In fact, it must be of just the right magnitude, and mathematicians have the task of finding the force by which a body can be kept exactly in any given orbit with a given velocity and, alternatively, to find the curvilinear path into which a body leaving any given place with a given velocity is deflected by a given force.

12 The quantity of centripetal force is of three kinds: absolute, accelerative, and motive.

11 通过增加发射速度，即可以随意增加它的抛射距离，减轻它的轨迹的弯曲度，直至它最终落在 10 度，30 度或 90 度的距离处①，甚至在落地之前环绕地球一周；或者，使它再也不返回地球，直入苍穹太空而去，作 infinitum(无限的)运动。运用同样的方法，抛射物在引力作用下，可以沿环绕整个地球的轨道运转。月球也是被引力，如果它有引力的话，或者别的力不断拉向地球，偏离其惯性力所遵循的直线路径，沿着其现在的轨道运转。如果没有这样的力，月球将不能保持在其轨道上。如果这个力太小，就将不足以使月球偏离直线路径；如果它太大，则将偏转太大，把月球由其轨道上拉向地球。这个力必须是一个适当的量，数学家的职责在于求出使一个物体以给定速度精确地沿着给定的轨道运转的力。反之，必须求出从一个给定处所，以给定速度抛射的物体，在给定力的作用下偏离其原来的直线路径所进入的曲线路径。

12 可以认为，任何一个向心力均有以下三种度量：绝对度量、加速度度量和运动度量。

① 此当指地球表面经度，因剑桥地处经度 0 度。——译者注

AXIOMS, OR THE LAWS OF MOTION

13 Law 1: Every body perseveres in its state of being at rest or of moving uniformly straight forward, except insofar as it is compelled to change its state by forces impressed.

14 Projectiles persevere in their motions, except insofar as they are retarded by the resistance of the air and are impelled downward by the force of gravity. A spinning hoop, which has parts that by their cohesion continually draw one another back from rectilinear motions, does not cease to rotate, except insofar as it is retarded by the air. And larger bodies—planets and comets—preserve for a longer time both their progressive and their circular motions, which take place in spaces having less resistance.

15 Law 2: A change in motion is proportional to the motive force impressed and takes place along the straight line in which that force is impressed.

16 If some force generates any motion, twice the force will generate twice the motion, and three times the force will generate three times the motion, whether the force is impressed all at once or successively by degrees. And if the body was previously moving, the new motion (since motion is always in the same direction as the generative force) is added to the original motion if that motion was in the same direction or is subtracted from the original motion if it was in the opposite direction or, if it was in an oblique direction, is combined obliquely and compounded with it according to the directions of both motions.

17 Law 3: To any action there is always an opposite and equal reaction; in other words, the actions of two bodies upon each other are always equal and

运动的公理或定律

13 定律Ⅰ：每个物体都保持其静止或匀速直线运动的状态，除非有外力作用于它迫使它改变那个状态。

14 抛射体如果没有空气阻力的阻碍或重力向下牵引，将维持其射出时的运动。陀螺各部分的凝聚力不断使之偏离直线运动，如果没有空气的阻碍，就不会停止旋转。行星和彗星一类较大物体，在自由空间中没有什么阻力，可以在很长时间里保持其向前的或圆周的运动。

15 定律Ⅱ：运动的变化正比于外力，变化的方向沿外力作用的直线方向。

16 如果某力产生一种运动，则加倍的力产生加倍的运动，三倍的力产生三倍的运动，无论这力是一次还是逐次施加的。而且如果物体原先是运动的，则它应加上原先的运动或是从中减去，这由它的方向与原先运动一致或相反来决定。如果它是斜向加入的，则它们之间有夹角，由二者的方向产生出新的复合运动。

17 定律Ⅲ：每一种作用都有一个相等的反作用；或者，两个物体间的相互作用总是相等的，而且指向相反。

always opposite in direction.

18 Whatever presses or draws something else is pressed or drawn just as much by it. If anyone presses a stone with a finger, the finger is also pressed by the stone. If a horse draws a stone tied to a rope, the horse will (so to speak) also be drawn back equally toward the stone, for the rope, stretched out at both ends, will urge the horse toward the stone and the stone toward the horse by one and the same endeavor to go slack and will impede the forward motion of the one as much as it promotes the forward motion of the other. If some body impinging upon another body changes the motion of that body in any way by its own force, then, by the force of the other body (because of the equality of their mutual pressure), it also will in turn undergo the same change in its own motion in the opposite direction. By means of these actions, equal changes occur in the motions, not in the velocities—that is, of course, if the bodies are not impeded by anything else. For the changes in velocities that likewise occur in opposite directions are inversely proportional to the bodies because the motions are changed equally. This law is valid also for attractions, as will be proved in the next scholium.

Scholium

19 Thus far it has seemed best to explain the senses in which less familiar words are to be taken in this treatise. Although time, space, place, and motion are very familiar to everyone, it must be noted that these quantities are popularly conceived solely with reference to the objects of sense perception. And this is the source of certain preconceptions; to eliminate them it is useful to distinguish these quantities into absolute and relative, true and apparent, mathematical and common.

(1) Absolute, true, and mathematical time, in and of itself and of its own

18 不论是拉或是压另一个物体，都会受到该物体同等的拉或是压。如果用手指压一块石头，则手指也受到石头的压。如果马拉一系于绳索上的石头，则马(如果可以这样说的话)也同等地被拉向石头，因为绷紧的绳索同样企图使自身放松，将像它把石头拉向马一样同样强地把马拉向石头，它阻碍马前进就像它拉石头前进一样强。如果某个物体撞击另一物体，并以其撞击力使后者的运动改变，则该物体的运动也(由于互压等同性)发生一个同等的变化，变化方向相反。这些作用造成的变化是相等的，但不是速度变化，而是指物体的运动变化，如果物体不受到任何其他阻碍的话。由于运动是同等变化的，向相反方向速度的变化反比于物体。本定律在吸引力情形也成立，我们将在附注中证明。

附注

19 至此，我已定义了这些鲜为人知的术语，解释了它们的意义，以便在以后的讨论中理解它们。我没有定义时间、空间、处所和运动，因为它们是人所共知的。唯一必须说明的是，一般人除了通过可感知客体外无法想象这些量，并会由此产生误解。为了消除误解，可方便地把这些量分为绝对的与相对的，真实的与表象的以及数学的与普通的。(1)绝对的、真实的和数学的时间，由其特性决定，自身均匀地流逝，与一切外在事物无关，又名延续；相对的、表象的和普通的时间是可感知和外在的(不论是精确的或是不均匀的)对运动之延续的量度，它常被用以代替真实时间，如一小时、一天、一个月、一年。

nature, without reference to anything external, flows uniformly and by another name is called duration. Relative, apparent, and common time is any sensible and external measure a (precise or imprecise) 3 of duration by means of motion; such a measure—for example, an hour, a day, a month, a year—is commonly used instead of true time.

(2) Absolute space, of its own nature without reference to anything external, always remains homogeneous and immovable. Relative space is any movable measure or dimension of this absolute space; such a measure or dimension is determined by our senses from the situation of the space with respect to bodies and is popularly used for immovable space, as in the case of space under the earth or in the air or in the heavens, where the dimension is determined from the situation of the space with respect to the earth. Absolute and relative space are the same in species and in magnitude, but they do not always remain the same numerically. For example, if the earth moves, the space of our air, which in a relative sense and with respect to the earth always remains the same, will now be one part of the absolute space into which the air passes, now another part of it, and thus will be changing continually in an absolute sense.

(3) Place is the part of space that a body occupies, and it is, depending on the space, either absolute or relative. I say the part of space, not the position of the body or its outer surface. For the places of equal solids are always equal, while their surfaces are for the most part unequal because of the dissimilarity of shapes; and positions, properly speaking, do not have quantity and are not so much places as attributes of places. The motion of a whole is the same as the sum of the motions of the parts; that is, the change in position of a whole from its place is the same as the sum of the changes in position of its parts from their places, and thus the place of a whole is the same as the sum of the places of the parts and therefore is internal and in the whole body.

(4) Absolute motion is the change of position of a body from one absolute

(2)绝对空间：其自身特性与一切外在事物无关，处处均匀，永不移动。相对空间是一些可以在绝对空间中运动的结构或是对绝对空间的量度，我们通过它与物体的相对位置来感知它；它一般被当做不可移动空间，如地表以下、大气中或天空中的空间，都是以其与地球的相互关系确定的。绝对空间与相对空间在形状与大小上相同，但在数值上并不总是相同。例如，地球在运动，大气的空间相对于地球总是不变，但在一个时刻大气通过绝对空间的一部分，而在另一时刻又通过绝对空间的另一部分，因此，在绝对的意义上看，它是连续变化的。

(3)处所是空间的一个部分，为物体占据着，它可以是绝对的或相对的，随空间的性质而定。我这里说的是空间的一部分，不是物体在空间中的位置，也不是物体的外表面。因为相等的固体其处所总是相等，但其表面却常常由于外形的不同而不相等。位置实在没有量可言，它们至多是处所的属性，绝非处所本身。整体的运动等同于各部分的运动的总和，即是说，整体离开其处所的迁移等同于其各部分离开各自的处所的迁移的总和，因此，总体的处所等同于部分处所的和，由于这个缘故，它是内在的，在整个物体内部。

(4)绝对运动是物体由一个绝对处所迁移到另一个绝对处所；相对运动是由一个相对处所迁移到另一个相对处所。一艘航行的船中，物体的相对处所是它所占据的船的一部分，或物体在船舱中充填的那一部

place to another; relative motion is change of position from one relative place to another. Thus, in a ship under sail, the relative place of a body is that region of the ship in which the body happens to be or that part of the whole interior of the ship which the body fills and which accordingly moves along with the ship, and relative rest is the continuance of the body in that same region of the ship or same part of its interior. But true rest is the continuance of a body in the same part of that unmoving space in which the ship itself, along with its interior and all its contents, is moving. Therefore, if the earth is truly at rest, a body that is relatively at rest on a ship will move trulyand absolutely with the velocity with which the ship is moving on the earth. But if the earth is also moving, the true and absolute motion of the body will arise partly from the true motion of the earth in unmoving space and partly from the relative motion of the ship on the earth. Further, if the body is also moving relatively on the ship, its true motion will arise partly from the true motion of the earth in unmoving space and partly from the relative motions both of the ship on the earth and of the body on the ship, and from these relative motions the relative motion of the body on the earth will arise. For example, if that part of the earth where the ship happens to be is truly moving eastward with a velocity of 10,010 units, and the ship is being borne westward by sails and wind with a velocity of 10 units, and a sailor is walking on the ship toward the east with a velocity of 1 unit, then the sailor will be moving truly and absolutely in unmoving space toward the east with a velocity of 10,001 units and relatively on the earth toward the west with a velocity of 9 units.

20 In astronomy, absolute time is distinguished from relative time by the equation of common time. For natural days, which are commonly considered equal for the purpose of measuring time, are actually unequal. Astronomers correct this inequality in order to measure celestial motions on the basis of a truer time. It is possible that there is no uniform motion by which time may

分，它与船共同运动：所谓相对静止，就是物体滞留在船或船舱的同一部分处。但实际上，绝对静止应是物体滞留在不动空间的同一部分处，船、船舱以及它携载的物品都已相对于它作了运动。所以，如果地球真的静止，那个相对于船静止的物体，将以等于船相对于地球的速度真实而绝对地运动。但如果地球也在运动，物体真正的绝对运动应当一部分是地球在不动空间中的运动，另一部分是船在地球上的运动：如果物体也相对于船运动，它的真实运动将部分来自地球在不动空间中的真实运动，部分来自船在地球上的相对运动，以及该物体相对于船的运动。这些相对运动决定物体在地球上的相对运动。例如，船所处的地球的那一部分，真实地向东运动，速度为 10010 等分，而船则在强风中扬帆向西航行，速度为 10 等分，水手在船上以 1 等分速度向东走，则水手在不动空间中实际上是向东运动，速度为 10001 等分，而他相对于地球的运动则是向西，速度为 9 等分。

20 天文学中，由表象时间的均差或勘误来区别绝对时间与相对时间，因为自然日并不真正相等，虽然一般认为它们相等，并用以度量时间。天文学家纠正这种不相等性，以便用更精确的时间测量天体的运动。能用以精确测定时间的等速运动可能是不存在的。所有运动都可能是加速或减速，但绝对时间的流逝并不迁就任何变化。事物的存在顽强地延续维持不变，无论运动是快是慢抑或停止，因此这种延续应当同

have an exact measure. All motions can be accelerated and retarded, but the flow of absolute time cannot be changed. The duration or perseverance of the existence of things is the same, whether their motions are rapid or slow or null; accordingly, duration is rightly distinguished from its sensible measures and is gathered from them by means of an astronomical equation. Moreover, the need for using this equation in determining when phenomena occur is proved by experience with a pendulum clock and also by eclipses of the satellites of Jupiter.

21 Just as the order of the parts of time is unchangeable, so, too, is the order of the parts of space. Let the parts of space move from their places, and they will move (so to speak) from themselves. For times and spaces are, as it were, the places of themselves and of all things. All things are placed in time with reference to order of succession and in space with reference to order of position. It is of the essence of spaces to be places, and for primary places to move is absurd. They are therefore absolute places, and it is only changes of position from these places that are absolute motions.

22 But since these parts of space cannot be seen and cannot be distinguished from one another by our senses, we use sensible measures in their stead. For we define all places on the basis of the positions and distances of things from some body that we regard as immovable, and then we reckon all motions with respect to these places, insofar as we conceive of bodies as being changed in position with respect to them. Thus, instead of absolute places and motions we use relative ones, which is not inappropriate in ordinary human affairs, although in philosophy abstraction from the senses is required. For it is possible that there is nobody truly at rest to which places and motions may be referred.

只能借着感官测量的时间区别开来，由此我们可以运用天文学时差把它推算出来。这种时差的必要性，在对现象作时间测定中已显示出来，如摆钟实验，以及木星卫星的食亏。

21 与时间间隔的顺序不可互易一样，空间部分的次序也不可互易。设想空间的一些部分被移出其处所，则它们将是(如果允许这样表述的活)移出其自身。因为时间和空间是，而且一直是它们自己以及一切其他事物的处所。所有事物置于时间中以列出顺序；置于空间中以排出位置。时间和空间在本质上或特性上就是处所，事物的基本处所可以移动的说法是不合理的。所以，这些是绝对处所，而离开这些处所的移动，是唯一的绝对运动。

22 但是，由于空间的这一部分无法看见，也不能通过感官把它与别的部分加以区分，所以我们代之以可感知的度量。由事物的位置及其到我们视为不动的物体的距离定义出所有处所，再根据物体由某些处所移向另一些处所，测出相对于这些处所的所有运动。这样，我们就以相对处所和运动取代绝对处所和运动，而且在一般情况下没有任何不便。但在哲学研究中，我们则应当从感官抽象出并且思考事物自身，把它们与单凭感知测度的表象加以区分。因为实际上借以标志其他物体的处所和运动的静止物体，可能是不存在的。

23 Moreover, absolute and relative rest and motion are distinguished from each other by their properties, causes, and effects. It is a property of rest that bodies truly at rest are at rest in relation to one another. And therefore, since it is possible that some body in the regions of the fixed stars or far beyond is absolutely at rest, and yet it cannot be known from the position of bodies in relation to one another in our regions whether or not any of these maintains a given position with relation to that distant body, true rest cannot be defined on the basis of the position of bodies in relation to one another.

24 It is a property of motion that parts which keep given positions in relation to wholes participate in the motions of such wholes. For all the parts of bodies revolving in orbit endeavor to recede from the axis of motion, and the impetus of bodies moving forward arises from the joint impetus of the individual parts. Therefore, when bodies containing others move, whatever is relatively at rest within them also moves. And thus true and absolute motion cannot be determined by means of change of position from the vicinity of bodies that are regarded as being at rest. For the exterior bodies ought to be regarded not only as being at rest but also as being truly at rest. Otherwise all contained bodies, besides being subject to change of position from the vicinity of the containing bodies, will participate in the true motions of the containing bodies and, if there is no such change of position, will not be truly at rest but only be regarded as being at rest. For containing bodies are to those inside them as the outer part of the whole to the inner part or as the shell to the kernel. And when the shell moves, the kernel also, without being changed in position from the vicinity of the shell, moves as a part of the whole.

25 A property akin to the preceding one is that when a place moves, whatever is placed in it moves along with it, and therefore a body moving away from a

23 不过我们可以由事物的属性、原因和效果把一事物与其他事物的静止与运动、绝对与相对区别开来。静止的属性在于，真正静止的物体相对于另一静止物体也是静止的，因此，在遥远的恒星世界，也许更为遥远的地方，有可能存在着某些绝对静止的物体，但却不可能由我们世界中物体间的相互位置知道这些物体是否保持着与遥远物体不变的位置，这意味着在我们世界中物体的位置不能确定绝对静止。

24 运动的属性在于，部分维持其在整体中的原有位置并参与整体的运动。转动物体的所有部分都有离开其转动轴的倾向，而向前行进的物体其力量来自所有部分的力量之和。所以，如果处于外围的物体运动了，处于其内原先相对静止的物体也将参与其运动。基于此项说明，物体真正的绝对运动，不能由它相对于只是看起来是静止的物体发生移动来确定，因为外部的物体不仅应看起来是静止的，而且还应是真正静止的。反过来，所有包含在内的物体，除了移开它们附近的物体外，同样也参与真正的运动，即使没有这项运动，它们也不是真正的静止，只是看起来静止而已。因为周围的物体与包含在内的物体的关系，类似于一个整体靠外的部分与其靠内的部分，或者类似于果壳与果仁，但如果果壳运动了，则果仁作为整体的一部分也将运动，而它与靠近的果壳之间并无任何移动。

25 与上述有关的一个属性是，如果处所运动了，则处于其中的物体也与之一同运动。所以，移开其运动处所的物体，也参与了其处所的运动。基于此项说明，一切脱离运动处所的运动，都只是整体和绝对运

place that moves participates also in the motion of its place. Therefore, all motions away from places that move are only parts of whole and absolute motions, and every whole motion is compounded of the motion of a body away from its initial place, and the motion of this place away from its place, and so on, until an unmoving place is reached, as in the abovementioned example of the sailor. Thus, whole and absolute motions can be determined only by means of unmoving places, and therefore in what has preceded I have referred such motions to unmoving places and relative motions to movable places. Moreover, the only places that are unmoving are those that all keep given positions in relation to one another from infinity to infinity and therefore always remain immovable and constitute the space that I call immovable.

26 The causes which distinguish true motions from relative motions are the forces impressed upon bodies to generate motion. True motion is neither generated nor changed except by forces impressed upon the moving body itself, but relative motion can be generated and changed without the impression of forces upon this body. For the impression of forces solely on other bodies with which a given body has a relation is enough, when the other bodies yield, to produce a change in that relation which constitutes the relative rest or motion of this body. Again, true motion is always changed by forces impressed upon a moving body, but relative motion is not necessarily changed by such forces. For if the same forces are impressed upon a moving body and also upon other bodies with which it has a relation, in such a way that the relative position is maintained, the relation that constitutes the relative motion will also be maintained. Therefore, every relative motion can be changed while the true motion is preserved, and can be preserved while the true one is changed, and thus true motion certainly does not consist in relations of this sort.

动的一部分。每个整体运动都由移出其初始处所的物体的运动和这个处所移出其原先位置的运动等构成，直至最终到达一不动的处所，如前面举过的航行的例子。所以，整体和绝对的运动，只能由不动的处所加以确定，正因为如此，我在前文里把绝对运动与不动处所相联系，而相对运动与相对处所相联系。所以，不存在不变的处所，只是那些从无限到无限的事物除外，它们全部保持着相互间既定的不变位置，必定永远不动，因而构成不动空间。

26 真实运动与相对运动之所以不同，原因在于施于物体上使之产生运动的力。真实运动，除非某种力作用于运动物体之上，是既不会产生也不会改变的，但相对运动在没有力作用于物体时也会产生或改变。因为，只要对与前者作比较的其他物体施加以某种力就足够了，其他物体的后退，使它们先前的相对静止或运动的关系发生改变，再者，当有力施于运动物体上时，真实的运动总是发生某种变化，而这种力却未必能使相对运动作同样变化。因为如果把相同的力同样施加在用作比较的其他物体上，相对的位置有可能得以维持，进而维持相对运动所需条件。因此，相对运动改变时，真实运动可维持不变，而相对运动得以维持时，真实运动却可能变化了。因此，这种关系决不包含真实运动。

27 The effects distinguishing absolute motion from relative motion are the forces of receding from the axis of circular motion. For in purely relative circular motion these forces are null, while in true and absolute circular motion they are larger or smaller in proportion to the quantity of motion. If a bucket is hanging from a very long cord and is continually turned around until the cord becomes twisted tight, and if the bucket is thereupon filled with water and is at rest along with the water and then, by some sudden force, is made to turn around in the opposite direction and, as the cord unwinds, perseveres for a while in this motion; then the surface of the water will at first be level, just as it was before the vessel began to move. But after the vessel, by the force gradually impressed upon the water, has caused the water also to begin revolving perceptibly, the water will gradually recede from the middle and rise up the sides of the vessel, assuming a concave shape (as experience has shown me), and, with an ever faster motion, will rise further and further until, when it completes its revolutions in the same times as the vessel, it is relatively at rest in the vessel. The rise of the water reveals its endeavor to recede from the axis of motion, and from such an endeavor one can find out and measure the true and absolute circular motion of the water, which here is the direct opposite of its relative motion. In the beginning, when the relative motion of the water in the vessel was greatest, that motion was not giving rise to any endeavor to recede from the axis; the water did not seek the circumference by rising up the sides of the vessel but remained level, and therefore its true circular motion had not yet begun. But afterward, when the relative motion of the water decreased, its rise up the sides of the vessel revealed its endeavor to recede from the axis, and this endeavor showed the true circular motion of the water to be continually increasing and finally becoming greatest when the water was relatively at rest in the vessel. Therefore, that endeavor does not depend on the change of position of the water with respect to surrounding bodies, and thus true circular motion

27 绝对运动与相对运动的效果的区别是飞离旋转运动轴的力。在纯粹的相对转动中不存在这种力，而在真实和绝对转动中，该力的大小取决于运动的量。如果将一悬在长绳之上的桶不断旋转，使绳拧紧，再向桶中注满水，并使桶与水都保持平静，然后通过另一个力的突然作用，桶沿相反方向旋转，同时绳自己放松，桶作这项运动会持续一段时间。开始时，水的表面是平坦的，因为桶尚未开始转动；但之后，桶通过逐渐把它的运动传递给水，将使水开始明显地旋转，一点一点地离开中间，并沿桶壁上升，形成一个凹形(我验证过)，而且旋转越快，水上升得越高，直至最后与桶同时转动，达到相对静止。水的上升表明它有离开转动轴的倾向，而水的真实和绝对的转动，在此与其相对运动直接矛盾，可以知道并由这种倾向加以度量。起初，当水在桶中的相对运动最大时，它并未表现出离开轴的倾向，也未显示出旋转的趋势，未沿桶壁上升，水面保持平坦，因此水的真正旋转并未开始。但在那之后，水的相对运动减慢，水沿桶壁上升表明它企图离开转轴，这种倾向说明水的真实的转动正逐渐加快，直到它获得最大量，这时水相对于桶静止。因此，水的这种倾向并不取决于水相对于其周围物体的移动，这种移动也不能说明真实的旋转运动。任何一个旋转的物体只存在一种真实的旋转运动，它只对应于一种企图离开运动轴的力，这才是其独特而恰当的后果。但在一个完全相同的物体中的相对运动，由其与外界物体的各种关系决定，多得不可胜数，而且与其他关系一样，都缺乏真实的效果，除非它们或许参与了那唯一的真实运动。因此，按这种见解，宇宙体系是：我们的天空在恒星天层之下携带着行星一同旋转，天空中的若干部分以及行星相对于它们的天空可能的确是静止的，但却实实在在地运动着。因为它们相互间变换着位置(真正静止的物体决不如此)，被“裹挟”在它们的天空中参与其运动，而且作为旋转整体的一部分，企图离开它们的运动轴。

cannot be determined by means of such changes of position. The truly circular motion of each revolving body is unique, corresponding to a unique endeavor as its proper and sufficient effect, while relative motions are innumerable in accordance with their varied relations to external bodies and, like relations, are completely lacking in true effects except insofar as they participate in that true and unique motion. Thus, even in the system of those who hold that our heavens revolve below the heavens of the fixed stars and carry the planets around with them, the individual parts of the heavens, and the planets that are relatively at rest in the heavens to which they belong, are truly in motion. For they change their positions relative to one another (which is not the case with things that are truly at rest), and as they are carried around together with the heavens, they participate in the motions of the heavens and, being parts of revolving wholes, endeavor to recede from the axes of those wholes.

28 Relative quantities, therefore, are not the actual quantities whose names they bear but are those sensible measures of them (whether true or erroneous) that are commonly used instead of the quantities being measured. But if the meanings of words are to be defined by usage, then it is these sensible measures which should properly be understood by the terms "time," "space," "place," and "motion," and the manner of expression will be out of the ordinary and purely mathematical if the quantities being measured are understood here. Accordingly those who there interpret these words as referring to the quantities being measured do violence to the Scriptures. And they no less corrupt mathematics and philosophy who confuse true quantities with their relations and common measures.

29 It is certainly very difficult to find out the true motions of individual bodies and actually to differentiate them from apparent motions, because the parts of that immovable space in which the bodies truly move make no impression on the senses. Nevertheless, the case is not utterly hopeless. For it is possible to

28 正因为如此，相对的量并不是负有其名的那些量本身，而是其可感知的度量(精确的或不精确的)，它通常用以代替量本身的度量。如果这些词的含义是由其用途决定的，则时间、空间、处所和运动这些词，其(可感知的)度量就能得到恰当的理解，而如果度量出的量意味着它们自身，则其表述就非同寻常，而且是纯数学的了。由此看来，有人在解释这些表示度量的量的同时，违背了本应保持准确的语言的精确性，他们混同了真实的量和与之有关的可感知的度量，这无助于减轻对数学和哲学真理的纯洁性的玷污。

29 要认识特定物体的真实运动，并切实地把它与表象的运动区分开，的确是一件极为困难的事，因为在其中发生运动的不动空间的那一部分，无法为我们的感官所感知，不过这件事也没有彻底绝望，我们还有若干见解作指导，一方面来自表象运动，它与真实运动有所差异；

draw evidence partly from apparent motions, which are the differences between the true motions, and partly from the forces that are the causes and effects of the true motions. For example, if two balls, at a given distance from each other with a cord connecting them, were revolving about a common center of gravity, the endeavor of the balls to recede from the axis of motion could be known from the tension of the cord, and thus the quantity of circular motion could be computed. Then, if any equal forces were simultaneously impressed upon the alternate faces of the balls to increase or decrease their circular motion, the increase or decrease of the motion could be known from the increased or decreased tension of the cord, and thus, finally, it could be discovered which faces of the balls the forces would have to be impressed upon for a maximum increase in the motion, that is, which were the posterior faces, or the ones that are in the rear in a circular motion. Further, once the faces that follow and the opposite faces that precede were known, the direction of the motion would be known. In this way both the quantity and the direction of this circular motion could be found in any immense vacuum, where nothing external and sensible existed with which the balls could be compared. Now if some distant bodies were set in that space and maintained given positions with respect to one another, as the fixed stars do in the regions of the heavens, it could not, of course, be known from the relative change of position of the balls among the bodies whether the motion was to be attributed to the bodies or to the balls. But if the cord was examined and its tension was discovered to be the very one which the motion of the balls required, it would be valid to conclude that the motion belonged to the balls and that the bodies were at rest, and then, finally, from the change of position of the balls among the bodies, to determine the direction of this motion. But in what follows, a fuller explanation will be given of how to determine true motions from their causes, effects, and apparent differences, and, conversely, of how to determine from motions, whether true or apparent, their causes and effects. For this was the purpose for which I composed the following treatise.

另一方面来自力，它是真实运动的原因与后果。例如，两只球由一根线连接并保持给定距离，围绕它们的公共重心旋转，则我们可以由线的张力发现球欲离开转动轴的倾向，进而可以计算出它们的转动量。如果用同等的力施加在球的两侧使其转动增加或减少，则由线的张力的增加或减少可以推知运动的增减，进而可以发现力应施加在球的什么面上才能使其运动有最大增加，即可以知道是它的最后面，或在转动中居后的一面。而知道了这后面的一面，以及与之对应的一面，也就同样可以知道其运动方向了。这样，我们就能知道这种转动的量和方向，即使在巨大的真空中，没有供球与之作比较的外界的可感知的物体存在，也能做到。但是，如果在那个空间里有一些遥远的物体，其相互间位置保持不变，就像我们世界中的恒星一样，我们就确实无法从球在那些物体中的相对移动来判定究竟这运动属于球还是属于那些物体。但如果我们观察绳子，发现其张力正是球运动时所需要的，就能断定运动属于球，而那些物体是静止的；最后，由球在物体间的运动，我们还能发现其运动的方向。但如何由其原因、效果及表象差异推知真正的运动，以及相反的推理，正是我要在随后的篇章中详细阐述的，这正是我写作本书的目的。

Rule of Reasoning in Philosophy

30 Rule 1 No more causes of natural things should be admitted than are both true and sufficient to explain their phenomena.

31 As the philosophers say: Nature does nothing in vain, and more causes are in vain when fewer suffice. For nature is simple and does not indulge in the luxury of superfluous causes.

32 Rule 2 Therefore, the causes assigned to natural effects of the same kind must be, so far as possible, the same.

33 Examples are the cause of respiration in man and beast, or of the falling of stones in Europe and America, or of the light of a kitchen fire and the sun, or of the reflection of light on our earth and the planets.

34 Rule 3 Those qualities of bodies that cannot be intended and remitted and that belong to all bodies on which experiments can be made should be taken as qualities of all bodies universally.

35 For the qualities of bodies can be known only through experiments; and therefore qualities that square with experiments universally are to be regarded as universal qualities; and qualities that cannot be diminished cannot be taken away from bodies. Certainly idle fancies ought not to be fabricated recklessly against the evidence of experiments, nor should we depart from the analogy of nature, since nature is always simple and ever consonant with itself. The extension of bodies is known to us only through our senses, and yet there are bodies beyond the range of these senses; but because extension is found in all

哲学中的推理规则

30 规则 1　寻求自然事物的原因，不得超出真实和足以解释其现象者。

31 为达此目的，哲学家们说，自然不做徒劳的事，解释多了白费口舌，言简意赅才见真谛；因为自然喜欢简单性，不会响应于多余原因的侈谈。

32 规则 2　因此对于相同的自然现象，必须尽可能地寻求相同的原因。

33 例如，人与野兽的呼吸；欧洲与美洲的石头下落；炊事用火的光亮与阳光；地球反光与行星反光。

34 规则 3　物体的特性，若其程度既不能增加也不能减少，且在实验所及范围内为所有物体所共有，则应视为一切物体的普遍属性。

35 因为，物体的特性只能通过实验为我们所了解，我们认为是普适的属性只能是实验上普适的；只能是既不会减少又绝不会消失的。我们当然不会因为梦幻和凭空臆想而放弃实验证据；也不会背弃自然的相似性，这种相似性应是简单的，首尾一致的。我们无法逾越感官而了解物体的广延，也无法由此而深入物体内部。但是，因为我们假设所有物体的广延是可感知的，所以也把这一属性普遍地赋予所有物体。我们由经验知道许多物体是硬的，而全体的硬度是由部分的硬度所产生的，所以我们恰当地推断，不仅我们感知的物体的粒子是硬的，而且

sensible bodies, it is ascribed to all bodies universally. We know by experience that some bodies are hard. Moreover, because the hardness of the whole arises from the hardness of its parts, we justly infer from this not only the hardness of the undivided particles of bodies that are accessible to our senses, but also of all other bodies. That all bodies are impenetrable we gather not by reason but by our senses. We find those bodies that we handle to be impenetrable, and hence we conclude that impenetrability is a property of all bodies universally. That all bodies are movable and persevere in motion or in rest by means of certain forces (which we call forces of inertia) we infer from finding these properties in the bodies that we have seen. The extension, hardness, impenetrability, mobility, and force of inertia of the whole arise from the extension, hardness, impenetrability, mobility, and force of inertia of each of the parts; and thus we conclude that every one of the least parts of all bodies is extended, hard, impenetrable, movable, and endowed with a force of inertia. And this is the foundation of all natural philosophy. Further, from phenomena we know that the divided, contiguous parts of bodies can be separated from one another, and from mathematics it is certain that the undivided parts can be distinguished into smaller parts by our reason. But it is uncertain whether those parts which have been distinguished in this way and not yet divided can actually be divided and separated from one another by the forces of nature. But if it were established by even a single experiment that in the breaking of a hard and solid body, any undivided particle underwent division, we should conclude by the force of this third rule not only that divided parts are separable but also that undivided parts can be divided indefinitely. Finally, if it is universally established by experiments and astronomical observations that all bodies on or near the earth gravitate [lit. are heavy] toward the earth, and do so in proportion to the quantity of matter in each body, and that the moon gravitates [is heavy] toward the earth in proportion to the quantity of its matter, and that our sea in turn

所有其他粒子都是硬的。说所有物体都是不可穿透的，这不是推理而来的结论，而是感知的。我们发现拿着的物体是不可穿透的，由此推断出不可穿透性是一切物体的普遍性质。说所有物体都能运动，并赋予它们在运动时或静止时具有某种保持其状态的能力(我们称之为惯性)，只不过是由我们曾见到过的物体中所发现的类似特性而推断出来的。全体的广延、硬度、不可穿透性、可运动性和惯性，都是由部分的广延、硬度、不可穿透性、可运动性和惯性所造成的；因而我们推断所有物体的最小粒子也都具有广延、硬度、不可穿透性、可运动性，并赋予它们以惯性性质。这是一切哲学的基础。此外，物体分离的但又相邻接的粒子可以相互分开，是观测事实；在未被分开的粒子内，我们的思维能区分出更小的部分，正如数学所证明的那样。但如此区分开的，以及未被分开的部分，能否确实由自然力分割并加以分离，我们尚不得而知。然而，只要有哪怕是一例实验证明，由坚硬的物体上取下的任何未分开的小粒子被分割开来了，我们就可以沿用本规则得出结论，已分开的和未分开的粒子实际上都可以分割为无限小。最后，如果实验和天文观测普遍发现，一方面，地球附近的物体都被吸引向地球，吸引力正比于物体所各自包含的物质；月球也根据其物质量被吸引向地球；而另一方面，我们的海洋被吸引向月球；所有的行星相互吸引；彗星以类似方式被吸引向太阳；则我们必须沿用本规则赋予一切物体以普遍相互吸引的原理。因为一切物体的普遍吸引是由现象得到的结论，它比物体的不可穿透性显得有说服力；后者在天体活动范围内无法由实验或任何别的观测手段加以验证。我肯定重力不是物体的基本属性；我说到固有的力时，只是指它们的惯性。这才是不会变更的。物体的重力会随其远离地球而减小。

gravitates [is heavy] toward the moon, and that all planets gravitate [are heavy] toward one another, and that there is a similar gravity [heaviness] of comets toward the sun, it will have to be concluded by this third rule that all bodies gravitate toward one another. Indeed, the argument from phenomena will be even stronger for universal gravity than for the impenetrability of bodies, for which, of course, we have not a single experiment, and not even an observation, in the case of the heavenly bodies. Yet I am by no means affirming that gravity is essential to bodies. By inherent force I mean only the force of inertia. This is immutable. Gravity is diminished as bodies recede from the earth.

36 Rule 4 In experimental philosophy, propositions gathered from phenomena by induction should be considered either exactly or very nearly true notwithstanding any contrary hypotheses, until yet other phenomena make such propositions either more exact or liable to exceptions.

37 This rule should be followed so that arguments based on induction may not be nullified by hypotheses.

36 规则 4 在实验哲学中，我们必须将由现象所归纳出的命题视为完全正确的或基本正确的，而不管想象所可能得到的与之相反的种种假说，直到出现了其他的或可排除这些命题或可使之变得更加精确的现象之时。

37 我们必须遵守这一规则，使不脱离假说归纳出的结论。

《狭义与广义相对论浅说》导引

阿尔伯特·爱因斯坦(Albert Einstein，1879—1955)是著名物理学家、思想家、哲学家。1879年他出生于德国乌尔姆市的一个犹太人家庭，那一年电磁学理论的集大成者麦克斯韦驾鹤西去。1900年爱因斯坦毕业于苏黎世联邦理工学院，1905年获苏黎世大学哲学博士学位，1955年4月18日去世，享年76岁。他提出的光量子假设成功解释了光电效应，因此获得1921年诺贝尔物理学奖。他对人类的贡献远不止于此，1905年创立的狭义相对论，1915年发表的广义相对论，为核能开发奠定了理论基础，开创了现代科学技术新纪元，被公认为是与牛顿相提并论的伟大的物理学家。爱因斯坦有多厉害呢？仅以1905年为例，年仅26岁的他就写出了5篇震惊世界的论文：《分子体积的新测定》证明了分子的存在，《热的分子运动论所要求的静止液体中悬浮小粒子的运动》论证了微粒的运动规则，《论动体的电动力学》创立了狭义相对论，《物体的惯性同它所含的能量有关吗?》阐明了“$E=mc^2$”质能方程，《关于光的产生和转化的一个试探性观点》解释了光电效应。所以，联合国教科文组织将2005年(这一年是狭义相对论创立100周年，也是爱因斯坦离世50周年)定为“世界物理年”以此纪念物理学上的这一“奇迹年”。杨振宁先生这样评价爱因斯坦，“20世纪物理学的三大贡献中，两个半都是爱因斯坦的。”这“三大贡献”指的是狭义相对论、广义相对论和量子力学。爱因斯坦把科学看做他的生活，但他的生活远远不止科学一个方面，比如他是一名音乐爱好者，小提琴拉得相当不错；他还是一位关心人类发展的和平主义者，“二战”之后一直致力于号召全世界科学家团结起来反对核战争。如果大家想进一步了解这位伟人，可以参看由他同事派斯撰写的荣获过美国国家图书奖的《上帝难以捉摸：爱因斯坦的科学与生平》。

谈及爱因斯坦的相对论，大家往往第一反应是难，是高深的数学，是晦涩的理论，难以理解。因为它讨论的概念和很多推论与我们大多数人的基本认知是不同的，比如我们对时间和空间的看法。为了“以最简单、最明了的方式来介绍相对论”，1916年爱因斯坦亲自为“具备相当于大学入学考试的知识水平”的人们，写了这本《狭义与广义相对论浅说》(以下简称《浅说》)。《浅说》由四个部分组成：狭义相对论、广义相对论和关于整个宇宙的一些考虑，另加两个附录，它的第1版由德文写成(*Über die spezielle und die allgemeine Relativitätstheorie*：(*Gemeinverständlich*))，出版于1917年，5年后已经有了第40版，它还陆续被译成10多种语言广为流传，中译本第1版在1922年发行。爱因斯坦在这本书中阐明了时间和空间的本质属性，论证了时间和空间的内在联系和协变性；发展和拓宽了牛顿力学，揭示了质量和能量之间的内在联系；对物质的存在和运动提出了全新的解释，使人们对所处世界和宇宙的认识提升到新的阶段。《浅说》是一本引发现代科学伟大变革，对整个人类思想发展产生深远影响的划时代的著作。请相信，你们完全能够读懂它，只需要多一点自信和坚持！

到底什么是相对论呢？爱因斯坦常常这样回答，“你和一个漂亮姑娘在公园长椅上坐一小时，觉得只过了一分钟；你紧挨着一个火炉坐一分钟，却觉得过了一个小时。这就是相对论。”事实上相对论的发展分为两个阶段，即狭义相对论和广义相对论。狭义相对论有两个前提，分别是惯性系相互等价和光速在真空中不变。只要彻底理解以上两个假设，加上一点并不复杂的数学知识，狭义相对论的所有推论都能够简单地推导出来。如时间和空间是一个整体；如能量与质量之间简单而本质的联系。值得一提的是，在狭义相对论创立过程中有两位先驱：一位是洛伦兹(Hendrik Antoon Lorentz，1853—1928，著名物理学家、数学家，1902年诺贝尔物理学奖获得者)，他提出了收缩假设；另一位是庞加莱(Jules Henri Poincaré，1854—1912，著名数学家、天文学家、理论物理学家)，他首次给出“相对性原理”并讨论了光速不变的问题。爱因斯坦在洛伦兹的葬礼上说洛伦兹的成就“对我产生了最伟大的影响”，而杨振宁先生的评价更为中肯：“庞加莱懂得了相对论的哲学，没有懂得它的物理；洛伦兹懂得了它的数学，也没有懂得它的物理。只有年轻的爱因斯坦能创建出此革命”。这是因为爱因斯坦对数学有着直觉性的欣赏能力，对物理既能近看又能远看，可以从不可能出问题的地方看出了问题，颠覆了人类对时间空间的看法。

这个世界真的存在一个优越的绝对的惯性系吗？这是爱因斯坦思索了十年的问题。在1915年发表的广义相对论中，他认为自然界不存在一个绝对的惯性系；所有的基本物理规律在任一参考系中具有相同的形式。爱因斯坦还告诉我们，对任意弯曲空间如何用任意形式的坐标系来表述时空距离，如何确定两点之间的最短路径。以上种种是对普通大众时空观的又一次巨大挑战。如果要用一句话高度概括广义相对论，爱因斯坦的合作者惠勒(John Wheeler，1911—2008，美国理论物理学家)说得非常好，他说“时空告诉物质如何运动；物质告诉时空如何弯曲(Spacetime tells matter how to move；matter tells spacetime how to curve)。”

从伽利略、牛顿建立现代科学以来，几乎所有的科学理论发展都遵循一个规律：先观测到新的现象，研究者们再对此提出相应的理论解释，接着用新的数据和实验来检验理论的真伪，如爱因斯坦的狭义相对论，如达尔文的进化论，如门捷列夫的元素周期表，除了广义相对论。对狭义相对论而言，如果没有爱因斯坦，它依然会出现在世人眼前，最多晚上几年；但这样的说法放在广义相对论上，就另当别论。广义相对论是爱因斯坦凭借一己之力在脑海中思考、孕育而得。它横空出世，在当时没有社会生产的需求，没有多少人能够真正理解，但其蕴藏着深奥的哲学、直观的物理学和精湛的数学。

从1915年爱因斯坦发表广义相对论，到1916年预言引力波的存在，直至2016年人类首次探测到它，刚好是100年。一百年的时光里，人们验证了所有广义相对论的经典预言，从光线偏折，到黑洞，再到引力波。正在阅读的你对这一切是否感到好奇？

时不我待，让我们赶快开启这趟时空之旅吧！

（祁宁）

Relativity: the Special and the General Theory

Albert Einstein

I. The Principle of Relativity (in the restricted sense)

1 In order to attain the greatest possible clearness, let us return to our example of the railway carriage supposed to be travelling uniformly. We call its motion a uniform translation ("uniform" because it is of constant velocity and direction, "translation" because although the carriage changes its position relative to the embankment yet it does not rotate in so doing). Let us imagine a raven flying through the air in such a manner that its motion, as observed from the embankment, is uniform and in a straight line. If we were to observe the flying raven from the moving railway carriage, we should find that the motion of the raven would be one of different velocity and direction, but that it would still be uniform and in a straight line. Expressed in an abstract manner we may say: If a mass m is moving uniformly in a straight line with respect to a co-ordinate system K, then it will also be moving uniformly and in a straight line relative to a second co-ordinate system K' provided that the latter is executing a uniform translatory motion with respect to K. In accordance with the discussion contained in the preceding section, it follows that:

2 If K is a Galileian co-ordinate system, then every other co-ordinate system K' is a Galileian one, when, in relation to K, it is in a condition of uniform motion of translation. Relative to K' the mechanical laws of Galilei-Newton hold good exactly as they do with respect to K.

本文选自阿尔伯特·爱因斯坦著 *Relativity: The Special and The General Theory*。

狭义与广义相对论浅说

阿尔伯特·爱因斯坦

一、相对性原理(狭义)

1　为了使我们的论述尽可能地清楚明确，让我们回到设想为匀速行驶中的火车车厢这个实例上来。我们称该车厢的运动为一种匀速平移运动(称为“匀速”是由于速度和方向是恒定的；称为“平移”是由于虽然车厢相对于路基不断改变其位置，但在这样的运动中并无转动)。设想一只大乌鸦在空中飞过，它的运动方式从路基上观察是匀速直线运动。如果我们在行驶着的车厢里观察这只飞鸟，我们就会发现这乌鸦是以另一种速度和方向在飞行，但仍然是匀速直线运动。用抽象的方式来表述，我们可以说：若一质量 m 相对于一坐标系 K 作匀速直线运动，只要第二个坐标系 K' 相对于 K 是在作匀速平移运动，则该质量相对于第二个坐标系 K' 亦作匀速直线运动。根据上节的论述可以推出：

2　若 K 为一伽利略坐标系，则其他每一个相对于 K 作匀速平移运动的坐标系 K' 亦为一伽利略坐标系。相对于 K'，正如相对于 K 一样，伽利略-牛顿力学定律也是成立的。

本文选自[美]阿尔伯特·爱因斯坦：《狭义与广义相对论浅说》(科学素养文库·科学元典丛书)，杨润殷译，北京大学出版社 2006 年版，第 12-14 页，第 15-23 页，第 47-49 页，第 52-54 页，第 55-59 页。

3 We advance a step farther in our generalisation when we express the tenet thus: If, relative to K, K' is a uniformly moving co-ordinate system devoid of rotation, then natural phenomena run their course with respect to K' according to exactly the same general laws as with respect to K. This statement is called the *principle of relativity* (in the restricted sense).

4 As long as one was convinced that all natural phenomena were capable of representation with the help of classical mechanics, there was no need to doubt the validity of this principle of relativity. But in view of the more recent development of electrodynamics and optics it became more and more evident that classical mechanics affords an insufficient foundation for the physical description of all natural phenomena. At this juncture the question of the validity of the principle of relativity became ripe for discussion, and it did not appear impossible that the answer to this question might be in the negative.

5 Nevertheless, there are two general facts which at the outset speak very much in favour of the validity of the principle of relativity. Even though classical mechanics does not supply us with a sufficiently broad basis for the theoretical presentation of all physical phenomena, still we must grant it a considerable measure of "truth," since it supplies us with the actual motions of the heavenly bodies with a delicacy of detail little short of wonderful. The principle of relativity must therefore apply with great accuracy in the domain of mechanics. But that a principle of such broad generality should hold with such exactness in one domain of phenomena, and yet should be invalid for another, is a priori not very probable.

6 We now proceed to the second argument, to which, moreover, we shall return later. If the principle of relativity (in the restricted sense) does not

3 如果我们把上面的推论作如下的表述，我们在推广方面就前进了一步：如果 K' 是相对于 K 作匀速运动而无转动的坐标系，那么，自然现象相对于坐标系 K' 的实际演变将与相对于坐标系 K 的实际演变一样依据同样的普遍定律。这个陈述称为相对性原理(狭义)。

4 只要人们确信一切自然现象都能够借助于经典力学来得到完善的表述，就没有必要怀疑这个相对性原理的正确性。但是由于晚近在电动力学和光学方面的发展，人们越来越清楚地看到，经典力学为一切自然现象的物理描述所提供的基础还是不够充分的。到这个时候，讨论相对性原理的正确性问题的时机就成熟了，而且在当时看来对这个问题作否定的答复并不是不可能的。

5 然而有两个普遍事实在一开始就给予相对性原理的正确性以很有力的支持。虽然经典力学对于一切物理现象的理论表述没有提供一个足够广阔的基础，但是我们仍然必须承认经典力学在相当大的程度上是“真理”，因为经典力学对天体的实际运动的描述，所达到的精确度简直是惊人的。因此，在力学的领域中应用相对性原理必然达到很高的准确度。一个具有如此广泛的普遍性的原理，在物理现象的一个领域中的有效性具有这样高的准确度，而在另一个领域中居然会无效，这从先验的观点来看是不大可能的。

6 现在我们来讨论第二个论据，这个论据以后还要谈到。如果相对性原理(狭义)不成立，那么，彼此作相对匀速运动的 K、K'、K'' 等一系列伽利

hold, then the Galileian co-ordinate systems K, K', K'', etc., which are moving uniformly relative to each other, will not be equivalent for the description of natural phenomena. In this case we should be constrained to believe that natural laws are capable of being formulated in a particularly simple manner, and of course only on condition that, from amongst all possible Galileian co-ordinate systems, we should have chosen one(K_0) of a particular state of motion as our body of reference. We should then be justified (because of its merits for the description of natural phenomena) in calling this system "absolutely at rest," and all other Galileian systems K'' in motion." If, for instance, our embankment were the system K_0 then our railway carriage would be a system K, relative to which less simple laws would hold than with respect to K_0. This diminished simplicity would be due to the fact that the carriage K would be in motion (i.e. "really") with respect to K_0. In the general laws of nature which have been formulated with reference to K, the magnitude and direction of the velocity of the carriage would necessarily play a part. We should expect, for instance, that the note emitted by an organpipe placed with its axis parallel to the direction of travel would be different from that emitted if the axis of the pipe were placed perpendicular to this direction. Now in virtue of its motion in an orbit round the sun, our earth is comparable with a railway carriage travelling with a velocity of about 30 kilometres per second. If the principle of relativity were not valid we should therefore expect that the direction of motion of the earth at any moment would enter into the laws of nature, and also that physical systems in their behaviour would be dependent on the orientation in space with respect to the earth. For owing to the alteration in direction of the velocity of revolution of the earth in the course of a year, the earth cannot be at rest relative to the hypothetical system K_0 throughout the whole year. However, the most careful observations have never revealed such anisotropic properties in terrestrial physical space, i. e. a physical non-equivalence of different directions. This is very powerful argument in favour of the principle of relativity.

略坐标系，对于描述自然现象就不是等效的。在这个情况下我们就不得不相信自然界定律能够以一种特别简单的形式来表述，这当然只有在下列条件下才能做到，即我们已经从一切可能有的伽利略坐标系中选定了一个具有特别的运动状态的坐标系(K_0)作为我们的参考系。这样，我们就会有理由(由于这个坐标系对描述自然现象具有优点)称这个坐标系是“绝对静止的”，而所有其他的伽利略坐标系 K 都是“运动的”。举例来说，设我们的铁路路基是坐标系 K_0，那么我们的火车车厢就是坐标系 K，相对于坐标系 K 成立的定律将不如相对于坐标系 K_0成立的定律那样简单。定律的简单性的此种减退是由于车厢 K 相对于 K_0而言是运动的(亦即“真正”是运动的)。在参照 K 所表述的普遍的自然界定律中，车厢速度的大小和方向必然是起作用的。例如，我们应该预料到，一个风琴管当它的轴与运动的方向平行时所发出的音调将不同于当它的轴与运动的方向垂直时所发出的音调。由于我们的地球是在环绕太阳的轨道上运行，因而我们可以把地球比作以每秒大约 30 公里的速度行驶的火车车厢。如果相对性原理是不正确的，我们就应该预料到，地球在任一时刻的运动方向将会在自然界定律中表现出来，而且物理系统的行为将与其相对于地球的空间取向有关。因为由于在一年中地球公转速度的方向的变化，地球不可能在全年中相对于假设的坐标系 K_0处于静止状态。但是，最仔细的观察也从来没有显示出地球物理空间的这种各向异性(即不同方向的物理不等效性)。这是一个支持相对性原理的十分强有力的论据。

Ⅱ. The Theorem of the Addition of Velocities Employed in Classical Mechanics

7 Let us suppose our old friend the railway carriage to be travelling along the rails with a constant velocity v, and that a man traverses the length of the carriage in the direction of travel with a velocity w. How quickly or, in other words, with what velocity W does the man advance relative to the embankment during the process? The only possible answer seems to result from the following consideration: If the man were to stand still for a second, he would advance relative to the embankment through a distance v equal numerically to the velocity of the carriage. As a consequence of his walking, however, he traverses an additional distance w relative to the carriage, and hence also relative to the embankment, in this second, the distance w being numerically equal to the velocity with which he is walking. Thus in total he covers the distance $W = v + w$ relative to the embankment in the second considered. We shall see later that this result, which expresses the theorem of the addition of velocities employed in classical mechanics, cannot be maintained; in other words, the law that we have just written down does not hold in reality. For the time being, however, we shall assume its correctness.

Ⅲ. The Apparent Incompatibility of the Law of Propagation of Light with the Principle of Relativity

8 There is hardly a simpler law in physics than that according to which light is propagated in empty space. Every child at school knows, or believes he knows, that this propagation takes place in straight lines with a velocity $c=$ 300,000 km./sec. At all events we know with great exactness that this velocity is the same for all colours, because if this were not the case, the

二、经典力学中所用的速度相加定理

7 假设我们的旧相识，火车车厢，在铁轨上以恒定速度 v 行驶；并假设有一个人在车厢里沿着车厢行驶的方向以速度 w 从车厢一头走到另一头。那么在这个过程中，对于路基而言，这个人向前走得有多快呢？换句话说，这个人前进的速度 W 有多大呢？唯一可能的解答似乎可以根据下列考虑而得：如果这个人站住不动一秒钟，在这一秒钟里他就相对于路基前进了一段距离 v，在数值上与车厢的速度相等。但是，由于他在车厢中向前运动，在这一秒钟里他相对于车厢向前走了一段距离 w，也就是相对于路基又多走了一段距离 w，这段距离在数值上等于这个人在车厢里走动的速度。这样，在所考虑的这一秒钟里他总共相对于路基走了距离 $W=v+w$。我们以后将会看到，表述了经典力学的速度相加定理的这一结果，是不能加以支持的；换句话说，我们刚才写下的定律实质上是不成立的。但目前我们暂时假定这个定理是正确的。

三、光的传播定律与相对性原理的表面抵触

8 在物理学中几乎没有比真空中光的传播定律更简单的定律了。学校里的每个儿童都知道，或者相信他知道，光在真空中沿直线以速度 $c=$ 300 000千米/秒传播。无论如何我们非常精确地知道，这个速度对于所有各色光线都是一样的。因为如果不是这样，则当一颗恒星为其邻近的黑暗星体所掩食时，其各色光线的最小发射值就不会同时被看到。荷兰天文学家德西特(De Sitter)根据对双星的观察，也以相似的理由指出，光的传播速度不能依赖于发光物体的运动速度。关于光的传播速度与其"在空间中"的方向有关的假定即就其本身而言也是难以成立的。

minimum of emission would not be observed simultaneously for different colours during the eclipse of a fixed star by its dark neighbour. By means of similar considerations based on observations of double stars, the Dutch astronomer De Sitter was also able to show that the velocity of propagation of light cannot depend on the velocity of motion of the body emitting the light. The assumption that this velocity of propagation is dependent on the direction "in space" is in itself improbable.

9 In short, let us assume that the simple law of the constancy of the velocity of light c (in vacuum) is justifiably believed by the child at school. Who would imagine that this simple law has plunged the conscientiously thoughtful physicist into the greatest intellectual difficulties? Let us consider how these difficulties arise.

10 Of course we must refer the process of the propagation of light (and indeed every other process) to a rigid reference-body (co-ordinate system). As such a system let us again choose our embankment. We shall imagine the air above it to have been removed. If a ray of light be sent along the embankment, we see from the above that the tip of the ray will be transmitted with the velocity c relative to the embankment. Now let us suppose that our railway carriage is again travelling along the railway lines with the velocity v, and that its direction is the same as that of the ray of light, but its velocity of course much less. Let us inquire about the velocity of propagation of the ray of light relative to the carriage. It is obvious that we can here apply the consideration of the previous section, since the ray of light plays the part of the man walking along relatively to the carriage. The velocity w of the man relative to the embankment is here replaced by the velocity of light relative to the embankment. w is the required velocity of light with respect to the carriage, and we have

9　总之，我们可以假定关于光(在真空中)的速度 c 是恒定的这一简单的定律已有充分的理由为学校里的儿童所确信。谁会想到这个简单的定律竟会使思想周密的物理学家陷入智力上的极大的困难呢？让我们来看看这些困难是怎样产生的。

10　当然我们必须参照一个刚体(坐标系)来描述光的传播过程(对于所有其他的过程而言确实也都应如此)。我们再次选取我们的路基作为这种参考系。我们设想路基上面的空气已经抽空。如果沿着路基发出一道光线，根据上面的论述我们可以看到，这道光线的前端将相对于路基以速度 c 传播。现在我们假定我们的车厢仍然以速度 v 在路轨上行驶，其方向与光线的方向相同，不过车厢的速度当然要比光的传播速度小得多。我们来研究一下这光线相对于车厢的传播速度问题。显然我们在这里可以应用前一节的推论，因为光线在这里就充当了相对于车厢走动的人。人相对于路基的速度 w 在这里由光相对于路基的速度代替。w 是所求的光相对于车厢的速度，我们得到：

$$w=c-v.$$

于是光线相对于车厢的传播速度就出现了小于 c 的情况。

$$w = c - v.$$

The velocity of propagation of a ray of light relative to the carriage thus comes out smaller than c.

11 But this result comes into conflict with the principle of relativity set forth in Section Ⅰ. For, like every other general law of nature, the law of the transmission of light in vacuo must, according to the principle of relativity, be the same for the railway carriage as reference-body as when the rails are the body of reference. But, from our above consideration, this would appear to be impossible. If every ray of light is propagated relative to the embankment with the velocity c, then for this reason it would appear that another law of propagation of light must necessarily hold with respect to the carriage—a result contradictory to the principle of relativity.

12 In view of this dilemma there appears to be nothing else for it than to abandon either the principle of relativity or the simple law of the propagation of light in vacuo. Those of you who have carefully followed the preceding discussion are almost sure to expect that we should retain the principle of relativity, which appeals so convincingly to the intellect because it is so natural and simple. The law of the propagation of light in vacuo would then have to be replaced by a more complicated law conformable to the principle of relativity. The development of theoretical physics shows, however, that we cannot pursue this course. The epoch-making theoretical investigations of H. A. Lorentz on the electrodynamical and optical phenomena connected with moving bodies show that experience in this domain leads conclusively to a theory of electromagnetic phenomena, of which the law of the constancy of the velocity of light in vacuo is a necessary consequence. Prominent theoretical physicists were therefore more inclined to reject the principle of relativity, in spite of the fact that no empirical data had been found which were contradictory to this principle.

11 但是这个结果与第一节所阐述的相对性原理是相抵触的。因为，根据相对性原理，真空中光的传播定律，就像所有其他普遍的自然界定律一样，不论以车厢作为参考物体还是以路轨作为参考物体，都必须是一样的。但是，从我们前面的论述看来，这一点似乎是不可能成立的。如果所有的光线相对于路基都以速度 c 传播，那么由于这个理由似乎光相对于车厢的传播就必然服从另一定律——这是一个与相对性原理相抵触的结果。

12 由于这种抵触，除了放弃相对性原理或放弃真空中光的传播的简单定律以外，其他办法似乎是没有的。仔细地阅读了以上论述的读者几乎都相信我们应该保留相对性原理，这是因为相对性原理如此自然而简单，在人们的思想中具有很大的说服力。因而，真空中光的传播定律就必须由一个能与相对性原理一致的比较复杂的定律所取代。但是，理论物理学的发展说明了我们不能遵循这一途径。具有划时代意义的洛伦兹对于与运动物体相关的电动力学和光学现象的理论研究表明，在这个领域中的经验无可争辩地导致了关于电磁现象的一个理论，而真空中光速恒定定律是这个理论的必然推论。因此，尽管不曾发现与相对性原理相抵触的实验数据，许多著名的理论物理学家还是比较倾向于舍弃相对性原理。

13 At this juncture the theory of relativity entered the arena. As a result of an analysis of the physical conceptions of time and space, it became evident that in reality there is not the least incompatibility between the principle of relativity and the law of propagation of light, and that by systematically holding fast to both these laws a logically rigid theory could be arrived at. This theory has been called the special theory of relativity to distinguish it from the extended theory, with which we shall deal later. In the following pages we shall present the fundamental ideas of the special theory of relativity.

Ⅳ. On the Idea of Time in Physics

14 Lightning has struck the rails on our railway embankment at two places *A* and *B* far distant from each other. I make the additional assertion that these two lightning flashes occurred simultaneously. If now I ask you whether there is sense in this statement, you will answer my question with a decided "Yes." But if I now approach you with the request to explain to me the sense of the statement more precisely, you find after some consideration that the answer to this question is not so easy as it appears at first sight.

15 After some time perhaps the following answer would occur to you: "The significance of the statement is clear in itself and needs no further explanation; of course it would require some consideration if I were to be commissioned to determine by observations whether in the actual case the two events took place simultaneously or not." I cannot be satisfied with this answer for the following reason. Supposing that as a result of ingenious considerations an able meteorologist were to discover that the lightning must always strike the places *A* and *B* simultaneously, then we should be faced with the task of testing whether or not this theoretical result is in accordance

13 相对论就是在这个关头产生的。由于分析了时间和空间的物理概念，人们开始清楚地看到，相对性原理和光的传播定律实际上丝毫没有抵触之处，如果系统地贯彻这两个定律，就能够得到一个逻辑严谨的理论。这个理论已被称为狭义相对论，以区别于推广了的理论，对于广义理论我们将留待以后再去讨论。下面我们将叙述狭义相对论的基本观念。

四、物理学的时间观

14 在我们的铁路路基上彼此相距相当远的两处 A 和 B，雷电击中了铁轨。我再补充一句，这两处的雷电闪光是同时发生的。如果我问你这句话有没有意义，你会很肯定地回答说，“有”。但是，如果我接下去请你更确切地向我解释一下这句话的意义，那么你在考虑一下以后就会感到回答这个问题并不像乍看起来那样容易。

15 经过一些时间的考虑之后，你或许会想出如下的回答：“这句话的意义本来就是清楚的，无需再加解释；当然，如果要我用观测的方法来确定在实际情况中这两个事件是否同时发生的，我就需要考虑考虑。”对于这个答复我不能感到满意，理由如下：假定有一位能干的气象学家经过巧妙的思考发现闪电必然同时击中 A 处和 B 处的话，那么我们就面对着这样的任务，即必须检验一下这个理论结果是否与实际相符。在一切物理陈述中凡是含有“同时”概念之处，我们都遇到了同样的困难。对于物理学家而言，在他有可能判断一个概念在实际情况中是否真被满足以前，这概念就还不能成立。因此我们需要有这样一个同时性定义，这定义必须能提供一个方法，以便在本例中使物理学家可以用这个方法通过实验来确定那两处雷击是否真正同时发生。如果在这个要求还没有得到满足

with the reality. We encounter the same difficulty with all physical statements in which the conception " simultaneous " plays a part. The concept does not exist for the physicist until he has the possibility of discovering whether or not it is fulfilled in an actual case. We thus require a definition of simultaneity such that this definition supplies us with the method by means of which, in the present case, he can decide by experiment whether or not both the lightning strokes occurred simultaneously. As long as this requirement is not satisfied, I allow myself to be deceived as a physicist (and of course the same applies if I am not a physicist) , when I imagine that I am able to attach a meaning to the statement of simultaneity. (I would ask the reader not to proceed farther until he is fully convinced on this point.)

16 After thinking the matter over for some time you then offer the following suggestion with which to test simultaneity. By measuring along the rails, the connecting line *AB* should be measured up and an observer placed at the mid-point *M* of the distance *AB*. This observer should be supplied with an arrangement (e.g. two mirrors inclined at 90°) which allows him visually to observe both places *A* and *B* at the same time. If the observer perceives the two flashes of lightning at the same time, then they are simultaneous.

17 I am very pleased with this suggestion, but for all that I cannot regard the matter as quite settled, because I feel constrained to raise the following objection: " Your definition would certainly be right, if I only knew that the light by means of which the observer at *M* perceives the lightning flashes travels along the length $A \rightarrow M$ with the same velocity as along the length $B \rightarrow M$. But an examination of this supposition would only be possible if we already had at our disposal the means of measuring time. It would thus appear as though we were moving here in a logical circle."

以前，我就认为我能够赋予同时性这个说法以某种意义，那么作为一个物理学家，这就是自欺欺人(当然，如果我不是物理学家也是一样)。(请读者完全搞通这一点之后再继续读下去。)

16 在经过一些时间的思考之后，你提出下列建议来检验同时性。沿着铁轨测量就可以量出连线 AB 的长度，然后把一位观察者安置在距离 AB 的中点 M。这位观察者应备有一种装置(例如，相互成 $90°$ 的两面镜子)，使他能够同时既观察到 A 处又能观察到 B 处。如果这位观察者的视神经在同一时刻感觉到这两束雷电闪光，那么这两束雷电闪光就必定是同时的。

17 对于这个建议我感到十分高兴，但是尽管如此我仍然不能认为问题已经完全解决，因为我感到不得不提出以下的不同意见："如果我能够知道，观察者站在 M 处赖以看到闪电的那些光，从 A 传播到 M 的速度与从 B 传播到 M 的速度确是相同，那么你的定义当然是对的。但是，要对这个假定进行验证，只有我们已经掌握测量时间的方法才有可能。因此从逻辑上看来我们好像尽是在这里兜圈子。"

18 After further consideration you cast a somewhat disdainful glance at me—and rightly so—and you declare: "I maintain my previous definition nevertheless, because in reality it assumes absolutely nothing about light. There is only one demand to be made of the definition of simultaneity, namely, that in every real case it must supply us with an empirical decision as to whether or not the conception that has to be defined is fulfilled. That my definition satisfies this demand is indisputable. That light requires the same time to traverse the path $A \rightarrow M$ as for the path $B \rightarrow M$ is in reality neither a supposition nor a hypothesis about the physical nature of light, but a stipulation which I can make of my own freewill in order to arrive at a definition of simultaneity."

19 It is clear that this definition can be used to give an exact meaning not only to two events, but to as many events as we care to choose, and independently of the positions of the scenes of the events with respect to the body of reference 1 (here the railway embankment)①. We are thus led also to a definition of "time" in physics. For this purpose we suppose that clocks of identical construction are placed at the points A, B and C of the railway line (co-ordinate system), and that they are set in such a manner that the positions of their pointers are simultaneously (in the above sense) the same. Under these conditions we understand by the "time" of an event the reading (position of the hands) of that one of these clocks which is in the immediate vicinity (in space) of the event. In this manner a time-value is associated with every event which is essentially capable of observation.

① Note 1. We suppose further that, when three events A, B and C take place in different places in such a manner that, if A is simultaneous with B, and B is simultaneous with C (simultaneous in the sense of the above definition), then the criterion for the simultaneity of the pair of events A, C is also satisfied. This assumption is a physical hypothesis about the law of propagation of light; it must certainly be fulfilled if we are to maintain the law of the constancy of the velocity of light in vacuo.

18 经过进一步考虑后，你带着些轻蔑的神气瞟我一眼（这是无可非议的），并宣称，“尽管如此我仍然维持我先前的定义，因为实际上这个定义完全没有对光作过任何假定。对于同时性的定义仅有一个要求，那就是在每一个实际情况中这个定义必须为我们提供一个实验方法来判断所规定的概念是否真被满足。我的定义已经满足这个要求是无可争辩的。光从 A 传播到 M 与从 B 传播到 M 所需时间相同，这实际上既不是关于光的物理性质的推断，也不是关于光的物理性质的假说，而仅是为了得出同时性的定义我按照我自己的自由意志所能作出的一种规定。”

19 显然这个定义不仅能够对两个事件的同时性，而且能够对我们愿意选定的任意多个事件的同时性规定出一个确切的意义，而与这些事件发生的地点相对于参考物体（在这里就是铁路路基）的位置无关①。由此我们也可以得出物理学的“时间”定义。为此，我们假定把构造完全相同的钟放在铁路线（坐标系）上的 A、B 和 C 诸点上，并这样校准它们，使它们的指针同时（按照上述意义来理解）指着相同的位置。在这些条件下，我们把一个事件的“时间”理解为放置在该事件的（空间）最邻近处的那个钟上的读数（指针所指位置）。这样，每一个本质上可以观测的事件都有一个时间数值与之相联系。

① 我们进一步推测，如果有三个事件 A、B 和 C 在不同地点按照下列方式发生：即 A 与 B 同时，而 B 又与 C 同时（同时的意义按照上述定义来理解），那么 A 和 C 这一对事件的同时性的判据就也得到了满足。这个是关于光的传播定律的一个物理假说；如果我们支持真空中光速恒定定律，这个假说就必然成立。

20 This stipulation contains a further physical hypothesis, the validity of which will hardly be doubted without empirical evidence to the contrary. It has been assumed that all these clocks go at the same rate if they are of identical construction. Stated more exactly: When two clocks arranged at rest in different places of a reference-body are set in such a manner that a particular position of the pointers of the one clock is simultaneous (in the above sense) with the same position of the pointers of the other clock, then identical "settings" are always simultaneous (in the sense of the above definition).

V. The Relativity of Simultaneity

21 Up to now our considerations have been referred to a particular body of reference, which we have styled a "railway embankment." We suppose a very long train travelling along the rails with the constant velocity v and in the direction indicated in Fig. 1. People travelling in this train will with advantage use the train as a rigid reference-body (co-ordinate system); they regard all events in reference to the train. Then every event which takes place along the line also takes place at a particular point of the train. Also the definition of simultaneity can be given relative to the train in exactly the same way as with respect to the embankment. As a natural consequence, however, the following question arises.

22 Are two events (e. g. the two strokes of lightning A and B) which are simultaneous with reference to the railway embankment also simultaneous relatively to the train? We shall show directly that the answer must be in the negative.

23 When we say that the lightning strokes A and B are simultaneous with respect to the embankment, we mean: the rays of light emitted at the places A and B, where the lightning occurs, meet each other at the mid-point M of the

20 这个规定还包含着另一个物理假说，如果没有相反的实验证据的话，这个假说的有效性是不大会被人怀疑的。这里已经假定，如果所有这些钟的构造完全一样，它们就以同样的时率走动。说得更确切些：如果我们这样校准静止在一个参考物体的不同地方的两个钟，使其中一个钟的指针指着某一个特定的位置的同时（按照上述意义来理解），另一个钟的指针也指着相同的位置，那么完全相同的"指针位置"就总是同时的（"同时"的意义按照上述定义来理解）。

五、同时性的相对性

21 到目前为止，我们的论述一直是参照我们称之为"铁路路基"的一个特定的参考物体来进行的。假设有一列很长的火车，以恒速 v 沿着图 1 所标明的方向在轨道上行驶。在这列火车上旅行的人们可以很方便地把火车当作刚性参考物体（坐标系）；他们参照火车来观察一切事件。因而，在铁路线上发生的每一个事件也在火车上某一特定的地点发生。而且完全和相对于路基所作的同时性定义一样，我们也能够相对于火车作出同时性的定义。但是，作为一个自然的推论，下述问题就随之产生：

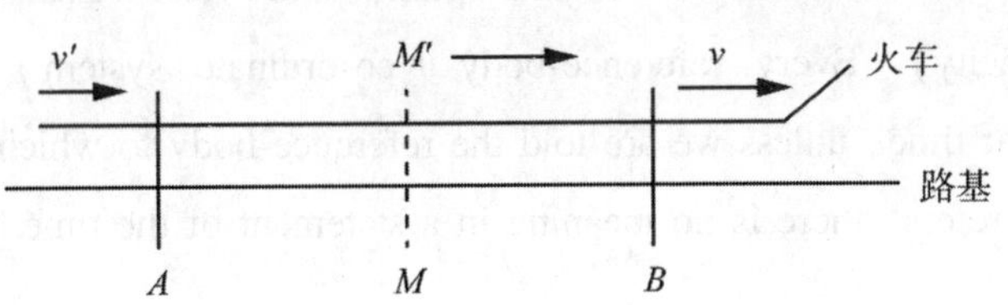

图 1

22 对于铁路路基来说是同时的两个事件（例如 A、B 两处雷击），对于火车来说是否也是同时的呢？我们将直接证明，回答必然是否定的。

23 当我们说 A、B 两处雷击相对于路基而言是同时的，我们的意思是：在发生闪电的 A 处和 B 处所发出的光，在路基 $A \rightarrow B$ 这段距离的中点 M 相遇。但是事件 A 和 B 也对应于火车上的 A 点和 B 点。令 M' 为在行驶中

length $A \rightarrow B$ of the embankment. But the events A and B also correspond to positions A and B on the train. Let M' be the mid-point of the distance $A \rightarrow B$ on the travelling train. Just when the flashes 1 of lightning occur, this point M' naturally coincides with the point M, but it moves towards the right in the diagram with the velocity v of the train. If an observer sitting in the position M' in the train did not possess this velocity, then he would remain permanently at M, and the light rays emitted by the flashes of lightning A and B would reach him simultaneously, i.e. they would meet just where he is situated. Now in reality (considered with reference to the railway embankment) he is hastening towards the beam of light coming from B, whilst he is riding on ahead of the beam of light coming from A. Hence the observer will see the beam of light emitted from B earlier than he will see that emitted from A. Observers who take the railway train as their reference-body must therefore come to the conclusion that the lightning flash B took place earlier than the lightning flash A. We thus arrive at the important result:

24 Events which are simultaneous with reference to the embankment are not simultaneous with respect to the train, and vice versa (relativity of simultaneity). Every reference-body (co-ordinate system) has its own particular time; unless we are told the reference-body to which the statement of time refers, there is no meaning in a statement of the time of an event.

25 Now before the advent of the theory of relativity it had always tacitly been assumed in physics that the statement of time had an absolute significance, i.e. that it is independent of the state of motion of the body of reference. But we have just seen that this assumption is incompatible with the most natural definition of simultaneity; if we discard this assumption, then the conflict between the law of the propagation of light in vacuo and the principle of relativity (developed in Section Ⅲ) disappears.

的火车上 $A \to B$ 这段距离的中点。正当雷电闪光发生的时候①，点 M' 自然与点 M 重合，但是点 M' 以火车的速度 v 向图中的右方移动。如果坐在火车上 M' 处的一个观察者并不具有这个速度，那么他就总是停留在 M 点，雷电闪光 A 和 B 所发出的光就同时到达他这里，也就是说正好在他所在的地方相遇。可是实际上(相对于铁路路基来考虑)这个观察者正在朝着来自 B 的光线急速行进，同时他又是在来自 A 的光线的前方向前行进。因此，这个观察者将先看见自 B 发出的光线，后看见自 A 发出的光线。所以，把列车当作参考物体的观察者就必然得出这样的结论，即雷电闪光 B 先于雷电闪光 A 发生。这样我们就得出以下的重要结果：

24 对于路基是同时的若干事件，对于火车并不是同时的，反之亦然(同时性的相对性)。每一个参考物体(坐标系)都有它本身的特殊的时间；除非我们讲出关于时间的陈述是相对于哪一个参考物体的，否则关于一个事件的时间的陈述就没有意义。

25 在相对论创立以前，在物理学中一直存在着一个隐含的假定，即时间的陈述具有绝对的意义，亦即时间的陈述与参考物体的运动状态无关。但是我们刚才看到，这个假定与最自然的同时性定义是不相容的；如果我们抛弃这个假定，那么真空中光的传播定律与相对性原理之间的抵触(详见第三节)就消失了。

① 从路基上判断。

26 We were led to that conflict by the considerations of Section Ⅱ, which are now no longer tenable. In that section we concluded that the man in the carriage, who traverses the distance *w* per second relative to the carriage, traverses the same distance also with respect to the embankment in each second of time. But, according to the foregoing considerations, the time required by a particular occurrence with respect to the carriage must not be considered equal to the duration of the same occurrence as judged from the embankment (as reference-body). Hence it cannot be contended that the man in walking travels the distance *w* relative to the railway line in a time which is equal to one second as judged from the embankment.

27 Moreover, the considerations of Section Ⅱ are based on yet a second assumption, which, in the light of a strict consideration, appears to be arbitrary, although it was always tacitly made even before the introduction of the theory of relativity.

Ⅵ. On the Relativity of the Conception of Distance

28 Let us consider two particular points on the train travelling along the embankment with the velocity *v*, and inquire as to their distance apart. We already know that it is necessary to have a body of reference for the measurement of a distance, with respect to which body the distance can be measured up. It is the simplest plan to use the train itself as reference-body (co-ordinate system). An observer in the train measures the interval by marking off his measuring-rod in a straight line (e.g. along the floor of the carriage) as many times as is necessary to take him from the one marked point to the other. Then the number which tells us how often the rod has to be laid down is the required distance.

26 这个抵触是根据第二节的论述推论出来的，这些论点现在已经站不住脚了。在该节我们曾得出这样的结论：在车厢里的人如果相对于车厢每秒走距离 w，那么在每一秒钟的时间里他相对于路基也走了相同的一段距离。但是，按照以上论述，相对于车厢发生一特定事件所需要的时间，绝不能认为就等于从路基(作为参考物体)上判断的发生同一事件所需要的时间。因此我们不能硬说在车厢里走动的人相对于铁路线走距离 w 所需的时间从路基上判断也等于一秒钟。

27 此外，第二节的论述还基于另一个假定。按照严格的探讨看来，这个假定是任意的，虽然在相对论创立以前人们一直在物理学中隐藏着这个假定。

六、距离概念的相对性

28 我们来考虑火车上的两个特定的点①，火车以速度 v 在铁路上行驶，现在要研究这两个点之间的距离。我们已经知道，测量一段距离，需要有一个参考物体，以便相对于这个物体量出这段距离的长度。最简单的办法是利用火车本身作为参考物体(坐标系)。在火车上的一个观察者测量这段间隔的方法是用他的量杆沿着一条直线(例如沿着车厢的地板)一下一下地量，从一个给定的点到另一个给定的点需要量多少下他就量多少下。那么，告诉我们这个量杆需要量多少下的那个数字就是所求的距离。

① 例如第 1 节车厢的中点和第 20 节车厢的中点。

29 It is a different matter when the distance has to be judged from the railway line. Here the following method suggests itself. If we call A' and B' the two points on the train whose distance apart is required, then both of these points are moving with the velocity v along the embankment. In the first place we require to determine the points A and B of the embankment which are just being passed by the two points A' and B' at a particular time t—judged from the embankment. These points A and B of the embankment can be determined by applying the definition of time given in Section Ⅳ. The distance between these points A and B is then measured by repeated application of thee measuring-rod along the embankment.

30 A priori it is by no means certain that this last measurement will supply us with the same result as the first. Thus the length of the train as measured from the embankment may be different from that obtained by measuring in the train itself. This circumstance leads us to a second objection which must be raised against the apparently obvious consideration of Section Ⅱ. Namely, if the man in the carriage covers the distance w in a unit of time—measured from the train—then this distance—as measured from the embankment—is not necessarily also equal to w.

Ⅶ. Special and General Principle of Relativity

31 The basal principle, which was the pivot of all our previous considerations, was the special principle of relativity, i.e. the principle of the physical relativity of all uniform motion. Let us once more analyse its meaning carefully.

32 It was at all times clear that, from the point of view of the idea it conveys to us, every motion must only be considered as a relative motion. Returning to

29 如果火车上的这段距离需要从铁路线上来判断，那就是另一回事了。这里可以考虑使用下述方法：如果我们把需要求出其距离的火车上的两个点称为 A' 和 B'，那么这两个点是以速度 v 沿着路基移动的。首先，我们需要在路基上确定两个对应点 A 和 B，使其在一特定时刻 t 恰好各为 A' 和 B' 所通过(由路基判断)。路基上的 A 点和 B 点可以引用第四节所提出的时间定义来确定。然后再用量杆沿着路基一段一段地量取 A、B 两点之间的距离。

30 从先验的观点来看，丝毫不能肯定这次测量的结果会与第一次在火车车厢中测量的结果完全一样。因此，在路基上量出的火车长度可能与在火车上量出的火车长度不同。这种情况使我们有必要对第二节中从表面上看来是明白的论述提出第二个不同意见。那就是，如果在车厢里的人在单位时间内走了一段距离 w(在火车上测量的)，那么这段距离如果在路基上测量并不一定也等于 w。

七、狭义和广义相对性原理

31 作为我们以前全部论述的中心的一个基本原理是狭义相对性原理，亦即一切匀速运动具有物理相对性的原理，让我们再一次仔细地分析它的意义。

32 从我们由狭义相对性原理所接受的观念来看，每一种运动都只能被认为是相对运动，这一点一直是很清楚的。回到我们经常引用的路基和车厢的例子，我们可以用下列两种方式来表述这里所发生的运动，这两种表述方式是同样合理的：

the illustration we have frequently used of the embankment and the railway carriage, we can express the fact of the motion here taking place in the following two forms, both of which are equally justifiable:

(1) The carriage is in motion relative to the embankment,

(2) The embankment is in motion relative to the carriage.

33 In (1) the embankment, in (2) the carriage, serves as the body of reference in our statement of the motion taking place. If it is simply a question of detecting or of describing the motion involved, it is in principle immaterial to what reference-body we refer the motion. As already mentioned, this is self-evident, but it must not be confused with the much more comprehensive statement called "the principle of relativity," which we have taken as the basis of our investigations.

34 The principle we have made use of not only maintains that we may equally well choose the carriage or the embankment as our reference-body for the description of any event (for this, too, is self-evident). Our principle rather asserts what follows: If we formulate the general laws of nature as they are obtained from experience, by making use of

(1) the embankment as reference-body,

(2) the railway carriage as reference-body.

35 Then these general laws of nature (e.g. the laws of mechanics or the law of the propagation of light in vacuo) have exactly the same form in both cases. This can also be expressed as follows: For the physical description of natural processes, neither of the reference-bodies K, K' is unique (lit. "specially marked out") as compared with the other. Unlike the first, this latter statement need not of necessity hold a priori; it is not contained in the conceptions of "motion" and "referencebody" and derivable from them; only experience can decide as to its correctness or incorrectness.

(1)车厢相对于路基而言是运动的；

(2)路基相对于车厢而言是运动的。

33 我们在表述所发生的运动时，在(1)中是把路基当作参考物体；在(2)中是把车厢当作参考物体。如果问题仅仅是要探测或者描述这个运动而已，那么我们相对于哪一个参考物体来考察这一运动在原则上是无关紧要的。前面已经提到，这一点是自明的，但是这一点绝不可以同我们已经用来作为研究的基础的、称之为“相对性原理”的更加广泛得多的陈述混淆起来。

34 我们所引用的原理不仅认为我们可以选取车厢也可选取路基作为我们描述任何事件的参考物(因为这也是自明的)，我们的原理所断言的是：如果我们表述从经验得来的普遍的自然界定律时，引用

(1)路基作为参考物体；

(2)车厢作为参考物体。

35 那么这些普遍的自然界定律(例如力学诸定律或真空中光的传播定律)在这两种情况中的形式完全一样。这一点也可以表述如下：对于自然过程的物理描述而言，在参考物体 K，K'中没有一个与另一个相比是唯一的(字面意义是“特别标出的”)。与第一个陈述不同，后一个陈述并不一定是根据推论必然成立的；这个陈述并不包含在“运动”和“参考物体”的概念中，也不能从这些概念推导出来；唯有经验才能确定这个陈述是正确的还是不正确的。

36 Up to the present, however, we have by no means maintained the equivalence of all bodies of reference K in connection with the formulation of natural laws. Our course was more on the following lines. In the first place, we started out from the assumption that there exists a reference-body K, whose condition of motion is such that the Galileian law holds with respect to it: A particle left to itself and sufficiently far removed from all other particles moves uniformly in a straight line. With reference to K (Galileian reference-body) the laws of nature were to be as simple as possible. But in addition to K, all bodies of reference K' should be given preference in this sense, and they should be exactly equivalent to K for the formulation of natural laws, provided that they are in a state of uniform rectilinear and non-rotary motion with respect to K; all these bodies of reference are to be regarded as Galileian reference-bodies. The validity of the principle of relativity was assumed only for these reference-bodies, but not for others (e.g. those possessing motion of a different kind). In this sense we speak of the special principle of relativity, or special theory of relativity.

37 In contrast to this we wish to understand by the "general principle of relativity" the following statement: All bodies of reference K, K', etc., are equivalent for the description of natural phenomena (formulation of the general laws of nature), whatever may be their state of motion. But before proceeding farther, it ought to be pointed out that this formulation must be replaced later by a more abstract one, for reasons which will become evident at a later stage.

38 Since the introduction of the special principle of relativity has been justified, every intellect which strives after generalisation must feel the temptation to venture the step towards the general principle of relativity. But a simple and apparently quite reliable consideration seems to suggest that, for the present at any rate, there is little hope of success in such an attempt. Let us imagine

36 但是，到目前为止，我们根本没有认定所有参考物体 K 在表述自然界定律方面具有等效性。我们的思路主要是沿着下列路线走的：首先我们从这样的假定出发，即存在着一个参考物体 K，它所具有的运动状态使伽利略定律对于它而言是成立的；一质点若不受外界作用并离所有其他质点足够远，则该质点沿直线作匀速运动。参照 K(伽利略参考物体)表述的自然界定律应该是最简单的。但是除 K 以外，参照所有参考物体 K' 表述的自然界定律也应该是最简单的，而且，只要这些参考物体相对于 K 是处于匀速直线无转动运动状态，这些参考物体对于表述自然界定律应该与 K 完全等效；所有这些参考物体都应认为是伽利略参考物体。以往我们假定相对性原理只是对于这些参考物体才是有效的，而对于其他参考物体(例如具有另一种运动状态的参考物体)则是无效的。在这个意义上我们说它是狭义相对性原理或狭义相对论。

37 与此对比，我们把“广义相对性原理”理解为下述陈述：所有参考物体 K、K'等不论它们的运动状态如何，对于描述自然现象(表述普遍的自然界定律)都是等效的。但是在我们继续谈下去以前应该指出，这一陈述在以后必须代之以一个更为抽象的陈述，其理由要等到以后才会明白。

38 由于已经证明引进狭义相对性原理是合理的，因而每一个追求普遍化结果的人必然很想朝着广义相对性原理探索前进。但是从一种简单而表面上颇为可靠的考虑看来，似乎至少就目前而论这样一种企图是没有多少成功的希望的。让我们转回到我们的旧相识——匀速向前行驶的火车车

ourselves transferred to our old friend the railway carriage, which is travelling at a uniform rate. As long as it is moving uniformly, the occupant of the carriage is not sensible of its motion, and it is for this reason that he can unreluctantly interpret the facts of the case as indicating that the carriage is at rest, but the embankment in motion. Moreover, according to the special principle of relativity, this interpretation is quite justified also from a physical point of view.

39 If the motion of the carriage is now changed into a non-uniform motion, as for instance by a powerful application of the brakes, then the occupant of the carriage experiences a correspondingly powerful jerk forwards. The retarded motion is manifested in the mechanical behaviour of bodies relative to the person in the railway carriage. The mechanical behaviour is different from that of the case previously considered, and for this reason it would appear to be impossible that the same mechanical laws hold relatively to the non-uniformly moving carriage, as hold with reference to the carriage when at rest or in uniform motion. At all events it is clear that the Galileian law does not hold with respect to the non-uniformly moving carriage. Because of this, we feel compelled at the present juncture to grant a kind of absolute physical reality to non-uniform motion, in opposition to the general principle of relativity. But in what follows we shall soon see that this conclusion cannot be maintained.

Ⅷ. The Equality of Inertial and Gravitational Mass as an Argument for the General Postulate of Relativity

40 We imagine a large portion of empty space, so far removed from stars and other appreciable masses, that we have before us approximately the conditions required by the fundamental law of Galilei. It is then possible to choose a Galileian reference-body for this part of space (world), relative to

厢，来设想一番。只要车厢作匀速运动，车厢里的人就不会感到车厢的运动。由于这个理由，他可以毫不勉强地作这样的解释，即这个例子表明车厢是静止的，而路基是运动的。而且，按照狭义相对性原理，这种解释从物理观点来看也是十分合理的。

39 如果车厢的运动变为非匀速运动，例如使用制动器猛然刹车，那么车厢里的人就体验到一种相应的朝向前方的猛烈冲动。这种减速运动由物体相对于车厢里的人的力学行为表现出来。这种力学行为与上述的例子里的力学行为是不同的；因此，对于静止的或作匀速运动的车厢能成立的力学定律，看来不可能对于作非匀速运动的车厢也同样成立。无论如何，伽利略定律对于作非匀速运动的车厢显然是不成立的。由于这个原因，我们感到在目前不得不暂时采取与广义相对性原理相反的做法而特别赋予非匀速运动以一种绝对的物理实在性。但是在下文我们就会看到，这个结论是不能成立的。

八、惯性质量和引力质量相等是广义相对性公理的一个论据

40 我们设想在一无所有的空间中有一个相当大的部分，这里距离众星及其他可以感知的质量非常遥远，可以说我们已经近似地有了伽利略基本定律所要求的条件。这样就有可能为这部分空间(世界)选取一个伽利略参考物体，使对之处于静止状态的点继续保持静止状态，而对之作相对运动的点永远继续作匀速直线运动。我们设想把一个像一间房子似的极宽大的箱子当作参考物体，里面安置一个配备有仪器的观察者。对于这个

which points at rest remain at rest and points in motion continue permanently in uniform rectilinear motion. As reference-body let us imagine a spacious chest resembling a room with an observer inside who is equipped with apparatus. Gravitation naturally does not exist for this observer. He must fasten himself with strings to the floor, otherwise the slightest impact against the floor will cause him to rise slowly towards the ceiling of the room.

41 To the middle of the lid of the chest is fixed externally a hook with rope attached, and now a "being" (what kind of a being is immaterial to us) begins pulling at this with a constant force. The chest together with the observer then begins to move "upwards" with a uniformly accelerated motion. In course of time their velocity will reach unheard of values—provided that we are viewing all this from another reference-body which is not being pulled with a rope.

42 But how does the man in the chest regard the process? The acceleration of the chest will be transmitted to him by the reaction of the floor of the chest. He must therefore take up this pressure by means of his legs if he does not wish to be laid out full length on the floor. He is then standing in the chest in exactly the same way as anyone stands in a room of a house on our earth. If he release a body which he previously had in his hand, the acceleration of the chest will no longer be transmitted to this body, and for this reason the body will approach the floor of the chest with an accelerated relative motion. The observer will further convince himself *that the acceleration of the body towards the floor of the chest is always of the same magnitude, whatever kind of body he may happen to use for the experiment.*

43 Relying on his knowledge of the gravitational field (as it was discussed in the preceding section), the man in the chest will thus come to the conclusion that

观察者而言引力当然并不存在。他必须用绳子把自己拴在地板上，否则他只要轻轻碰一下地板就会朝着箱子的天花板慢慢地浮起来。

41 在箱子盖外面的当中，安装了一个钩子，钩子上系有缆索。现在又设想有一“生物”(是何种生物对我们来说无关紧要)开始以恒力拉这根缆索。于是箱子连同观察者就要开始作匀加速运动“上升”。经过一段时间，它们的速度将会达到前所未闻的高值——倘若我们从另一个未用绳牵的参考物体来继续观察这一切的话。

42 但是箱子里的人会如何看待这个过程呢？箱子的加速度要通过箱子地板的反作用才能传给他。所以，如果他不愿意整个人卧倒在地板上，他就必须用他的腿来承受这个压力。因此，他站立在箱子里实际上与站立在地球上的一个房间里完全一样。如果他松手放开原来拿在手里的一个物体，箱子的加速度就不会再传到这个物体上，因而这个物体就必然作加速相对运动而落到箱子的地板上。观察者将会进一步断定：物体朝向箱子的地板的加速度总是有相同的量值，不论他碰巧用来做实验的物体为何。

43 依靠他对引力场的知识(如同在前节①所讨论的)，箱子里的人将会得出这样一个结论：他自己以及箱子是处在一个引力场中，而且该引力场对于时间而言是恒定不变的。当然他会一时感到迷惑不解为什么箱子在这

① 这里的“前节”指的是[美]阿尔伯特·爱因斯坦所著，杨润殷译，北京大学出版社2006年出版的《狭义与广义相对论浅说》的第19节引力场内容，而非本导引第19节。

he and the chest are in a gravitational field which is constant with regard to time. Of course he will be puzzled for a moment as to why the chest does not fall in this gravitational field. Just then, however, he discovers the hook in the middle of the lid of the chest and the rope which is attached to it, and he consequently comes to the conclusion that the chest is suspended at rest in the gravitational field.

44 Ought we to smile at the man and say that he errs in his conclusion? I do not believe we ought to if we wish to remain consistent; we must rather admit that his mode of grasping the situation violates neither reason nor known mechanical laws. Even though it is being accelerated with respect to the "Galileian space" first considered, we can nevertheless regard the chest as being at rest. We have thus good grounds for extending the principle of relativity to include bodies of reference which are accelerated with respect to each other, and as a result we have gained a powerful argument for a generalised postulate of relativity.

45 We must note carefully that the possibility of this mode of interpretation rests on the fundamental property of the gravitational field of giving all bodies the same acceleration, or, what comes to the same thing, on the law of the equality of inertial and gravitational mass. If this natural law did not exist, the man in the accelerated chest would not be able to interpret the behaviour of the bodies around him on the supposition of a gravitational field, and he would not be justified on the grounds of experience in supposing his reference-body to be "at rest."

46 Suppose that the man in the chest fixes a rope to the inner side of the lid, and that he attaches a body to the free end of the rope. The result of this will be to stretch the rope so that it will hang "vertically" downwards. If we ask for

个引力场中并不降落。但是正在这个时候他发现箱盖的当中有一个钩子，钩上系着缆索；因此他就得出结论，箱子是静止地悬挂在引力场中的。

44 我们是否应该讥笑这个人，说他的结论错了呢？如果我们要保持前后一致的话，我认为我们不应该这样说他；我们反而必须承认，他的思想方法既不违反理性，也不违反已知的力学定律。虽然我们先认定为箱子相对于“伽利略空间”在作加速运动，但是也仍然能够认定箱子是在静止中。因此我们确有充分理由可以将相对性原理推广到把相互作加速运动的参考物体也包括进去的地步，因而对于相对性公理的推广也就获得了一个强有力的论据。

45 我们必须充分注意到，这种解释方式的可能性是以引力场使一切物体得到同样的加速度这一基本性质为基础的；这也就等于说，是以惯性质量和引力质量相等的这一定律为基础的。如果这个自然规律不存在，处在作加速运动的箱子里的人就不能先假定出一个引力场来解释他周围物体的行为，他就没有理由根据经验假定他的参考物体是“静止的”。

46 假定箱子里的人在箱子盖内面系一根绳子，然后在绳子的自由端拴上一个物体。结果绳子受到伸张，“竖直地”悬垂着该物体。如果我们问一下绳子上产生张力的原因，箱子里的人就会说：“悬垂着的物体在引力场

an opinion of the cause of tension in the rope, the man in the chest will say: "The suspended body experiences a downward force in the gravitational field, and this is neutralised by the tension of the rope; what determines the magnitude of the tension of the rope is the *gravitational mass* of the suspended body." On the other hand, an observer who is poised freely in space will interpret the condition of things thus: "The rope must perforce take part in the accelerated motion of the chest, and it transmits this motion to the body attached to it. The tension of the rope is just large enough to effect the acceleration of the body. That which determines the magnitude of the tension of the rope is the *inertial mass* of the body." Guided by this example, we see that our extension of the principle of relativity implies the *necessity* of the law of the equality of inertial and gravitational mass. Thus we have obtained a physical interpretation of this law.

47 From our consideration of the accelerated chest we see that a general theory of relativity must yield important results on the laws of gravitation. In point of fact, the systematic pursuit of the general idea of relativity has supplied the laws satisfied by the gravitational field. Before proceeding farther, however, I must warn the reader against a misconception suggested by these considerations. A gravitational field exists for the man in the chest, despite the fact that there was no such field for the co-ordinate system first chosen. Now we might easily suppose that the existence of a gravitational field is always only an *apparent* one. We might also think that, regardless of the kind of gravitational field which may be present, we could always choose another reference-body such that *no* gravitational field exists with reference to it. This is by no means true for all gravitational fields, but only for those of quite special form. It is, for instance, impossible to choose a body of reference such that, as judged from it, the gravitational field of the earth (in its entirety) vanishes.

中受到一向下的力，此力为绳子的张力所平衡；决定绳子张力的大小的是悬垂着的物体的引力质量。”另一个角度，自由地稳定在空中的一个观察者将会这样解释这个情况：“绳子势必参与箱子的加速运动，并将此运动传给拴在绳子上的物体。绳子的张力的大小恰好足以引起物体的加速度。决定绳子张力大小的是物体的惯性质量。”我们从这个例子看到，我们对相对性原理的推广隐含着惯性质量和引力质量相等这一定律的必然性。这样我们就得到了这个定律的一个物理解释。

47 根据对作加速运动的箱子的讨论，我们看到，一个广义的相对论必然会对引力诸定律产生重要的结果。事实上，对广义相对性观念的系统研究已经补充了好些定律为引力场所满足。但是，在继续谈下去以前，我必须提醒读者不要接受这些论述中所隐含的一个错误概念。对于箱子里的人而言存在着一个引力场，尽管对于最初选定的坐标系而言并没有这样的场。于是我们可能会轻易地假定，引力场的存在永远只是一种表观的存在。我们也可能认为，不论存在着什么样的引力场，我们总是能够这样选取另外一个参考物体，使得对于该参考物体而言没有引力场存在。这绝对不是对于所有的引力场都是真实的，这仅仅是对于那些具有十分特殊的形式的引力场才是真实的。例如，我们不可能这样选取一个参考物体，使得由该参考物体来判断地球的引力场(就其整体而言)会等于零。

48 We can now appreciate why that argument is not convincing, which we brought forward against the general principle of relativity at the end of Section Ⅷ. It is certainly true that the observer in the railway carriage experiences a jerk forwards as a result of the application of the brake, and that he recognises in this the non-uniformity of motion (retardation) of the carriage. But he is compelled by nobody to refer this jerk to a "real" acceleration (retardation) of the carriage. He might also interpret his experience thus: "My body of reference (the carriage) remains permanently at rest. With reference to it, however, there exists (during the period of application of the brakes) a gravitational field which is directed forwards and which is variable with respect to time. Under the influence of this field, the embankment together with the earth moves nonuniformly in such a manner that their original velocity in the backwards direction is continuously reduced."

Ⅸ. In What Respects Are the Foundations of Classical Mechanics and of the Special Theory of Relativity Unsatisfactory?

49 We have already stated several times that classical mechanics starts out from the following law: Material particles sufficiently far removed from other material particles continue to move uniformly in a straight line or continue in a state of rest. We have also repeatedly emphasised that this fundamental law can only be valid for bodies of reference K which possess certain unique states of motion, and which are in uniform translational motion relative to each other. Relative to other reference-bodies K' the law is not valid. Both in classical mechanics and in the special theory of relativity we therefore differentiate between reference-bodies K relative to which the recognised "laws of nature" can be said to hold, and reference-bodies K' relative to which these laws do not hold.

48 现在我们可以认识到，为什么我们在第七节末尾所叙述的反对广义相对性原理的论据是不能令人信服的。车厢里的观察者由于刹车而体验到一种朝向前方的冲动，并由此察觉车厢的非匀速运动(阻滞)，这一点当然是真实的。但是谁也没有强迫他把这种冲动归因于车厢的“实在的”加速度(阻滞)。他也可以这样解释他的经验：“我的参考物体(车厢)一直保持静止。但是，对于这个参考物体存在着(在刹车期间)一个方向向前而且对于时间而言是可变的引力场。在这个场的影响下，路基连同地球以这样的方式作非匀速运动，即它们的向后的原有速度是在不断地减小下去。”

九、经典力学的基础和狭义相对论的基础在哪些方面不能令人满意

49 我们已经说过几次，经典力学是从下述定律出发的：离其他质点足够远的质点继续作匀速直线运动或继续保持静止状态。我们也曾一再强调，这个基本定律只有对于这样一些参考物体 K 才有效，这些参考物体具有某些特别的运动状态并相对作匀速平移运动。相对于其他参考物体 K'，这个定律就失效。所以，我们在经典力学中和在狭义相对论中都把参考物体 K 和参考物体 K' 区分开；相对于参考物体 K，公认的“自然界定律”可以说是成立的，而相对于参考物体 K' 则这些定律并不成立。

50 But no person whose mode of thought is logical can rest satisfied with this condition of things. He asks: "How does it come that certain reference-bodies (or their states of motion) are given priority over other reference-bodies (or their states of motion)? What is the reason for this preference?" In order to show clearly what I mean by this question, I shall make use of a comparison.

51 I am standing in front of a gas range. Standing alongside of each other on the range are two pans so much alike that one may be mistaken for the other. Both are half full of water. I notice that steam is being emitted continuously from the one pan, but not from the other. I am surprised at this, even if I have never seen either a gas range or a pan before. But if I now notice a luminous something of bluish colour under the first pan but not under the other, I cease to be astonished, even if I have never before seen a gas flame. For I can only say that this bluish something will cause the emission of the steam, or at least possibly it may do so. If, however, I notice the bluish something in neither case, and if I observe that the one continuously emits steam whilst the other does not, then I shall remain astonished and dissatisfied until I have discovered some circumstance to which I can attribute the different behaviour of the two pans.

52 Analogously, I seek in vain for a real something in classical mechanics (or in the special theory of relativity) to which I can attribute the different behaviour of bodies considered with respect to the reference-systems K and K'.① Newton saw this objection and attempted to invalidate it, but without success. But E. Mach recognised it most clearly of all, and because of this

① Note 1. The objection is of importance more especially when the state of motion of the reference-body is of such a nature that it does not require any external agency for its maintenance, e.g. in the case when the reference-body is rotating uniformly.

50 但是，凡是思想方法合乎逻辑的人谁也不会满足于此种情形。他要问："为什么要认定某些参考物体(或它们的运动状态)比其他参考物体(或它们的运动状态)优越呢？此种偏爱的理由何在？"为了讲清楚我提出这个问题是什么意思，我来打一个比方。

51 比方我站在一个煤气灶前面。灶上并排放着两个平底锅。这两个锅非常相像，常常会认错。里面都盛着半锅水。我注意到一个锅不断地冒出蒸汽，而另一个锅则没有蒸汽冒出。即使我以前从来没有见过煤气灶或者平底锅，我也会对这种情况感到奇怪。但是如果在这个时候我注意到在第一个锅底下有一种蓝色的发光的东西，而在另一个锅底下则没有，那么我就不会再感到惊奇，即使以前我从来没有见过煤气的火焰。因为我只要说是这种蓝色的东西使得锅里冒出蒸汽，或者至少可以说有这种可能。但是如果我注意到这两个锅底下都没有什么蓝色的东西，而且如果我还观察到其中一个锅不断冒出蒸汽，而另外一个锅则没有蒸汽，那么我就总是感到惊奇和不满足，直到我发现某种情况能够用来说明为什么这两个锅有不同的表现为止。

52 与此类似，我在经典力学中(或在狭义相对论中)找不到什么实在的东西能够用来说明为什么相对于参考系 K 和 K' 来考虑时物体会有不同的表现①。牛顿看到了这个缺陷，并曾试图消除它，但没有成功。只有马赫对它看得最清楚，由于这个缺陷他宣称必须把力学放在一个新的基础

① 这个缺陷在下述情况尤为严重，即当参考物体的运动状态无需任何外力来维持时，例如在参考物体作匀速转动时。

objection he claimed that mechanics must be placed on a new basis. It can only be got rid of by means of a physics which is conformable to the general principle of relativity, since the equations of such a theory hold for every body of reference, whatever may be its state of motion.

X. A Few Inferences from the General Principle of Relativity

53 The considerations of Section Ⅷ show that the general theory of relativity puts us in a position to derive properties of the gravitational field in a purely theoretical manner. Let us suppose, for instance, that we know the space-time "course" for any natural process whatsoever, as regards the manner in which it takes place in the Galileian domain relative to a Galileian body of reference *K*. By means of purely theoretical operations (i. e. simply by calculation) we are then able to find how this known natural process appears, as seen from a reference-body K' which is accelerated relatively to *K*. But since a gravitational field exists with respect to this new body of reference K', our consideration also teaches us how the gravitational field influences the process studied.

54 For example, we learn that a body which is in a state of uniform rectilinear motion with respect to *K* (in accordance with the law of Galilei) is executing an accelerated and in general curvilinear motion with respect to the accelerated reference-body K' (chest). This acceleration or curvature corresponds to the influence on the moving body of the gravitational field prevailing relatively to K'. It is known that a gravitational field influences the movement of bodies in this way, so that our consideration supplies us with nothing essentially new.

上。只有借助于与广义相对性原理一致的物理学才能消除这个缺陷，因为这样的理论的方程，对于一切参考物体，不论其运动状态如何，都是成立的。

十、广义相对性原理的几个推论

53 第八节的论述表明，广义相对性原理能够使我们以纯理论方式推出引力场的性质。例如，假定我们已经知道任一自然过程在伽利略区域中相对于一个伽利略参考物体 K 如何发生，亦即已经知道该自然过程的空时"进程"，借助于纯理论运算（亦即单凭计算），我们就能够断定这个已知自然过程从一个相对于 K 作加速运动的参考物体 K' 去观察，是如何表现的。但是由于对于这个新的参考物体 K' 而言存在着一个引力场，所以以上的考虑也告诉我们引力场如何影响所研究的过程。

54 例如，我们知道，相对于 K（按照伽利略定律）作匀速直线运动的一个物体，它相对于作加速运动的参考物体 K'（箱子）是在作加速运动的，一般还是在作曲线运动的。此种加速度或曲率相当于相对于 K' 存在的引力场对运动物体的影响。引力场以此种方式影响物体的运动是大家已经知道的，因此以上的考虑并没有为我们提供任何本质上新的结果。

55 However, we obtain a new result of fundamental importance when we carry out the analogous consideration for a ray of light. With respect to the Galileian reference-body K, such a ray of light is transmitted rectilinearly with the velocity c. It can easily be shown that the path of the same ray of light is no longer a straight line when we consider it with reference to the accelerated chest (reference-body K'). From this we conclude, that, in general, rays of light are propagated curvilinearly in gravitational fields. In two respects this result is of great importance.

56 In the first place, it can be compared with the reality. Although a detailed examination of the question shows that the curvature of light rays required by the general theory of relativity is only exceedingly small for the gravitational fields at our disposal in practice, its estimated magnitude for light rays passing the sun at grazing incidence is nevertheless 1.7 seconds of arc. This ought to manifest itself in the following way. As seen from the earth, certain fixed stars appear to be in the neighbourhood of the sun, and are thus capable of observation during a total eclipse of the sun. At such times, these stars ought to appear to be displaced outwards from the sun by an amount indicated above, as compared with their apparent position in the sky when the sun is situated at another part of the heavens. The examination of the correctness or otherwise of this deduction is a problem of the greatest importance, the early solution of which is to be expected of astronomers.①

57 In the second place our result shows that, according to the general theory of relativity, the law of the constancy of the velocity of light in vacuo, which constitutes one of the two fundamental assumptions in the special theory of

① Note 1. By means of the star photographs of two expeditions equipped by a Joint Committee of the Royal and Royal Astronomical Societies, the existence of the deflection of light demanded by theory was confirmed during the solar eclipse of 29th May, 1919. (Cf. Appendix III.)

55 但是，如果我们对一道光线进行类似的考虑就得到一个新的具有基本重要性的结果。相当于伽利略参考物体 K，这样的一道光线是沿直线以速度 c 传播的。不难证明，当我们相对于作加速运动的箱子（参考物体 K'）来考察这同一道光线时，它的路线就不再是一条直线。由此我们得出结论，光线在引力场中一般沿曲线传播。这个结果在两个方面具有重大意义。

56 首先这个结果可以同实际比较。虽然对这个问题的详细研究表明，按照广义相对论，光线穿过我们在实践中能够加以利用的引力场时，只有极其微小的曲率；但是，以掠入射方式经过太阳的光线，其曲率的估计值达到 1.7″。这应该以下述方式表现出来。从地球上观察，某些恒星看起来是在太阳的邻近处，因此这些恒星能够在日全食时加以观测。当日全食时这些恒星在天空的视位置与当太阳位于天空的其他部位时这些恒星的视位置相比较应该偏离太阳，偏离的数值同上。检验这个推断正确与否是一个极其重要的问题，希望天文学家能够早日予以解决①。

57 其次，我们的结果表明，按照广义相对论，我们时常提到的作为狭义相对论中两个基本假定之一的真空中光速恒定定律，就不能被认为具有无限的有效性。光线的弯曲只有在光的传播速度随位置而改变时才能发

① 理论所要求的光线偏转的存在，首次于 1919 年 5 月 29 日的日食期间，借助于英国皇家学会和英国皇家天文学会的一个联合委员会所装备的两个远征观测队的摄影星图得到证实。

relativity and to which we have already frequently referred, cannot claim any unlimited validity. A curvature of rays of light can only take place when the velocity of propagation of light varies with position. Now we might think that as a consequence of this, the special theory of relativity and with it the whole theory of relativity would be laid in the dust. But in reality this is not the case. We can only conclude that the special theory of relativity cannot claim an unlimited domain of validity; its result hold only so long as we are able to disregard the influences of gravitational fields on the phenomena (e.g. of light).

58 Since it has often been contended by opponents of the theory of relativity that the special theory of relativity is overthrown by the general theory of relativity is overthrown by the general theory of relativity, it is perhaps advisable to make the facts of the case clearer by means of an appropriate comparison. Before the development of electrodynamics the laws of electrostatics and the laws of electricity were regarded indiscriminately. At the present time we know that electric fields can be derived correctly from electrostatic considerations only for the case, which is never strictly realised, in which the electrical masses are quite at rest relatively to each other, and to the co-ordinate system. Should we be justified in saying that for this reason electrostatics is overthrown by the field-equations of Maxwell in electrodynamics? Not in the least. Electrostatics is contained in electrodynamics as a limiting case; the laws of the latter lead directly to those of the former for the case in which the fields are invariable with regard to time. No fairer destiny could be allotted to any physical theory, than that it should of itself point out the way to the introduction of a more comprehensive theory, in which it lives on as a limiting case.

生。我们或许会想，由于这种情况，狭义相对论以及随之整个相对论，都要化为灰烬了。但实际上并不是这样。我们只能作这样的结论：不能认为狭义相对论的有效性是无止境的；只有在我们能够不考虑引力场对现象(例如光的现象)的影响时，狭义相对论的结果才能成立。

58 由于反对相对论的人时常说狭义相对论被广义相对论推翻了，因此用一个适当的比方来把这个问题的实质弄得更清楚些也许是允当的。在电动力学发展前，静电学定律被看作是电学定律。现在我们知道，只有在电质量相互之间并相对于坐标系完全保持静止的情况下(这种情况是永远不会严格实现的)，才能够从静电学的考虑出发正确地推导出电场。我们是否可以说，由于这个理由，静电学被电动力学的麦克斯韦场方程推翻了呢？绝对不可以。静电学作为一个极限情况包含在电动力学中；在场不随时间而改变的情况下，电动力学的定律就直接得出静电学的定律。任何物理理论都不会获得比这更好的命运了，即一个理论本身指出创立一个更为全面的理论的道路，而在这个更为全面的理论中，原来的理论作为一个极限情况继续存在下去。

59 In the example of the transmission of light just dealt with, we have seen that the general theory of relativity enables us to derive theoretically the influence of a gravitational field on the course of natural processes, the laws of which are already known when a gravitational field is absent. But the most attractive problem, to the solution of which the general theory of relativity supplies the key, concerns the investigation of the laws satisfied by the gravitational field itself. Let us consider this for a moment.

60 We are acquainted with space-time domains which behave (approximately) in a "Galileian" fashion under suitable choice of reference-body, i. e. domains in which gravitational fields are absent. If we now refer such a domain to a reference-body K' possessing any kind of motion, then relative to K' there exists a gravitational field which is variable with respect to space and time.① The character of this field will of course depend on the motion chosen for K'. According to the general theory of relativity, the general law of the gravitational field must be satisfied for all gravitational fields obtainable in this way. Even though by no means all gravitational fields can be produced in this way, yet we may entertain the hope that the general law of gravitation will be derivable from such gravitational fields of a special kind. This hope has been realised in the most beautiful manner. But between the clear vision of this goal and its actual realisation it was necessary to surmount a serious difficulty, and as this lies deep at the root of things, I dare not withhold it from the reader. We require to extend our ideas of the space-time continuum still farther.

① Note 1. This follows from a generalisation of the discussion in Section Ⅷ.

59 在刚才讨论的关于光的传播的例子中，我们已经看到，广义相对论使我们能够从理论上推导引力场对自然过程的进程的影响，这些自然过程的定律在没有引力场时是已知的。但最吸引人的问题是，广义相对论提供了关键的解，它涉及引力场本身所满足的定律的研究。让我们对此稍微考虑一下。

60 我们已经熟悉了经过适当选取参考物体后处于(近似地)“伽利略”形式的那种空时区域，亦即没有引力场的区域。如果我们相对于一个不论作何种运动的参考物体 K' 来考察这样的一个区域，那么相对于 K' 就存在着一个引力场，该引力场对于空间和时间是可变的①。这个场的特性当然取决于为 K' 选定的运动。按照广义相对论，普遍的引力场定律对于所有能够按这一方式得到的引力场都必须被满足。虽然绝不是所有的引力场都能够如此产生，我们仍然可以希望普遍的引力定律能够从这样的一些特殊的引力场推导出来。这个希望已经以极其美妙的方式实现了；但是从认清这个目标到完全实现它，是经过克服了一个严重的困难之后才达到的。由于这个问题具有很深刻的意义，我不敢对读者略而不谈。我们需要进一步推广我们对于空时连续区的观念。

① 这一点可由第八节的讨论推广得出。

三、生命领域

《物种起源》导引

如果说起生物学领域最为人所熟知的理论，进化论必是不二之选；如果评选最容易理解的生物学法则，进化论会再次问鼎。160年来，无数科学家从达尔文的理论中吸取灵感，极大地推动了生物学领域研究的范围和深度，但也有不少人为自己的理论披上进化论的外衣并广泛传播，这使进化理论得到了最有效的推广，同时也产生了最多误读。当所有人都以为自己理解进化论的时候，也正是我们要打开书本，认真重读160年前出版的惊世骇俗的《物种起源》的时候。让我们通过阅读原文来了解达尔文本人在他的著作中表达了什么思想，如何对这些思想进行表述，还原达尔文和真实的《物种起源》。同时，我们也许能做如下思考：我从哪里来？我在自然界中占有什么样的位置？我与自然有什么关系？人类的发展与自然有什么关系？

让我们先从《物种起源》的作者谈起。达尔文出生于1809年，从少年时代开始，他就显示出对自然历史的强烈求知欲，这种求知欲不仅表现为对标本采集的狂热爱好，更体现为敏锐的观察力和深入的思考能力。1831年，达尔文参加了英国海军"小猎犬号"的环球旅行，谁也没想到，5年环球旅行的经历对这位初登船时虔诚的基督徒产生了无比深远的影响。这期间的博物采集工作以及回国后的资料整理和科学交流对达尔文产生了巨大的影响，他的思想发生巨大转变。通过严密的推理论证，达尔文提出了以自然选择学说为框架的进化理论，颠覆了“上帝创造万物并主宰万物”的宗教神话，成为进化论的斗士。达尔文一生著述颇丰，尽管常年饱受病痛侵扰，达尔文仍始终保持强烈的科学探索精神，他的著作除最著名的《物种起源》外，还有大量基于自己博物采集工作和生物学实验研究工作的论文和专著，涉及领域包括植物学、动物学、地质学等各个方面。

达尔文在学术上的巨大成功让我们不由得要追寻他的思想奥秘。他在自传中这样写道“作为一个科学工作者，我的成功取决于我复杂的心理素质。其中最重要的是：热爱科学、善于思索，勤于观察和搜集资料，具有相当的发现能力和广博的常识。这些看起来的确令人奇怪，凭借这些极平常的能力，我居然在一些重要地方影响了科学家们的信仰。”正是通过对前人思想财富的批判式继承、在孜孜以求的科学实践中大量积累资料并加以细致的观察和分析，再加上与同行的学术交流获得的灵感，才成就了这一伟大的思想理论体系。

《物种起源》共 15 章，在引言和绪论中，达尔文对近代进化理论的渊源进行了梳理和点评，对前人的思想进行了思辨、批判和继承，更重要的是，他对自己生命中最重要的科学实践——跟随皇家“小猎犬号”进行的科考远行以及在自家进行的动植物人工驯养实验进行了回顾和总结。达尔文正是通过这些科学实践搜集整理了大量素材，并通过对这些材料的观察、理解和总结，才提出了伟大的进化论思想。在第 1~5 章，达尔文开宗明义提出了他最重要的学术观点：自然选择和生命之树；在第 6~10 章，达尔文提出了进化理论面临的难题，分别进行回答，并提出他关于变异如何传递的遗传学理论；在第 11~14 章，达尔文从地质年代演化和地理分布两个维度，纵览时空，再次讨论物种间的亲缘关系以及这些亲缘关系对进化论思想的支持；在最后一章，达尔文对自己的学说进行了概括总结，最后达尔文以“认为生命及其种种力量是由‘造物主’(这里指‘大自然’，而非宗教上的造物主——译者注)注入到少数几个或仅仅一个类型中去的，而且认为地球这个行星按照地球的引力法则，旋转不息，并从最简单的无形物体演化出如此美丽和令人惊叹的生命体，而且这一演化过程仍在持续，这才是一种真正伟大的思想理念！”这样恢弘的陈词为《物种起源》画上了华丽的句号。英国著名博物学家赫胥黎曾这样评价：“我认为《物种起源》这本书的格调是再好也没有的，它可以感动那些对这个问题一无所知的人们。至于达尔文的理论，我准备即使赴汤蹈火也要支持。”

《物种起源》的完整标题是《论通过自然选择的物种起源，或生存竞争中优赋族群之保存》(*On the Origin of Species By Means of Natural Selection*, *or*, *the Preservation of Favoured Races in the Struggle for Life*)，达尔文在标题中清晰地表达出“自然选择”学说在他的学术思想和进化理论中的重要性，我们的节选也来自于最能体现他思想精华的第4章——自然选择即适者生存。

第4章共72段，包含了《物种起源》中最重要的两条进化生物学原理，即“自然选择理论”和“生命之树理论”，我们略去部分篇幅较长的、对自然选择进行实例分析的段落，使读者可以较为清晰地了解这两条重要原理提出的背景知识、原理内涵以及达尔文的解读。

自然选择理论是达尔文进化思想的精髓，这一理论是在对家养状态下的变异以及自然状态下的变异进行对比，并引入生存斗争的概念和实例后提出的。生存斗争的结果是适者生存，即自然对于变异进行选择，选择的结果是能利于生物体适应环境变化的性状得以保留并传递给后代，在这一过程中，变异的积累和代间传递最终导致新物种的形成。在自然选择的过程中，新物种携带变异不断形成，旧有物种也在经历不断的灭绝，从共同祖先开始，经过自然选择繁衍出形形色色以不同方式适应环境的后代，开枝散叶，繁衍生息，生命之树就这样形成了繁复的谱系演化，见证着地球上壮丽的生命图景。在阅读过程中，让我们尝试寻找出这一章节的思想主线，更加深入地理解达尔文是如何提出、证明并捍卫自己的思想的。

达尔文首先提出变异普遍存在并可以遗传，接着他提出生物界普遍存在的现象“繁殖出来的个体比能够生存下来的个体要多得多”，因此生存斗争广泛存在，那些“具有任何优势的个体，无论其优势多么微小，都将比其他个体有更多的生存和繁殖的机会”，于是水到渠成，作者明确提出“我把这种有利于生物个体的差异或变异的保存，以及有害变异的毁灭，称为‘自然选择’或‘适者生存’”。

自然选择与家养变异选择主体和被选择的性状特质方面均有不同，自然选择的选择主体是生物赖以生存的外部环境，而家养变异的选择主体是人类。“人类仅为自己的利益去选择，而‘自然’却是为保护生物的利益去选择”。

自然选择在生存斗争中表现为“适者生存”，在自然的作用下，生物表现出对环境的适应，而且适应既表现在形态结构方面，也表现在行为和习性方面。不仅如此，“自然选择会根据亲体使子体的构造发生变异，也能根据子体使亲体的结构发生变异。在群居的动物中，如果选择出来的变异有利于群体，自然选择就会为了整体的利益改变个体的构造。”同时，文中列举了大量实例来说明自然选择和适者生存，其中狼和鹿相互选择，共同进化的例子最广为人知，在这一实例中，达尔文还表达出环境造就不同适应方向的思想，例如同在卡茨基尔山脉，狼群既有形状略似长嘴猎狗的捕鹿变种，也有躯干较粗而腿较短的捕羊变种。

自然选择在繁殖过程中表现为“性选择”，选择的结果为“适者繁衍”，这是相较生存斗争而言比较温和的一种选择。

在对物种的繁衍特点、自然选择的作用方式和地质条件的客观描述之后，达尔文推断出原有物种的绝灭与新物种的诞生都是自然选择作用的必然结果。“在每一变种或物种形成的过程中，它们给最近源种类造成了最大的威胁，以至于往往最终消灭它们”。

达尔文对性状趋异这一原理进行了充分阐述，在比较了物种间的显著差异和变种间的细微差异后，他提出“能造成差别的所谓趋异原理的作用，它使最初难以觉察的差异逐渐扩大，使品种彼此间以及品种与其亲体间的性状发生分异”，不仅如此，“任何一个物种的后代越是在结构、体质和习性上分异，它就越能占据自然体系中的不同位置，因而数量会大大增加”。

在构建适者生存、物种绝灭和性状趋异等思想的基础上，达尔文提出了他为当代进化论提供的最宝贵的思想财富：生命之树思想。通过对本书唯一的一张插图进行的详细说明，我们可以看到，自然选择下物种保留有利的可遗传变异，新物种的发生与旧物种的绝灭交替反复出现在物种变迁的历史中，这一过程在我们眼前勾画出了一幅壮丽的物种演化图景：所有生物共享同一类远古祖先，随着变异性状的出现和代际遗传，繁衍出的大量后代新物种也在不断各自形成不同的谱系演化树，直至最后构成属于地球的独一无二的生命之树。虽然达尔文在追寻变异遗传的原理时错失与现代遗传学相遇的良机，但现代分子生物学揭示的生命共享的生物密码对他的生命之树理论提供了强有力的支持。

宗教将人类定位为万物之灵，而《物种起源》则打破了这一神话：人类并不是天之骄子，而是与自然界的生物有共同的祖先，共享地球家园。作为有着出众智力的自然大家庭成员，我们应该为自然界的和谐发展作出自己独特的贡献。

今日重读《物种起源》，我惊叹于达尔文极具前瞻性的科学理念，也感佩于他坚持追寻科学真理、挑战神学权威的勇气。

最后，想和各位读者分享阅读原文《物种起源》的一点感受，达尔文的写作沿袭了弥尔顿以降的典雅学院风格，与我们今天读到的现代英文差别甚大，如果多花一点时间和耐心尝试读完一整段文字，相信你也会像我一样折服于这些对自然规律精准细致又富于文学色彩的描述与比喻，进一步加深对《物种起源》的理解和对这位科学巨擘的崇敬。

（杜润蕾）

On the Origin of Species

Charles Darwin

Chapter IV Natural Selection

1 Natural Selection—its power compared with man's selection—its power on characters of trifling importance—its power at all ages and on both sexes—Sexual Selection—On the generality of intercrosses between individuals of the same species—Circumstances favourable and unfavourable to Natural Selection, namely, intercrossing, isolation, number of individuals—Slow action—Extinction caused by Natural Selection—Divergence of Character, related to the diversity of inhabitants of any small area, and to naturalisation—Action of Natural Selection, through Divergence of Character and Extinction, on the descendants from a common parent—Explains the Grouping of all organic beings.

2 How will the struggle for existence, discussed too briefly in the last chapter, act in regard to variation? Can the principle of selection, which we have seen is so potent in the hands of man, apply in nature? I think we shall see that it can act most effectually. Let it be borne in mind in what an endless number of strange peculiarities our domestic productions, and, in a lesser degree, those under nature, vary; and how strong the hereditary tendency is. Under domestication, it may be truly said that the whole organisation becomes in some degree plastic. Let it be borne in mind how infinitely complex and close-fitting are the mutual relations of all organic beings to each other and to their physical conditions of life. Can it, then, be thought

---

本文选自查尔斯·达尔文著 *On the Origin of Species*。

物种起源

查尔斯·达尔文

第四章　自然选择即适者生存

1 自然选择——自然选择与人工选择的比较——自然选择对次要性状的作用——自然选择对不同年龄和不同性别生物的作用——性选择——同种个体间杂交的普遍性——杂交、隔离、个体数量等对自然选择结果的有利和不利因素——自然选择的缓慢作用——由自然选择造成的生物绝灭——与小区域生物的分异和驯化作用相关的性状趋异——自然选择通过性状趋异和绝灭对共祖之后裔的作用——自然选择对生物分类的解释。

2 上一章所简要讨论过的生存斗争，到底对物种的变异有什么影响呢？在人类手中产生巨大作用的选择原理，能适用于自然界吗？回答是肯定的。我们将会看到，在自然状态下，选择的原理能够极其有效地发生作用。我们得记住，在自然状态下的生物也会产生如家养生物那样无数的微小变异和个体差异，只是程度稍小些而已。此外，还应记住的是遗传倾向的力量。在家养状况下，整个身体构造都具有了某种程度的可塑性。但是，正如胡克和阿沙·格雷所说，在家养生物中，我们普遍看到的变异，并不是由人类作用直接产生出来的；人类既不能创造变异，也不能阻止变异发生，人类只是保存和积累已发生的变异。当人类无意识地把生物置于新的、变化着的生活条件中时，变异就产生了；但类似的生活条件的变化，在自然状态下确实也可能发生。我们还应记住，一切生物彼此之间及生物与其自然生活条件之间有着多么复杂密切的关系。因而，构造上那些无穷尽的变异，对于每一生物，在变动的环境下生

本文选自[英]达尔文：《物种起源》(科学素养文库·科学元典丛书)，舒德干等译，北京大学出版社 2005 年版，第 53 页，第 55-62 页，第 63-72 页，第 77-78 页。

improbable, seeing that variations useful to man have undoubtedly occurred, that other variations useful in some way to each being in the great and complex battle of life, should sometimes occur in the course of thousands of generations? If such do occur, can we doubt (remembering that many more individuals are born than can possibly survive) that individuals having any advantage, however slight, over others, would have the best chance of surviving and of procreating their kind? On the other hand, we may feel sure that any variation in the least degree injurious would be rigidly destroyed. This preservation of favorable variations and the rejection of injurious variations, I call Natural Selection. Variations neither useful nor injurious would not be affected by natural selection, and would be left a fluctuating element, as perhaps we see in the species called polymorphic.

3 We shall best understand the probable course of natural selection by taking the case of a country undergoing some physical change, for instance, of climate. The proportional numbers of its inhabitants would almost immediately undergo a change, and some species might become extinct. We may conclude, from what we have seen of the intimate and complex manner in which the inhabitants of each country are bound together, that any change in the numerical proportions of some of the inhabitants, independently of the change of climate itself, would most seriously affect many of the others. If the country were open on its borders, new forms would certainly immigrate, and this also would seriously disturb the relations of some of the former inhabitants. Let it be remembered how powerful the influence of a single introduced tree or mammal has been shown to be. But in the case of an island, or of a country partly surrounded by barriers, into which new and better adapted forms could not freely enter, we should then have places in the economy of nature which would assuredly be better filled up, if some of the original inhabitants were in some manner modified; for, had the area been open to immigration, these same places would have been seized on by intruders. In such case, every slight modification, which in the course of

存，可能是很有用处的。既然家养生物肯定发生了对人类有益的变异，难道在广泛复杂的生存斗争中，对每一个生物本身有益的变异，在许多世代相传的历程中就不会发生了吗？由于繁殖出来的个体比能够生存下来的个体要多得多，我们可以毫不怀疑地说，如果上述情况的确曾发生，那么具有任何优势的个体，无论其优势多么微小，都将比其他个体有更多的生存和繁殖的机会。另一方面，我们也确信，任何轻微的有害变异，都必然招致绝灭。我把这种有利于生物个体的差异或变异的保存，以及有害变异的毁灭，称为"自然选择"或"适者生存"。无用也无害的变异，则不受自然选择作用的影响，它们或者成为不固定的性状，如在某些多型物种里所看到的性质一样，或者根据生物本身和外界生存环境的情况，最终成为生物固定的性状。

3 为使我们完全明白自然选择的大概过程，最好研究一下某个地区在自然条件轻微变化下发生的事情。例如：在气候变化的时候，当地各种生物的比例数，几乎立刻也会发生变化，有些物种很可能会绝灭。我们知道，任何一个地区的生物，都是由密切复杂的相互关系连接在一起的，即使不因气候的变化，仅仅是某些生物比例数的变化，就会严重影响到其他生物。如果一个地区的边界是开放的，新的生物类型必然要迁入，这就会严重扰乱原有生物间的关系。我们曾指出，从外地引进一种树或一种哺乳动物会引起多么大的影响。如果是在一个岛上，或是在一部分边界被障碍物环绕的地方，新的善于适应环境的生物不能自由进入这里，原有自然生态中出现空隙，必然会被当地善于发生变异的种类所充填。而这些位置，在迁入方便的情况下早就被外来生物所侵占了。在此种情况下，凡是有利于生物个体的任何微小变异，都能使此个体更好地去适应改变了的生活条件，这些变异就可能被保存下来，而自然选择就有充分的机会去进行改良生物的工作了。

ages chanced to arise, and which in any way favoured the individuals of any of the species, by better adapting them to their altered conditions, would tend to be preserved; and natural selection would thus have free scope for the work of improvement.

4 We have reason to believe, as stated in the first chapter, that a change in the conditions of life, by specially acting on the reproductive system, causes or increases variability; and in the foregoing case the conditions of life are supposed to have undergone a change, and this would manifestly be favourable to natural selection, by giving a better chance of profitable variations occurring; and unless profitable variations do occur, natural selection can do nothing. Not that, as I believe, any extreme amount of variability is necessary; as man can certainly produce great results by adding up in any given direction mere individual differences, so could Nature, but far more easily, from having incomparably longer time at her disposal. Nor do I believe that any great physical change, as of climate, or any unusual degree of isolation to check immigration, is actually necessary to produce new and unoccupied places for natural selection to fill up by modifying and improving some of the varying inhabitants. For as all the inhabitants of each country are struggling together with nicely balanced forces, extremely slight modifications in the structure or habits of one inhabitant would often give it an advantage over others; and still further modifications of the same kind would often still further increase the advantage. No country can be named in which all the native inhabitants are now so perfectly adapted to each other and to the physical conditions under which they live, that none of them could anyhow be improved; for in all countries, the natives have been so far conquered by naturalised productions, that they have allowed foreigners to take firm possession of the land. And as foreigners have thus everywhere beaten some of the natives, we may safely conclude that the natives might have been modified with advantage, so as to have better resisted such intruders.

4 正如第1章①所指出的，我们有足够的理由相信，自然条件的变化可能使变异性增加。外界环境条件发生变化时，有益变异的机会便会增加，这对于自然选择显然是有利的。如果没有有益变异的产生，自然选择也就无所作为。我相信不必积累大量变异，既然人类能在一定方向上积累个体差异，而且在家养的动植物中效果显著，那么，自然选择也能够而且更容易做到这一点，因为它可以在比人工选择长久得多的时间内发生作用。我认为不必通过巨大的自然变化，如气候的变化，或通过高度隔绝限制生物迁移，才可使自然生态系统中出现某些空白位置，以使自然选择去改进某些生物性状，使它们填补进去。因为每一地区的各种生物是以极微妙的均衡力量在进行竞争，当一种生物的构造或习性发生微小变化时，就会具有超过其他生物的优势，只要此种生物继续生活在同样的环境条件下，以同样的生存和防御方式获得利益，则同样的变异将继续发展，此物种的优势就越来越大。可以说没有一个地方，那里的生物与生物之间，生物与其生活的自然地理条件之间，已达适应的完美程度，以至于任何生物都不需要继续变异以适应得更好一些了，因为在许多地区，都可以看到外地迁入的生物迅速战胜土著生物，而在当地获得立足之地的事实。根据外来生物在各地仅能征服某些种类的土著生物的事实，我们可以断定，土著生物也曾产生过有利的变异以抵抗入侵者。

① 此处的"第1章"指的是[英]达尔文著，舒德干等译，北京大学出版社2005年出版的《物种起源》的第一章，而非本导引第1章。本文中的"第××章"皆指《物种起源》中的第××章，而非本导引中的。

5 As man can produce and certainly has produced a great result by his methodical and unconscious means of selection, what may not nature effect? Man can act only on external and visible characters: nature cares nothing for appearances, except in so far as they may be useful to any being. She can act on every internal organ, on every shade of constitutional difference, on the whole machinery of life. Man selects only for his own good; Nature only for that of the being which she tends. Every selected character is fully exercised by her; and the being is placed under well-suited conditions of life. Man keeps the natives of many climates in the same country; he seldom exercises each selected character in some peculiar and fitting manner; he feeds a long and a short beaked pigeon on the same food; he does not exercise a long-backed or long-legged quadruped in any peculiar manner; he exposes sheep with long and short wool to the same climate. He does not allow the most vigorous males to struggle for the females. He does not rigidly destroy all inferior animals, but protects during each varying season, as far as lies in his power, all his productions. He often begins his selection by some half-monstrous form; or at least by some modification prominent enough to catch his eye, or to be plainly useful to him. Under nature, the slightest difference of structure or constitution may well turn the nicely-balanced scale in the struggle for life, and so be preserved. How fleeting are the wishes and efforts of man! How short his time! And consequently how poor will his products be, compared with those accumulated by nature during whole geological periods. Can we wonder, then, that nature's productions should be far "truer" in character than man's productions; that they should be infinitely better adapted to the most complex conditions of life, and should plainly bear the stamp of far higher workmanship?

5 通过有计划的或无意识的选择方法，人类能够产生并确实已经产生极大的成果，那么自然选择为什么就不能产生如此效力呢？人类仅就生物的外部和可见的性状加以选择，而“自然”（请允许我把“自然保存”或“适者生存”拟人化）并不关心外表，除非是对生物有用的外表。“自然”可以作用到每一内部器官、每一体质的细微差异及整个生命机制。人类仅为自己的利益去选择，而“自然”却是为保护生物的利益去选择。从选择的事实可以看出，每一个被选择的性状，都充分受到“自然”的陶冶；而人类把许多不同气候的产物，畜养于同一地区，很少用特殊、合适的方式去增强每一选择出的性状。人类用同样的食料饲养长喙鸽和短喙鸽，也不用特殊的方法，去训练长背的或长脚的哺乳动物，人类把长毛羊和短毛羊畜养在同一种气候下，也不让最强壮的雄性动物通过争斗获得雌性配偶。人类也不严格地把所有劣等动物淘汰掉，反而在各个不同的季节里，利用人类的能力不分良莠地保护一切生物。人类往往根据半畸形的生物，或至少根据能引起他注意的显著变异，或根据对他非常有用的某些性状去进行选择。在自然状态下，任何生物在构造上和体质上的微小差异，都能改变生存斗争中的微妙平衡关系，并把差异保存下来。与自然选择在整个地质时期内的成果比较起来，人类的愿望与努力，只是瞬息间的事，人类的生命是多么短暂，所获得的成果也是多么贫乏！“自然”产物的性状，比人工产物的性状更加“实用”，它们能更好地适应极其复杂的生活条件，能更明显地表现出选择优良性状的高超技巧，对此，难道我们还会感到惊奇吗？

6 It may be said that natural selection is daily and hourly scrutinising, throughout the world, every variation, even the slightest; rejecting that which is bad, preserving and adding up all that is good; silently and insensibly working, whenever and wherever opportunity offers, at the improvement of each organic being in relation to its organic and inorganic conditions of life. We see nothing of these slow changes in progress, until the hand of time has marked the long lapse of ages, and then so imperfect is our view into long past geological ages, that we only see that the forms of life are now different from what they formerly were.

7 Natural selection will modify the structure of the young in relation to the parent, and of the parent in relation to the young. In social animals it will adapt the structure of each individual for the benefit of the community; if each in consequence profits by the selected change. What natural selection cannot do, is to modify the structure of one species, without giving it any advantage, for the good of another species; and though statements to this effect may be found in works of natural history, I cannot find one case which will bear investigation. A structure used only once in an animal's whole life, if of high importance to it, might be modified to any extent by natural selection; for instance, the great jaws possessed by certain insects, and used exclusively for opening the cocoon—or the hard tip to the beak of nestling birds, used for breaking the egg. It has been asserted, that of the best short-beaked tumbler-pigeons more perish in the egg than are able to get out of it; so that fanciers assist in the act of hatching. Now, if nature had to make the beak of a full-grown pigeon very short for the bird's own advantage, the process of modification would be very slow, and there would be simultaneously the most rigorous selection of the young birds within the egg, which had the most powerful and hardest beaks, for all with weak

6 打个比方说吧，在世界范围内，自然选择每日每时都在对变异进行检查，去掉差的，保存、积累好的。不论何时何地，只要一有机会，它就默默地不知不觉地工作，去改进各种生物与有机的和无机的生活条件的关系。除非标志出时代的变迁，岁月的流逝，否则人们很难看出这种缓慢的变化，而人们对于远古的地质时代所知甚少，所以我们现在所看到的，只是现在的生物与以前的生物不同而已。

7 自然选择可以根据亲体使子体的构造发生变异，也能根据子体使亲体的构造发生变异。在群居的动物中，如果选择出来的变异有利于群体，自然选择就会为了整体的利益改变个体的构造。自然选择不可能在改变一个物种的构造时不是为了对这一物种有利而是为了对另一物种有利。虽然在自然史著作中有对此种作用的记载，但是我们没有见到一个能经得起检验的实例。自然选择可以使动物一生中仅用一次的重要构造发生极大变化，例如：某些昆虫专门用于破茧的大颚或雏鸟破卵壳用的坚硬的喙尖。有人说，优良的短喙翻飞鸽死在蛋壳里的数量比能破壳孵出的多，所以养鸽者必须帮助它们孵出。假如为了这种鸽自身的利益，自然选择使这种成年鸽具有极短的喙，必定是一个非常缓慢的变异过程，而在这个严格选择的过程中，那些在蛋壳内具有强有力喙的雏鸟将被选择出来，因为弱喙的雏鸟必然死在蛋壳内，或者蛋壳较脆弱易破碎的，也可能被选择出来，因为和其他构造一样，蛋壳也是能够变异的。

beaks would inevitably perish: or, more delicate and more easily broken shells might be selected, the thickness of the shell being known to vary like every other structure.

Sexual Selection

8 In as much as peculiarities often appear under domestication in one sex and become hereditarily attached to that sex, the same fact probably occurs under nature, and if so, natural selection will be able to modify one sex in its functional relations to the other sex, or in relation to wholly different habits of life in the two sexes, as is sometimes the case with insects. And this leads me to say a few words on what I call Sexual Selection. This depends, not on a struggle for existence, but on a struggle between the males for possession of the females; the result is not death to the unsuccessful competitor, but few or no offspring. Sexual selection is, therefore, less rigorous than natural selection. Generally, the most vigorous males, those which are best fitted for their places in nature, will leave most progeny. But in many cases, victory will depend not on general vigour, but on having special weapons, confined to the male sex. A hornless stag or spurless cock would have a poor chance of leaving offspring. Sexual selection by always allowing the victor to breed might surely give indomitable courage, length to the spur, and strength to the wing to strike in the spurred leg, as well as the brutal cock-fighter, who knows well that he can improve his breed by careful selection of the best cocks. How low in the scale of nature this law of battle descends, I know not; male alligators have been described as fighting, bellowing, and whirling round, like Indians in a war-dance, for the possession of the females; male salmons have been seen fighting all day long; male stag-beetles often bear wounds from the huge mandibles of other males. The war is, perhaps, severest between the males of polygamous animals, and these seem oftenest provided with special weapons. The males of carnivorous animals are already well armed; though to them and to others, special means

性选择

8 在家养状态下，有些特征往往只见于一个性别，并由这个性别遗传；在自然状态下无疑也有这种情况。因此，通过自然选择作用，有时雌雄两性个体在不同生活习性方面都能发生变异，或者更常见的是这一性别对另一性别的关系发生变异。这促使我要谈一下所谓性选择的问题。性选择的形式，并不是一种生物和其他生物为了生存或和外界自然条件进行的斗争，而是在同一物种的同一性别的个体间，一般是雄性之间，为了获得雌性配偶而发生的斗争。这种斗争的结果，不是让失败的一方死掉，而是让失败的一方不留或少留下后代，所以性的选择不如自然选择那样激烈。一般来说，最强壮的雄性，是自然界中最适应的个体，它们留下的后代也最多。但往往胜利并不全靠体格的强壮，而是靠雄性特有的武器。如无角雄鹿和无距(spur)(雄鸡爪后面突出像脚趾似的部分——译者注)公鸡就不可能留下很多后代。由于性选择可以使获胜者得到更多繁殖的机会，所以和残忍的斗鸡者挑选善斗的公鸡一样，性选择可以赋予公鸡不屈不挠斗争的勇气、增加距的长度和在争斗时拍击翅膀加强距的攻击力量。我不知道在动物的分类中，直到哪一类动物才没有性选择的作用。但是有人曾描述说，雄性鳄鱼(alligator)在争取雌性鳄鱼时会像美洲印第安人跳战斗舞蹈那样吼叫并旋绕转身；雄鲑鱼(salmon)整天彼此争斗；雄性锹形虫(slag-beetle)的大颚常被其他雄虫咬伤。这种争斗，可能在“多妻”的雄性动物中最为激烈，而这种雄性常有特殊的武器。食肉动物原来就已具备良好的战斗武器，性选择又使它们和别的动物一样，又具备了更特殊的防御手段，例如，雄狮的鬃毛，雄鲑鱼的钩形上颚等。要知道，为了在战斗中取胜，盾的作用和矛、剑是同等重要的。

of defence may be given through means of sexual selection, as the mane to the lion, the shoulder-pad to the boar, and the hooked jaw to the male salmon; for the shield may be as important for victory, as the sword or spear.

9 Amongst birds, the contest is often of a more peaceful character. All those who have attended to the subject, believe that there is the severest rivalry between the males of many species to attract by singing the females. The rock-thrush of Guiana, birds of Paradise, and some others, congregate; and successive males display their gorgeous plumage and perform strange antics before the females, which standing by as spectators, at last choose the most attractive partner. Those who have closely attended to birds in confinement well know that they often take individual preferences and dislikes: thus Sir R. Heron has described how one pied peacock was eminently attractive to all his hen birds. It may appear childish to attribute any effect to such apparently weak means; I cannot here enter on the details necessary to support this view; but if man can in a short time give elegant carriage and beauty to his bantams, according to his standard of beauty, I can see no good reason to doubt that female birds, by selecting, during thousands of generations, the most melodious or beautiful males, according to their standard of beauty, might produce a marked effect. I strongly suspect that some well-known laws with respect to the plumage of male and female birds, in comparison with the plumage of the young, can be explained on the view of plumage having been chiefly modified by sexual selection, acting when the birds have come to the breeding age or during the breeding season; the modifications thus produced being inherited at corresponding ages or seasons, either by the males alone, or by the males and females; but I have not space here to enter on this subject.

10 Thus it is, as I believe, that when the males and females of any animal have the same general habits of life, but differ in structure, colour, or ornament,

9 就鸟类来说，这类争斗要平和得多。研究过这一问题的人都相信，许多鸟类的雄性间最激烈的斗争，是用歌唱去吸引雌鸟。圭亚那(Guiana)的岩鸫(rook-thrush)、极乐鸟(birds of paradise)及其他鸟类常常聚集一处，雄鸟一个个精心地以最殷勤的态度展示它们艳丽的羽毛，在雌鸟面前做出种种奇特的姿态，而雌鸟在一旁观赏，最后选择最有吸引力的雄鸟做配偶。仔细观察过笼养鸟的人，都知道鸟有各自的爱憎。赫龙爵士(Sir R. Heron)曾描述他养的斑纹孔雀是如何极为成功地吸引了所有的雌孔雀。认为这样明显微弱的方式就能产生任何效果也许显得幼稚，这里虽不能叙述详情，但是可以说人类能在很短时间内按自己的审美标准，使矮脚鸡具有美丽、优雅的姿态。毫无疑问，在数千代的相传中，雌鸟一定会根据它们的审美标准，选择出声调最动听、羽毛最美丽的雄鸟，并产生了显著的性选择效果。在生命不同时期出现的变异，会在相应时期单独出现在雌性后代或者雄、雌两性后代身上，性选择会对这些变异起作用；用这种性选择的作用，可以在一定程度上解释关于雄鸟和雌鸟的羽毛不同于雏鸟羽毛的著名法则，在此就不详细讨论这个题目了。

10 因此任何动物的雌雄两体，如果它们的生活习性相同而构造、颜色或装饰不同，可以说这些差异主要是性选择造成的，即在世代遗传中，雄性

such differences have been mainly caused by sexual selection; that is, individual males have had, in successive generations, some slight advantage over other males, in their weapons, means of defence, or charms; and have transmitted these advantages to their male offspring. Yet, I would not wish to attribute all such sexual differences to this agency: for we see peculiarities arising and becoming attached to the male sex in our domestic animals (as the wattle in male carriers, horn-like protuberances in the cocks of certain fowls, etc.), which we cannot believe to be either useful to the males in battle, or attractive to the females. We see analogous cases under nature, for instance, the tuft of hair on the breast of the turkey-cock, which can hardly be either useful or ornamental to this bird—indeed, had the tuft appeared under domestication, it would have been called a monstrosity.

Illustrations of the Action of Natural Selection

11 In order to make it clear how, as I believe, natural selection acts, I must beg permission to give one or two imaginary illustrations. Let us take the case of a wolf, which preys on various animals, securing some by craft, some by strength, and some by fleetness; and let us suppose that the fleetest prey, a deer for instance, had from any change in the country increased in numbers, or that other prey had decreased in numbers, during that season of the year when the wolf is hardest pressed for food. I can under such circumstances see no reason to doubt that the swiftest and slimmest wolves would have the best chance of surviving, and so be preserved or selected—provided always that they retained strength to master their prey at this or at some other period of the year, when they might be compelled to prey on other animals. I can see no more reason to doubt this, than that man can improve the fleetness of his greyhounds by careful and methodical selection, or by that unconscious selection which results from each man trying to keep the best dogs without any thought of modifying the breed.

个体把稍优于其他雄性的攻击武器、防御手段或漂亮雄壮的外形等特点，遗传给它们的雄性后代。不过，我们不应该把所有性别间的差异都归因于性选择，因为在家养动物中，有些雄性专有的特征并不能通过人工选择而扩大(如雄信鸽的肉垂，特定种类雄性禽类的角状突)。野生雄火鸡(turkey-cock)胸间的丛毛，并无什么用处，而在雌火鸡眼里，也很难说这是一种装饰；说实在的，如果这丛毛出现在家养动物身上，就会被视为畸形。

自然选择，即适者生存作用的实例

11 让我举一两个假想的例子，来说明自然选择是如何起作用的吧。以狼为例，在捕食各种动物时，狼有时用技巧，有时用力量，有时则用速度。假设一个地区由于某种变化，狼所捕食的动物中，跑得最快的鹿数量增加或其他动物数量减少，这是狼捕食最困难的时期，在这种情况下，当然只有跑动最敏捷、体型最灵巧的狼才能获得充分的生存机会，从而被选择和保存，当然它们还必须在各个时期总能保存足够的力量去征服和捕食其他动物。人类为了保存最优良的个体(并非为了改变品种)，在进行仔细有计划的或无意识的选择时，能提高长嘴猎狗(灵猩)的敏捷性。毫无疑问，自然选择也会产生如此效果。

12 I may add, that, according to Mr. Pierce, there are two varieties of the wolf inhabiting the Catskill Mountains in the United States, one with a light greyhound-like form, which pursues deer, and the other more bulky, with shorter legs, which more frequently attacks the shepherd's flocks.

13 Let us now take a more complex case. Certain plants excrete a sweet juice, apparently for the sake of eliminating something injurious from their sap: this is effected by glands at the base of the stipules in some Leguminosae, and at the back of the leaf of the common laurel. This juice, though small in quantity, is greedily sought by insects. Let us now suppose a little sweet juice or nectar to be excreted by the inner bases of the petals of a flower. In this case insects in seeking the nectar would get dusted with pollen, and would certainly often transport the pollen from one flower to the stigma of another flower. The flowers of two distinct individuals of the same species would thus get crossed; and the act of crossing, we have good reason to believe (as will hereafter be more fully alluded to), would produce very vigorous seedlings, which consequently would have the best chance of flourishing and surviving. Some of these seedlings would probably inherit the nectar-excreting power. Those individual flowers which had the largest glands or nectaries, and which excreted most nectar, would be oftenest visited by insects, and would be oftenest crossed; and so in the long-run would gain the upper hand. Those flowers, also, which had their stamens and pistils placed, in relation to the size and habits of the particular insects which visited them, so as to favour in any degree the transportal of their pollen from flower to flower, would likewise be favoured or selected. We might have taken the case of insects visiting flowers for the sake of collecting pollen instead of nectar; and as pollen is formed for the sole object of fertilisation, its destruction appears a simple loss to the plant; yet if a little pollen were carried, at first occasionally and then habitually, by the pollen-devouring insects from flower

12 顺便提一下，根据皮尔斯先生(Mr. Pierce)所说，在美国的卡茨基尔山脉(Catskill Mountains)栖息着两种狼的变种，一种形状略似长嘴猎狗，逐鹿为食，另一种躯干较粗腿较短，常常袭击牧人的羊群。

13 再举一个较复杂的例子来说明自然选择的作用吧。有些植物分泌甜汁，这显然是为了排除体液内的有害物质。例如，某些豆科植物(Leguminosae)从托叶基部的腺体排出分泌物，普通月桂树(laurel)从叶背分泌液体。这种甜汁量虽少，却被昆虫贪婪地寻求着，然而这些昆虫的来访，对植物本身并无任何益处。假如甜汁是从一种植物的若干植株的花里分泌出来的，寻找这种甜汁(花蜜)的昆虫会沾上花粉，并把花粉从一朵花传到另一朵花上去，这样同种的两个不同个体就可以进行杂交，从而产生强壮的幼苗，并使幼苗得到更好的生存和繁殖机会，这些情况都可以得到充分证明。那些花蜜腺体最大的植株，分泌的花蜜最多，最常受到昆虫的光顾，获得杂交机会也就最多。长此以往，它们就会占有优势并形成一个地方变种。有些花的雄蕊和雌蕊所处的位置能适合前来采蜜昆虫的大小和习性，这在一定程度上有利于昆虫传授花粉，这样的花同样也会受益。如果一只来往于花间的昆虫并不采蜜而专采花粉，这种对花粉的破坏显然是植物的一种损失，因为花粉是专为受精用的。可是如果因这一昆虫的媒介作用，少量的花粉由一朵花传到另一朵花，最初可能出于偶然，尔后就可能形成习惯，这种情况促进植物的杂交，即使9/10的花粉损失掉了，对于花粉被盗的植物来说，结果仍然是非常有利的。因而，那些产花粉较多的、粉囊较大的个体将被选择出来。

to flower, and a cross thus effected, although nine-tenths of the pollen were destroyed, it might still be a great gain to the plant; and those individuals which produced more and more pollen, and had larger and larger anthers, would be selected.

14 When our plant, by this process of the continued preservation or natural selection of more and more attractive flowers, had been rendered highly attractive to insects, they would, unintentionally on their part, regularly carry pollen from flower to flower; and that they can most effectually do this, I could easily show by many striking instances. I will give only one—not as a very striking case, but as likewise illustrating one step in the separation of the sexes of plants, presently to be alluded to. Some holly-trees bear only male flowers, which have four stamens producing rather a small quantity of pollen, and a rudimentary pistil; other holly-trees bear only female flowers; these have a full-sized pistil, and four stamens with shrivelled anthers, in which not a grain of pollen can be detected. Having found a female tree exactly sixty yards from a male tree, I put the stigmas of twenty flowers, taken from different branches, under the microscope, and on all, without exception, there were pollen-grains, and on some a profusion of pollen. As the wind had set for several days from the female to the male tree, the pollen could not thus have been carried. The weather had been cold and boisterous, and therefore not favourable to bees, nevertheless every female flower which I examined had been effectually fertilised by the bees, accidentally dusted with pollen, having flown from tree to tree in search of nectar. But to return to our imaginary case: as soon as the plant had been rendered so highly attractive to insects that pollen was regularly carried from flower to flower, another process might commence. No naturalist doubts the advantage of what has been called the "physiological division of labour;" hence we may believe that it would be advantageous to a plant to produce stamens alone in one

14 如果上述过程长期继续下去，植物就将变得很能吸引昆虫，昆虫也就不自觉地在花间规律而有效地传递花粉。这方面突出的例子很多。现举一例，这个例子同时还能说明植物雌雄分株的步骤。有些冬青树(holly-tree)只生雄花，每花有含少量花粉的四枚雄蕊和一枚不发育的雌蕊；另一雌冬青树只生雌花，每花有一枚发育完全的雌蕊和四个粉囊萎缩的雄蕊，且雄蕊上无一粒花粉。在距一株雄冬青树60码的地方，我找到一株雌冬青树并从不同枝干上采下20朵花，当我把雌花柱头放在显微镜下观察时，发现所有柱头上毫无例外地都沾有几粒花粉，有的还相当多。那几天风是从雌树的方向吹往雄树，所以这些花粉不是由风力传送的；虽然天气很冷并有暴风雨(这对蜂类不利)，但是我检查的所有雌花都因在花间寻找花蜜的蜂而有效地受精了。现在再回过头来谈一下我们想象的情况：一旦植物变得很能吸引昆虫，以致昆虫在花间规则地传递花粉，另一个步骤可能就开始了。博物学者们都不怀疑所谓生理分工的益处。因此我们相信，一树或一花只生雄蕊而另一树或另一花只生雌蕊对植物是有利的。栽培的植物和被置于新的生活环境的植物雄性器官，有时是雌性器官的功能会有所减退，假定在自然状态下，这种情况也会发生，即使程度极其轻微。既然花粉已经能在花间有规律地传递，既然“生理分工”的原理显示，更完全的性别分离对植物更为有利，那么雌雄分离的倾向越显明的个体，将会不断受益并被选择，直到雌雄两体最终完全分离。

flower or on one whole plant, and pistils alone in another flower or on another plant. In plants under culture and placed under new conditions of life, sometimes the male organs and sometimes the female organs become more or less impotent; now if we suppose this to occur in ever so slight a degree under nature, then as pollen is already carried regularly from flower to flower, and as a more complete separation of the sexes of our plant would be advantageous on the principle of the division of labour, individuals with this tendency more and more increased, would be continually favoured or selected, until at last a complete separation of the sexes would be effected.

15 Let us now turn to the nectar-feeding insects in our imaginary case: we may suppose the plant of which we have been slowly increasing the nectar by continued selection, to be a common plant; and that certain insects depended in main part on its nectar for food. I could give many facts, showing how anxious bees are to save time; for instance, their habit of cutting holes and sucking the nectar at the bases of certain flowers, which they can, with a very little more trouble, enter by the mouth. Bearing such facts in mind, I can see no reason to doubt that an accidental deviation in the size and form of the body, or in the curvature and length of the proboscis, etc., far too slight to be appreciated by us, might profit a bee or other insect, so that an individual so characterised would be able to obtain its food more quickly, and so have a better chance of living and leaving descendants. Its descendants would probably inherit a tendency to a similar slight deviation of structure. The tubes of the corollas of the common red and incarnate clovers (Trifolium pratense and incarnatum) do not on a hasty glance appear to differ in length; yet the hive-bee can easily suck the nectar out of the incarnate clover, but not out of the common red clover, which is visited by humble-bees alone; so that whole fields of the red clover offer in vain an abundant supply of precious nectar to the hive-bee. Thus it might be a great advantage to the hive-bee to

15 现在谈谈吃花蜜的昆虫。假如一种普通的植物因连续的选择作用而使花蜜逐渐增加，而某种昆虫又是以这种花蜜为食的。我能举出多种例子说明蜂是如何急于采蜜而设法节省时间的。例如，一些蜂习惯于在花的基部咬一口来吸食花蜜，而本来它们稍费点劲就能从花的开口部位钻到花里去。想到这些情况，我们就会相信，那些容易被忽视的微小个体差异，如口吻的长度、弯曲度等，在一定条件下，对于蜂和其他昆虫是有利的。因此有些个体能比其他个体更快地获得食物，它们所属的群体能够繁盛，而从它们分出去的许多蜂群也都继承了同样的性状。普通红三叶草和肉色三叶草（T. incarnatum）的管形花冠，粗看上去长度并无差异，但蜜蜂可以轻易地吸取肉色三叶草的花蜜却不能吸到红三叶草的花蜜，能采红三叶草花蜜的只有野蜂。蜜蜂不能享受遍布田野的红三叶草花蜜，一方面可以说，在长满红三叶草的地方，具有略长或不同形状吻的蜜蜂能够获得好处。从另一方面来说，由于红三叶草完全靠能来采花蜜的蜂受精，如果一个地区的野蜂少了，则花管较短或分裂较深的植株将会得到好处，而蜜蜂也就可以采这种红三叶草的花蜜了。现在，我们理解了蜂与花是如何通过不断保存结构上互相有利的微小差异，同时或先后发生变异以达到完美的相互适应了。

have a slightly longer or differently constructed proboscis. On the other hand, I have found by experiment that the fertility of clover greatly depends on bees visiting and moving parts of the corolla, so as to push the pollen on to the stigmatic surface. Hence, again, if humble-bees were to become rare in any country, it might be a great advantage to the red clover to have a shorter or more deeply divided tube to its corolla, so that the hive-bee could visit its flowers. Thus I can understand how a flower and a bee might slowly become, either simultaneously or one after the other, modified and adapted in the most perfect manner to each other, by the continued preservation of individuals presenting mutual and slightly favourable deviations of structure.

16 I am well aware that this doctrine of natural selection, exemplified in the above imaginary instances, is open to the same objections which were at first urged against Sir Charles Lyell's noble views on "the modern changes of the earth, as illustrative of geology;" but we now very seldom hear the action, for instance, of the coast-waves, called a trifling and insignificant cause, when applied to the excavation of gigantic valleys or to the formation of the longest lines of inland cliffs. Natural selection can act only by the preservation and accumulation of infinitesimally small inherited modifications, each profitable to the preserved being; and as modern geology has almost banished such views as the excavation of a great valley by a single diluvial wave, so will natural selection, if it be a true principle, banish the belief of the continued creation of new organic beings, or of any great and sudden modification in their structure.

Extinction

17 This subject will be more fully discussed in our chapter on Geology; but it

16 我知道，用上述想象的例子来说明自然选择的原理，是会遭到反对的，正如莱伊尔爵士最初“用地球近代的变迁来解释地质学”时遇到反对一样。不过现在再运用仍然活跃的一些地质作用来解释深谷和内陆崖壁的形成时，很少再有人说是微不足道或毫无意义了。自然选择的作用，仅在于把每个有益的微小遗传变异保存和积累起来。近代地质学已经抛弃了那种一次大洪水就能凿出一个大山谷来的观点，同样地，自然选择学说也将排除那种以为新生物类型能连续被创生，或者生物的构造能够突然发生大变异的观点。

自然选择造成的绝灭

17 在地质学一章里将充分讨论这个问题，但因它与自然选择很有关系，所以在此很有必要提一下。自然选择的作用只是通过保存在某些方面的有

must be here alluded to from being intimately connected with natural selection. Natural selection acts solely through the preservation of variations in some way advantageous, which consequently endure. But as from the high geometrical powers of increase of all organic beings, each area is already fully stocked with inhabitants, it follows that as each selected and favoured form increases in number, so will the less favoured forms decrease and become rare. Rarity, as geology tells us, is the precursor to extinction. We can, also, see that any form represented by few individuals will, during fluctuations in the seasons or in the number of its enemies, run a good chance of utter extinction. But we may go further than this; for as new forms are continually and slowly being produced, unless we believe that the number of specific forms goes on perpetually and almost indefinitely increasing, numbers inevitably must become extinct. That the number of specific forms has not indefinitely increased, geology shows us plainly; and indeed we can see reason why they should not have thus increased.

18 Furthermore, the species which are most numerous in individuals will have the best chance of producing within any given period favourable variations. We have evidence of this, in the facts given in the second chapter, showing that it is the common species which afford the greatest number of recorded varieties, or incipient species. Hence, rare species will be less quickly modified or improved within any given period, and they will consequently be beaten in the race for life by the modified descendants of the commoner species.

19 From these several considerations I think it inevitably follows, that as new species in the course of time are formed through natural selection, others will become rarer and rarer, and finally extinct. The forms which stand in closest

利变异，使这些变异能持续下去。由于一切生物都以几何级数增加，致使每一地区都充满了生物；随着优势类型个体数目的增加，劣势类型的个体数就要减少以致稀少。地质学告诉我们，稀少就是绝灭的前奏。我们知道，任何个体数量少的类型在季节气候发生重大变动时，或在敌害数量暂时增多时，极有可能遭到灭顶之灾。进一步说，如果我们承认物种类型不可能无限地增多的话，那么，随着新类型的产生，必然导致众多旧类型的消亡。地质学明确显示，物种类型的数目没有无限地增加过。现在我们来说明一下，为什么世界上的物种数量没有无限地增加。

18 我们知道，在任何时期，个体最多的物种可获得最好的机会以产生有利变异，对此我们是有证据的。第 2 章提到的事实显示，正是那些常见的、广泛分布的或占优势的物种，产生了最多的有据可考的变种。因此，在任何时期，个体稀少的物种，发生变异和改良的速度较慢，在生存斗争中，它们很容易被已变异和改良过的常见物种的后代击败。

19 根据这些分析，结果必然是：随着时间的推移，通过自然选择，新物种产生了，而其他物种将变得越来越稀少以至于最终绝灭，并且，那些与正在变异和改良的类型进行最激烈的斗争的类型将最先灭亡。在生存斗

competition with those undergoing modification and improvement, will naturally suffer most. And we have seen in the chapter on the *Struggle for Existence* that it is the most closely-allied forms—varieties of the same species, and species of the same genus or of related genera—which, from having nearly the same structure, constitution, and habits, generally come into the severest competition with each other. Consequently, each new variety or species, during the progress of its formation, will generally press hardest on its nearest kindred, and tend to exterminate them. We see the same process of extermination amongst our domesticated productions, through the selection of improved forms by man. Many curious instances could be given showing how quickly new breeds of cattle, sheep, and other animals, and varieties of flowers, take the place of older and inferior kinds. In Yorkshire, it is historically known that the ancient black cattle were displaced by the long-horns, and that these "were swept away by the short-horns" (I quote the words of an agricultural writer) "as if by some murderous pestilence."

Divergence of Character

20 The principle, which I have designated by this term, is of high importance on my theory, and explains, as I believe, several important facts. In the first place, varieties, even strongly-marked ones, though having somewhat of the character of species—as is shown by the hopeless doubts in many cases how to rank them—yet certainly differ from each other far less than do good and distinct species. Nevertheless, according to my view, varieties are species in the process of formation, or are, as I have called them, incipient species. How, then, does the lesser difference between varieties become augmented into the greater difference between species? That this does habitually happen, we must infer from most of the innumerable species throughout nature

争一章中我们已经知道，由于具有近似的结构、体质及习性，最近缘的类型(同种的各变种，同属或近属的各物种)之间竞争最为激烈。结果是，在每一变种或物种形成的过程中，它们给最近缘种类造成最大的威胁，以至于往往最终消灭它们。在家养动物中，通过人类对改良类型的选择，也会出现同样的消亡过程。许多奇特的例子可以说明，牛、羊及其他动物的新品种和花草的变种，是多么迅速地取代了旧的低劣种类的。在约克郡人们都知道，古代的黑牛被长角牛取代，长角牛又被短角牛所排挤。用一老农的话说："简直就像被残酷的瘟疫一扫而光。"

性状趋异

20 这个术语所包含的原理是很重要的，一些现象可以用它来解释。首先，即便是那些多少已经具有了物种特征的显著变种，与明确的物种比起来，彼此间的差异还是小得多，因而在很多场合很难对它们进行分类；但我还是认为，变种是形成过程中的物种，我称之为初期物种。那么，变种间的较小差异是如何扩大为物种间的较大差异的呢？自然界中无数的物种呈现显著的差异，而变种是未来显著物种的原型和亲体，在它们之间仅呈现出微小而不确定的差异，由此，我们可以推论，较小差异向较大差异变化的过程是经常发生的。可以说，仅仅出于偶然变异，变种才与亲体之间有了某种性状的不同，后来这一发生变异的后代又与亲体在这同一性状方面产生了更大的不同。然而，仅此并不能解释同属异种间常见的巨大差异。

presenting well-marked differences; whereas varieties, the supposed prototypes and parents of future well-marked species, present slight and ill-defined differences. Mere chance, as we may call it, might cause one variety to differ in some character from its parents, and the offspring of this variety again to differ from its parent in the very same character and in a greater degree; but this alone would never account for so habitual and large an amount of difference as that between varieties of the same species and species of the same genus.

21 As has always been my practice, let us seek light on this head from our domestic productions. We shall here find something analogous. A fancier is struck by a pigeon having a slightly shorter beak; another fancier is struck by a pigeon having a rather longer beak; and on the acknowledged principle that "fanciers do not and will not admire a medium standard, but like extremes," they both go on (as has actually occurred with tumbler-pigeons) choosing and breeding from birds with longer and longer beaks, or with shorter and shorter beaks. Again, we may suppose that at an early period one man preferred swifter horses; another stronger and more bulky horses. The early differences would be very slight; in the course of time, from the continued selection of swifter horses by some breeders, and of stronger ones by others, the differences would become greater, and would be noted as forming two sub-breeds; finally, after the lapse of centuries, the sub-breeds would become converted into two well-established and distinct breeds. As the differences slowly become greater, the inferior animals with intermediate characters, being neither very swift nor very strong, will have been neglected, and will have tended to disappear. Here, then, we see in man's productions the action of what may be called the principle of divergence, causing differences, at first barely appreciable, steadily to increase, and the breeds to diverge in character both from each other and from their common parent.

21 按照惯常的做法，我到家养动植物中去寻求对此事的解释，因为在它们当中可找到一些相似的情况。在实践中，有的养鸽者喜爱短喙鸽，有的却喜欢长喙鸽。一个公认的事实是，鸽迷们喜欢极端的类型而不喜欢中间类型。他们继续选择和饲养那些喙较长的或较短的鸽，就像人们培育翻飞鸽的亚品种那样。此外，我们可以设想，在历史初期，一个国家或地区的人需要快跑的马，另一个国家或地区的人需要强壮的高头大马，随着时间的流逝，一方不断地选择快马，另一方则不断地选择壮硕的马，这样原来差异甚小的两种马会逐渐变为两个差异很大的亚品种，最终在几个世纪之后，亚品种就转变为两个界线分明的不同品种了。当两者之间的差异增大时，那些既不太快又不太壮的中间性状的劣等马就不会被选来配种，因此这种马也就逐渐消亡了。从这些人工选择产物中可以看到能造成差别的所谓趋异原理的作用，它使最初难以觉察的差异逐渐扩大，使品种彼此间以及品种与其亲体间的性状发生分异。

22 But how, it may be asked, can any analogous principle apply in nature? I believe it can and does apply most efficiently, from the simple circumstance that the more diversified the descendants from any one species become in structure, constitution, and habits, by so much will they be better enabled to seize on many and widely diversified places in the polity of nature, and so be enabled to increase in numbers.

23 We can clearly see this in the case of animals with simple habits. Take the case of a carnivorous quadruped, of which the number that can be supported in any country has long ago arrived at its full average. If its natural powers of increase be allowed to act, it can succeed in increasing (the country not undergoing any change in its conditions) only by its varying descendants seizing on places at present occupied by other animals: some of them, for instance, being enabled to feed on new kinds of prey, either dead or alive; some inhabiting new stations, climbing trees, frequenting water, and some perhaps becoming less carnivorous. The more diversified in habits and structure the descendants of our carnivorous animal became, the more places they would be enabled to occupy. What applies to one animal will apply throughout all time to all animals—that is, if they vary—for otherwise natural selection can do nothing. So it will be with plants. It has been experimentally proved, that if a plot of ground be sown with one species of grass, and a similar plot be sown with several distinct genera of grasses, a greater number of plants and a greater weight of dry herbage can thus be raised. The same has been found to hold good when first one variety and then several mixed varieties of wheat have been sown on equal spaces of ground. Hence, if any one species of grass were to go on varying, and those varieties were continually selected which differed from each other in at all the same manner as distinct species and genera of grasses differ from each other, a greater number of individual plants of this species of grass, including its modified descendants, would succeed in living on the same piece of ground. And we

22 那么能否把类似的原理应用于自然界呢？我认为这一原理可以非常有效地应用于自然界（虽然经过很长时间我才弄清楚该怎样应用），因为任何一个物种的后代越是在结构、体质和习性上分异，它就越能占据自然体系中的不同位置，因而数量大大增加。

23 从习性简单的动物中，我们可以清楚地看到这种情况。以食肉的哺乳动物为例，在任何可以容身的地方，哺乳动物的数量早就达到了饱和平均数。如果在一个地区，生活条件不发生改变而任其自然发展，只有那些发生变异了的子孙们，才能获取目前被其他动物占据着的一些位置。例如，它们有的能获取新的猎物，不管是死的还是活的；有的能生活在不同的场所，能上树或能下水；有的能减少食肉习性，等等。总之，食肉动物的后代越能在身体构造和习性方面产生分异，它们能占据的位置就越多。如果这一原理可应用于一种动物，那么，它就可以应用于一切时期的一切动物。也就是说，只要它们变异，自然选择就会起作用；如果不变异，选择就无所作为。植物界的情况也是如此。试验证明，如果一块地上只播种一种草，另一块大小相等的地上播种多种草，则后一块地上所得的植株数量和干草的重量都比前者要多。如果把小麦的变种分成单种和多种两组，并在同样大小的土地上种植，也会得到相同的结果。因为只要任何一种草仍在继续变异，哪怕差异十分微小，这些彼此不同的变种就能像不同的物种或属那样，以同样的方式继续被选择。于是，这一物种的大量个体，连同它的变种，都将成功地在同一块土地上存活下去。每年草类的各个物种和变种都要撒下无数的种子，在追求最大限度地增加个体的数量方面，可以说是竭尽全力；因此，在万千世代相传的过程中，只有那些草类最显著的变种，能够有机会增加个体数目并排除那些变异不够显著的变种。当各个变种变异得彼此截然不同时，它们就跻身于物种之列了。

well know that each species and each variety of grass is annually sowing almost countless seeds; and thus, as it may be said, is striving its utmost to increase its numbers. Consequently, I cannot doubt that in the course of many thousands of generations, the most distinct varieties of any one species of grass would always have the best chance of succeeding and of increasing in numbers, and thus of supplanting the less distinct varieties; and varieties, when rendered very distinct from each other, take the rank of species.

24 The truth of the principle, that the greatest amount of life can be supported by great diversification of structure, is seen under many natural circumstances. In an extremely small area, especially if freely open to immigration, and where the contest between individual and individual must be severe, we always find great diversity in its inhabitants. For instance, I found that a piece of turf, three feet by four in size, which had been exposed for many years to exactly the same conditions, supported twenty species of plants, and these belonged to eighteen genera and to eight orders, which shows how much these plants differed from each other. So it is with the plants and insects on small and uniform islets; and so in small ponds of fresh water. Farmers find that they can raise most food by a rotation of plants belonging to the most different orders: nature follows what may be called a simultaneous rotation. Most of the animals and plants which live close round any small piece of ground, could live on it (supposing it not to be in any way peculiar in its nature), and may be said to be striving to the utmost to live there; but, it is seen, that where they come into the closest competition with each other, the advantages of diversification of structure, with the accompanying differences of habit and constitution, determine that the inhabitants, which thus jostle each other most closely, shall, as a general rule, belong to what we call different genera and orders.

24 身体构造上的多样性，可使生物获得最大限度的生活空间，许多自然环境中的情况，都显示出这一原理的正确性。在一个对外开放、可以自由迁入的极小地区，个体之间的生存斗争一定非常激烈，生物间的分异也是非常之大。例如，一块多年的生活条件相同的 3 英尺宽 4 英尺长的草地，在这里生长的 20 种植物属于 8 个目的 18 个属，可见这些植物之间的差异有多么大。在地质构造一致的小岛上或是小小的淡水池塘里，植物与昆虫的情况也是如此。农民发现轮种不同科目的作物收获最多，自然界所遵循的则是所谓的同时轮种。假设一个没有什么特殊情况的小地方，密集在这里的大部分植物都可以生存，或者说是在为生活挣扎。按照一般规律，我们应该看到在生存斗争最激烈的地方，由于构造分异和习性、体质趋异，必然导致这样一种情况，即彼此倾轧最激烈的，正是那些所谓异属和异目的生物。

25 After the foregoing discussion, which ought to have been much amplified, we may, I think, assume that the modified descendants of any one species will succeed by so much the better as they become more diversified in structure, and are thus enabled to encroach on places occupied by other beings. Now let us see how this principle of great benefit being derived from divergence of character, combined with the principles of natural selection and of extinction, will tend to act.

26 The accompanying diagram will aid us in understanding this rather perplexing subject. Let A to L represent the species of a genus large in its own country; these species are supposed to resemble each other in unequal degrees, as is so generally the case in nature, and as is represented in the diagram by the letters standing at unequal distances. I have said a large genus, because we have seen in the second chapter, that on an average more of the species of large genera vary than of small genera; and the varying species of the large genera present a greater number of varieties. We have, also, seen that the species, which are the commonest and the most widely-diffused, vary more than rare species with restricted ranges. Let (A) be a common, widely-diffused, and varying species, belonging to a genus large in its own country. The little fan of diverging dotted lines of unequal lengths proceeding from (A), may represent its varying offspring. The variations are supposed to be extremely slight, but of the most diversified nature; they are not supposed all to appear simultaneously, but often after long intervals of time; nor are they all supposed to endure for equal periods. Only those variations which are in some way profitable will be preserved or naturally selected. And here the importance of the principle of benefit being derived from divergence of character comes in; for this will generally lead to the most different or divergent variations (represented by the outer dotted lines) being preserved and accumulated by natural selection. When a dotted line reaches one of the

25 通过以上简要的讨论，我们可以认为，某一物种的后代越变异，就越能成功地生存，因为它们在构造上越分异，就越能侵入其他生物所占据的位置。现在让我们看一看，这种从性状分异中获利的原理，与自然选择原理及绝灭原理，是如何结合起来发挥作用的。

26 下面的图表，可以帮助我们理解这个复杂的问题。图中从 A 到 L 代表某地一个大属的各个物种，它们彼此之间有不同程度的相似(自然界的情况普遍如此)，所以在图中各字母之间的距离不相等。在第 2 章我们

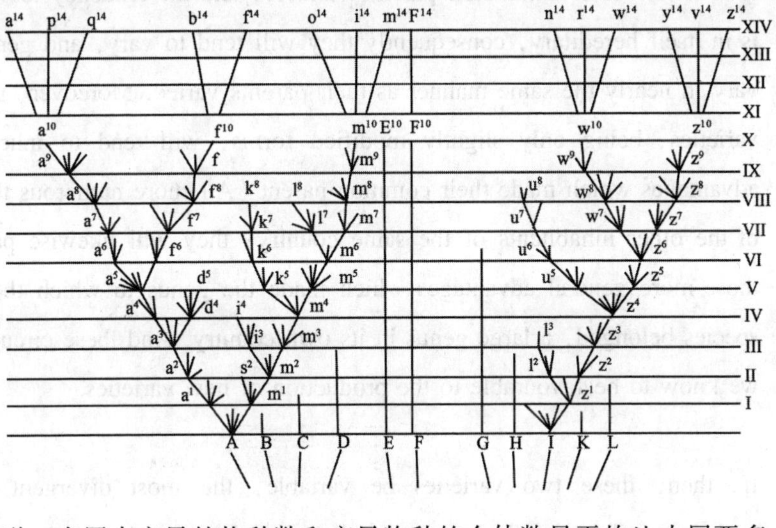

知道，大属中变异的物种数和变异物种的个体数量平均比小属要多；我们还知道，最常见、分布最广泛的物种，比罕见且分布范围狭小的物种产生的变异多。假设图中 A 代表大属中一个常见的、广泛分布的、正在变异着的物种，从 A 发出的长短不一的、呈树枝形状的虚线是它的后代。假设变异的分异度极高但程度甚微，而且变异并非同时发生或是常常间隔很长时间才发生，发生后能持续的时间也各不相同，那么，只有那些有利变异才能被保存下来，即被自然所选择。这时就显示出性状分异在形成物种上的重要性，因为只有性状分歧最大的变异(由图中外侧

horizontal lines, and is there marked by a small numbered letter, a sufficient amount of variation is supposed to have been accumulated to have formed a fairly well-marked variety, such as would be thought worthy of record in a systematic work.

27 The intervals between the horizontal lines in the diagram, may represent each a thousand generations; but it would have been better if each had represented ten thousand generations. After a thousand generations, species (A) is supposed to have produced two fairly well-marked varieties, namely a^1 and m^1. These two varieties will generally continue to be exposed to the same conditions which made their parents variable, and the tendency to variability is in itself hereditary, consequently they will tend to vary, and generally to vary in nearly the same manner as their parents varied. Moreover, these two varieties, being only slightly modified forms, will tend to inherit those advantages which made their common parent (A) more numerous than most of the other inhabitants of the same country; they will likewise partake of those more general advantages which made the genus to which the parent-species belonged, a large genus in its own country. And these circumstances we know to be favourable to the production of new varieties.

28 If, then, these two varieties be variable, the most divergent of their variations will generally be preserved during the next thousand generations. And after this interval, variety a^1 is supposed in the diagram to have produced variety a^2, which will, owing to the principle of divergence, differ more from (A) than did variety a^1. Variety m^1 is supposed to have produced two varieties, namely m^2 and s^2, differing from each other, and more considerably from their common parent (A). We may continue the process by similar steps for any length of time; some of the varieties, after each thousand generations, producing only a single variety, but in a more and

的虚线表示)才能通过自然选择被保存和积累。图中虚线与标有小写字母和数字的横线相遇，说明充分积累的变异已经形成了一个能在分类志上记载的显著变种。

27 图中每两条横线间的距离，代表一千代或更多的世代。假定一千代以后，(A)物种产生了两个显著的变种，即 a^1 和 m^1，这两个变种处于与它们的亲代变异时相同的生活条件中，它们本身具有遗传得来的变异倾向，所以它们很可能以它们亲代变异的方式继续产生变异。此外，这两个稍微变异的变种，还继承了它们亲代的和亲代所在属的优点，那些优点曾使它们的亲代(A)具有更多的个体，曾使它们所在的属成为大属，所有这些条件无疑都是有利于产生新变种的。

28 如果这两个变种仍然发生变异，最显著的性状变异将在下一个千代中被保存，这段时间过后，假定由图中的 a^1 产生出 a^2，由趋异原理可知，a^2 与(A)的差异一定大于 a^1 与(A)的差异。设想 m^1 产生了两个变种：m^2 和 s^2，它们彼此不同，与它们的共同祖先(A)更不同。按照同样的步骤，这个过程可以无限地延续下去，每经过一千代，有的变种仅产生一个变种，随着自然条件的变化，有的可产生二至三个变种，有的也许不能产生变种。这样，由共同祖先(A)所产生的变种，即改变了的后代的种类数目不断增加，性状会不断变异，从图中可看到，这个过程仅列到第一万代，再往后则用虚线简略表示直到一万四千代。

more modified condition, some producing two or three varieties, and some failing to produce any. Thus the varieties or modified descendants, proceeding from the common parent (A), will generally go on increasing in number and diverging in character. In the diagram the process is represented up to the ten-thousandth generation, and under a condensed and simplified form up to the fourteen-thousandth generation.

29 But I must here remark that I do not suppose that the process ever goes on so regularly as is represented in the diagram, though in itself made somewhat irregular. I am far from thinking that the most divergent varieties will invariably prevail and multiply: a medium form may often long endure, and may or may not produce more than one modified descendant; for natural selection will always act according to the nature of the places which are either unoccupied or not perfectly occupied by other beings; and this will depend on infinitely complex relations. But as a general rule, the more diversified in structure the descendants from any one species can be rendered, the more places they will be enabled to seize on, and the more their modified progeny will be increased. In our diagram the line of succession is broken at regular intervals by small numbered letters marking the successive forms which have become sufficiently distinct to be recorded as varieties. But these breaks are imaginary, and might have been inserted anywhere, after intervals long enough to have allowed the accumulation of a considerable amount of divergent variation.

30 As all the modified descendants from a common and widely-diffused species, belonging to a large genus, will tend to partake of the same advantages which made their parent successful in life, they will generally go on multiplying in number as well as diverging in character: this is represented in the diagram by the several divergent branches proceeding from (A). The modified

29 但是我必须指出，变异过程并非如图所示这样规则地(图表本身已能反映一些不规则)或连续地进行，而很有可能一个变种长时间内保持不变，而后又发生变化。我也不能断言，最分异的变种必然会被保存下来，有时中间类型也能持续很长时间，并能产生一种以上的后代，因为自然选择是按照自然体系中未占据或占据不完全位置的性质来发挥作用的，而且也是与许多复杂的因素相互关联的。不过按照一般规律，任何物种的后代性状越分异，它们所能占据的位置越多，所拥有的变异后代也就越多。在我们的图中，连续的系统每隔一定距离就规则地被一个小写字母所中断，那是表示此类型已经发生了充分变异，可以标记为一个变种。但这种间断完全是想象的，实际上只要间隔的时间长度足够使变异大量地积累起来，这种表示变种形成的间断是可以出现在任何位置上的。

30 大属内广泛分布的常见物种所产生的变异后代，大多从亲代那里继承了相同的优势，这种优势使它们的亲代成功地生存，一般也会使这些后代继续增加个体数量和性状变异的程度。图中从(A)延伸出来的数条分支虚线就表示了这一情况。图中几条位置较低没有达到上端线的分支虚线，表示早期的改进较小的后代，它们已被较晚产生的、图上位置较高

offspring from the later and more highly improved branches in the lines of descent, will, it is probable, often take the place of, and so destroy, the earlier and less improved branches: this is represented in the diagram by some of the lower branches not reaching to the upper horizontal lines. In some cases I do not doubt that the process of modification will be confined to a single line of descent, and the number of the descendants will not be increased; although the amount of divergent modification may have been increased in the successive generations. This case would be represented in the diagram, if all the lines proceeding from (A) were removed, excepting that from a^1 to a^{10}. In the same way, for instance, the English race-horse and English pointer have apparently both gone on slowly diverging in character from their original stocks, without either having given off any fresh branches or races.

31 After ten thousand generations, species (A) is supposed to have produced three forms, a^{10}, f^{10}, and m^{10}, which, from having diverged in character during the successive generations, will have come to differ largely, but perhaps unequally, from each other and from their common parent. If we suppose the amount of change between each horizontal line in our diagram to be excessively small, these three forms may still be only well-marked varieties; or they may have arrived at the doubtful category of sub-species; but we have only to suppose the steps in the process of modification to be more numerous or greater in amount, to convert these three forms into well-defined species: thus the diagram illustrates the steps by which the small differences distinguishing varieties are increased into the larger differences distinguishing species. By continuing the same process for a greater number of generations (as shown in the diagram in a condensed and simplified manner), we get eight species, marked by the letters between a^{14} and m^{14}, all descended from (A). Thus, as I believe, species are multiplied and genera are formed.

的、更为改进的后代所取代并绝灭。在某些情况下，变异仅限于一条支线，这样，虽然分歧变异的量在不断扩大，而变异后代的个体数量却没有增加。如果把图中 a^1 至 a^{10} 的支线留下，而去掉其他由 A 发出的各条虚线，这种情况就清楚地反映出来了。英国赛跑马和向导狗显然就属于这一情况，它们的性状慢慢地改进了，可是并没有增加新品种。

31 假定一万代后，由物种(A)产生出三个类型：a^{10}、f^{10} 和 m^{10}，由于历代性状的分异，它们之间以及它们与祖代之间的差异虽不相等，但一定非常之大。假定在图中每两条横线之间的变异量是极其微小的，这三个类型仅仅是三个显著的变种，假定在变异的过程中，步骤很多且变异量很大，这三个变种就会转为可疑物种，进而成为明确的物种。这样，此图就把区别变种的较小差异是如何上升为区别物种的较大差异的各个步骤清楚地表示出来了。如果这一过程如图中简略部分所示，以同一方式继续进行下去的话，那么，更多世代后便可得到如图上所标出的 a^{14} 和 m^{14} 之间的几个物种，它们都是由(A)传衍下来的后代。我相信物种就是这样增加的，属也是这样形成的。

32 In a large genus it is probable that more than one species would vary. In the diagram I have assumed that a second species (I) has produced, by analogous steps, after ten thousand generations, either two well-marked varieties (w^{10} and z^{10}) or two species, according to the amount of change supposed to be represented between the horizontal lines. After fourteen thousand generations, six new species, marked by the letters n^{14} to z^{14}, are supposed to have been produced. In each genus, the species, which are already extremely different in character, will generally tend to produce the greatest number of modified descendants; for these will have the best chance of filling new and widely different places in the polity of nature: hence in the diagram I have chosen the extreme species (A), and the nearly extreme species (I), as those which have largely varied, and have given rise to new varieties and species. The other nine species (marked by capital letters) of our original genus, may for a long period continue transmitting unaltered descendants; and this is shown in the diagram by the dotted lines not prolonged far upwards from want of space.

33 But during the process of modification, represented in the diagram, another of our principles, namely that of extinction, will have played an important part. As in each fully stocked country natural selection necessarily acts by the selected form having some advantage in the struggle for life over other forms, there will be a constant tendency in the improved descendants of any one species to supplant and exterminate in each stage of descent their predecessors and their original parent. For it should be remembered that the competition will generally be most severe between those forms which are most nearly related to each other in habits, constitution, and structure. Hence all the intermediate forms between the earlier and later states, that is between the less and more improved state of a species, as well as the original parent-species itself, will generally tend to become extinct. So it probably will be

32 在大属内，可能发生变异的物种不止一个，假设图表上的物种(I)以同样的步骤，在万代以后也产生了两个显著的变种，或根据图中横线所代表的变异量，产生了两个物种(w^{10}和z^{10})，而一万四千代以后，便可获得如图所示的由n^{14}到z^{14}的六个物种。某一属里具有极大差异的物种，可能产生的变异后代也会更多，因为它们有最好的机会去占据自然体系中新的不同位置。所以，在图中，我选择了一个极端的物种(A)和另一个近乎极端的物种(I)，因为它们已经大量变异并已产生了新变种和新物种。而同属内的其他九个物种(图中用大写字母表示)也能在长度不等的时间内，继续繁育它们的无变化的后代，对此情况，在图中用向上的长度不等的虚线来表示。

33 此外，如图所示，在变异过程中还有另一个原理，即绝灭的原理也起着重要的作用。在充满生物的地方，自然选择的作用体现在被选取的类型，具有在生存斗争中有超出其他类型的优势。任何物种的变异后代在繁衍发展的各个阶段都可能取代并排除它们的前辈或原始祖代。因为我们知道，在那些习性、体质和构造上最近似的类型中，生存斗争最为激烈。因此介于早期和后期的中间类型，即处于改进较少和改进较多之间的类型，以及原始亲种本身，都可能逐渐趋向消亡，甚至生物系统中有的整个分支的物种，都会被后起的改进类型排除而至绝灭。不过，如果变异的后代进入另一个区域并迅速适应了新的环境，则后代与祖代之间没有竞争，二者可各自生存下去。

with many whole collateral lines of descent, which will be conquered by later and improved lines of descent. If, however, the modified offspring of a species get into some distinct country, or become quickly adapted to some quite new station, in which child and parent do not come into competition, both may continue to exist.

34 If then our diagram be assumed to represent a considerable amount of modification, species (A) and all the earlier varieties will have become extinct, having been replaced by eight new species (a^{14} to m^{14}); and (I) will have been replaced by six (n^{14} to z^{14}) new species.

35 But we may go further than this. The original species of our genus were supposed to resemble each other in unequal degrees, as is so generally the case in nature; species (A) being more nearly related to B, C, and D, than to the other species; and species (I) more to G, H, K, L, than to the others. These two species (A) and (I), were also supposed to be very common and widely diffused species, so that they must originally have had some advantage over most of the other species of the genus. Their modified descendants, fourteen in number at the fourteen-thousandth generation, will probably have inherited some of the same advantages: they have also been modified and improved in a diversified manner at each stage of descent, so as to have become adapted to many related places in the natural economy of their country. It seems, therefore, to me extremely probable that they will have taken the places of, and thus exterminated, not only their parents (A) and (I), but likewise some of the original species which were most nearly related to their parents. Hence very few of the original species will have transmitted offspring to the fourteen-thousandth generation. We may suppose that only one (F), of the two species which were least closely related to the other nine original species, has transmitted descendants to this late stage of descent.

34 假定图中表示的变异量相当大，物种(A)和它的早期变种都会绝灭，代之而起的是 a^{14} 至 m^{14} 的八个新物种，物种 I 将被 n^{14} 至 z^{14} 的六个新物种所取代。

35 进一步说，假如原来同一属的物种彼此相似的程度不同(这是自然界的普遍情况)，物种(A)与 B、C 和 D 之间的关系比它与其他种的关系更近；物种(I)与 G、H、K 和 L 之间的关系之密切超出它与其他物种的关系；假如(A)与(I)是两个广泛分布的常见物种，它们本身就具备超越大部分其他物种的优势，那么，它们的变异后代，如前文图所示，即一万四千代之后产生的那十四个新物种，很可能既继承了它们祖代的优势，又在发展的不同阶段进行不同程度的分异和改进，已经适应了这个地区自然体系中的许多新位置，因此，很有可能它们不但取代并消灭它们的祖代(A)和(I)，而且可能消灭与它们祖代近缘的那些原始种。所以说，只有极少数原始物种能够传到一万四千代。在原始物种中那两个与其他九个种最疏远的物种，如图所示即(E)和(F)中，假设只有一个(F)可以将后代延续到最后阶段。

36 The new species in our diagram descended from the original eleven species, will now be fifteen in number. Owing to the divergent tendency of natural selection, the extreme amount of difference in character between species a^{14} and z^{14} will be much greater than that between the most different of the original eleven species. The new species, moreover, will be allied to each other in a widely different manner. Of the eight descendants from (A) the three marked a^{14}, q^{14}, p^{14}, will be nearly related from having recently branched off from a^{10}; b^{14} and f^{14}, from having diverged at an earlier period from a^{5}, will be in some degree distinct from the three first-named species; and lastly, o^{14}, i^{14}, and m^{14}, will be nearly related one to the other, but from having diverged at the first commencement of the process of modification, will be widely different from the other five species, and may constitute a sub-genus or even a distinct genus.

37 The six descendants from (I) will form two sub-genera or even genera. But as the original species (I) differed largely from (A), standing nearly at the extreme points of the original genus, the six descendants from (I) will, owing to inheritance, differ considerably from the eight descendants from (A); the two groups, moreover, are supposed to have gone on diverging in different directions. The intermediate species, also (and this is a very important consideration), which connected the original species (A) and (I), have all become, excepting (F), extinct, and have left no descendants. Hence the six new species descended from (I), and the eight descended from (A), will have to be ranked as very distinct genera, or even as distinct sub-families.

38 Thus it is, as I believe, that two or more genera are produced by descent, with modification, from two or more species of the same genus. And the two or more parent-species are supposed to have descended from some one species of an earlier genus. In our diagram, this is indicated by the broken

36 在图表中，由原来 11 个物种传下来的新物种已达到 15 个。由于自然选择的分异倾向，新种中两个极端物种 a^{14} 与 z^{14} 间的差异量，比原始种两个极端物种间的差异量要大得多。新种间的亲缘关系的远近程度也很不相同。在(A)的八个后代中，a^{14}、q^{14} 和 p^{14} 之间的关系较近。因为它们是近期由 a^{10} 分出来的；b^{14} 和 f^{14} 是较早期由 a^{5} 分出，所以它们与前面三个物种有一定程度的差别；最后的三个种，o^{14}、i^{14} 和 m^{14} 彼此之间亲缘关系很近，由于它们是从变异开始时期就分化出来，所以它们与以上五个物种差异非常大，它们可形成一个亚属或者形成一个特征显著的属。

37 从(I)传下来的六个后代，可形成两个亚属或两个属。因为原始种(I)与(A)本来就很不相同。在原属中，(I)几乎处于另一个极端，仅仅由于遗传的原因，(I)的六个后代与(A)的八个后代，就会有相当大的差异；再说，我们可以设想两组生物变异的方向也不相同。还有一个重要因素是，那些曾连接这些原始种(A)和(I)的中间种，设想除了(F)以外，全部绝灭了，而且没有留下后代，这样，(I)的六个新物种，以及(A)的八个后代物种，就势必被列为不同的属，甚至被列为不同的亚科。

38 由此我认为，两个或更多的属是通过后代的变异，由同属的两个或更多的物种中产生出来的；而这两个或更多的亲种，可以假定是由较早的属里某一个物种传衍下来的。图中各大写字母下面的虚线，就表示这种情况，这些虚线聚集为几个支群，再向下归结到一点。假定这一点代表一个物种，那么它就可能是上面提到的新亚属或者新属的祖先。

lines, beneath the capital letters, converging in sub-branches downwards towards a single point; this point representing a single species, the supposed single parent of our several new sub-genera and genera.

39 It is worth while to reflect for a moment on the character of the new species F^{14}, which is supposed not to have diverged much in character, but to have retained the form of (F), either unaltered or altered only in a slight degree. In this case, its affinities to the other fourteen new species will be of a curious and circuitous nature. Having descended from a form which stood between the two parent-species (A) and (I), now supposed to be extinct and unknown, it will be in some degree intermediate in character between the two groups descended from these species. But as these two groups have gone on diverging in character from the type of their parents, the new species (F^{14}) will not be directly intermediate between them, but rather between types of the two groups; and every naturalist will be able to bring some such case before his mind.

Summary of Chapter

40 If during the long course of ages and under varying conditions of life, organic beings vary at all in the several parts of their organisation, and I think this cannot be disputed; if there be, owing to the high geometrical powers of increase of each species, at some age, season, or year, a severe struggle for life, and this certainly cannot be disputed; then, considering the infinite complexity of the relations of all organic beings to each other and to their conditions of existence, causing an infinite diversity in structure, constitution, and habits, to be advantageous to them, I think it would be a most extraordinary fact if no variation ever had occurred useful to each

39 新物种 F^{14} 的特性很值得一提。假定这个物种保持着(F)的形态(即使改变也非常轻微)，性状也没有大的分异，这样它与其他 14 个新物种之间，就有了一种奇特而间接的亲缘关系。假定这个物种的祖先，是位于已绝灭而又不被人知道的两个早期物种(A)和(I)之间的一个类型，那么，这个物种的性状，就属于(A)和(I)的两组后代的中间类型。由于这两种后代在性状上与亲种已经分异，新物种 F^{14} 并不是直接介于各新物种之间，而且介于两个大组之间的中间类型，每一个博物学家都应该想到这样的情况。

摘　要

40 在变化的生活条件下，几乎生物构造的每一部分都会表现出个体差异；由于生物以几何级数增加，在生命的某一年龄，在某一年、某一季节都会出现激烈的生存斗争，这些都是无可争辩的事实。生物相互之间，生物与生活条件之间，无比复杂的关系会引起在构造、体质和习性上对生物有利的无限变异；所以，如果有人说，有利于生物本身的变异，从未像人类本身经历过许多有利变异那样地发生过，那就太奇怪了。如果对生物有利的变异的确曾发生，具有此特性的生物个体，就会在生存斗争中获得保存自己的最佳机会；根据强有力的遗传原理，它们便会产生具有相似特性的后代。我把这种保存有利变异的原理，或者适者生存的原理，称为自然选择。自然选择导致生物与其有机和无机的生活条件之间

being's own welfare, in the same way as so many variations have occurred useful to man. But if variations useful to any organic being do occur, assuredly individuals thus characterised will have the best chance of being preserved in the struggle for life; and from the strong principle of inheritance they will tend to produce offspring similarly characterised. This principle of preservation, I have called, for the sake of brevity, Natural Selection. Natural selection, on the principle of qualities being inherited at corresponding ages, can modify the egg, seed, or young, as easily as the adult. Amongst many animals, sexual selection will give its aid to ordinary selection, by assuring to the most vigorous and best adapted males the greatest number of offspring. Sexual selection will also give characters useful to the males alone, in their struggles with other males.

41 Whether natural selection has really thus acted in nature, in modifying and adapting the various forms of life to their several conditions and stations, must be judged of by the general tenour and balance of evidence given in the following chapters. But we already see how it entails extinction; and how largely extinction has acted in the world's history, geology plainly declares. Natural selection, also, leads to divergence of character; for more living beings can be supported on the same area the more they diverge in structure, habits, and constitution, of which we see proof by looking at the inhabitants of any small spot or at naturalised productions. Therefore during the modification of the descendants of any one species, and during the incessant struggle of all species to increase in numbers, the more diversified these descendants become, the better will be their chance of succeeding in the battle of life. Thus the small differences distinguishing varieties of the same species, will steadily tend to increase till they come to equal the greater differences between species of the same genus, or even of distinct genera.

关系不断得以改善，在许多情况下，能使生物体制进化。但如果低等、简单的类型能很好地适应其生活环境，它们也可以长久地生存下去。根据生物的特性在适当年龄期遗传的原理，自然选择可以像改变成体那样，很容易地改变卵、种子和幼体。在许多动物中，性选择有助于普通选择，它保证最健壮、适应力最强的雄性可以多留后代。性选择还能使雄体获得与其他雄体斗争、对抗的有用性状。

41 根据下一章的内容和例证，可以去判断自然选择是否真的能使各类生物适应它们的生活条件和场所。目前我们已经知道自然选择是怎样引起绝灭的，地质学可以显示，在世界历史上绝灭所起的作用是多么巨大。自然选择导致性状分异，因为生物在构造、习性和体质上越分异，一个地区所能容纳的生物数量就越多，对任何小地区的生物和在外地归化的生物进行观察，就可以证明这一点。因此，在任何物种后代的变异过程中，在一切生物都努力增加个体数量的斗争中，后代的性状越分异，它们就越能获得更好的生存机会。于是在同种内，用以区别变种的小差异，一定会逐渐扩大为区别同属内物种的大差异，甚至发展为区别属的更大的差异。

42 We have seen that it is the common, the widely-diffused, and widely-ranging species, belonging to the larger genera, which vary most; and these will tend to transmit to their modified offspring that superiority which now makes them dominant in their own countries. Natural selection, as has just been remarked, leads to divergence of character and to much extinction of the less improved and intermediate forms of life. On these principles, I believe, the nature of the affinities of all organic beings may be explained. It is a truly wonderful fact—the wonder of which we are apt to overlook from familiarity—that all animals and all plants throughout all time and space should be related to each other in group subordinate to group, in the manner which we everywhere behold—namely, varieties of the same species most closely related together, species of the same genus less closely and unequally related together, forming sections and sub-genera, species of distinct genera much less closely related, and genera related in different degrees, forming sub-families, families, orders, sub-classes, and classes. The several subordinate groups in any class cannot be ranked in a single file, but seem rather to be clustered round points, and these round other points, and so on in almost endless cycles. On the view that each species has been independently created, I can see no explanation of this great fact in the classification of all organic beings; but, to the best of my judgment, it is explained through inheritance and the complex action of natural selection, entailing extinction and divergence of character, as we have seen illustrated in the diagram.

43 The affinities of all the beings of the same class have sometimes been represented by a great tree. I believe this simile largely speaks the truth. The green and budding twigs may represent existing species; and those produced during each former year may represent the long succession of extinct species. At each period of growth all the growing twigs have tried to branch out on all

42 每一纲的大属中，那些分散的、广泛分布的常见物种最易变异，而且能把使它们在本土占优势的长处传衍给变异了的后代。如前所述，自然选择可导致性状趋异和使改良不大的中间类型大量绝灭。世界上各纲内无数生物之间的亲缘关系和它们彼此间所具有的明显差异，都可以用以上的原理加以解释。令人称奇的是，一切时间、空间内的动植物，竟然都可以归属于不同的类群，在群内又相互关联。也就是说，同一物种内的变种间关系最密切，同一属内各物种间的关系较疏远且关联程度不等，它们构成生物的组(section)和亚属；异属物种间的关系则更疏远些；各属之间的关系亲疏不等，它们形成亚科、科、目、亚纲和纲。上述情况随处可见，看惯了就不会感到新奇了。任何纲内总有若干附属类群，不能形成单独的行列，它们围绕着某些点，这些点又环绕另一些点继续下去，以至无穷。如果物种都是单个被上帝创造出来的，上述分类情况则无法解释；这种情况，只有通过遗传，通过图表中所显示的造成绝灭和性状分异的自然选择的复杂作用，才能得以解释。

43 同一纲内生物间的亲缘关系，可以用一株大树来表示，我认为这个比喻很能说明问题。绿色生芽的树枝，代表现存的物种；过去年代所生的枝条，代表那些长期的、先后继承的绝灭物种。在每一生长期内，发育的枝条竭力向各个方向生长延伸，去遮盖周围的枝条并使它们枯萎，这就像在任何时期的生存斗争中，一些物种和物种群征服其他物种的情况一

sides, and to overtop and kill the surrounding twigs and branches, in the same manner as species and groups of species have tried to overmaster other species in the great battle for life. The limbs divided into great branches, and these into lesser and lesser branches, were themselves once, when the tree was small, budding twigs; and this connexion of the former and present buds by ramifying branches may well represent the classification of all extinct and living species in groups subordinate to groups. Of the many twigs which flourished when the tree was a mere bush, only two or three, now grown into great branches, yet survive and bear all the other branches; so with the species which lived during long-past geological periods, very few now have living and modified descendants. From the first growth of the tree, many a limb and branch has decayed and dropped off; and these lost branches of various sizes may represent those whole orders, families, and genera which have now no living representatives, and which are known to us only from having been found in a fossil state. As we here and there see a thin straggling branch springing from a fork low down in a tree, and which by some chance has been favoured and is still alive on its summit, so we occasionally see an animal like the Ornithorhynchus or Lepidosiren, which in some small degree connects by its affinities two large branches of life, and which has apparently been saved from fatal competition by having inhabited a protected station. As buds give rise by growth to fresh buds, and these, if vigorous, branch out and overtop on all sides many a feebler branch, so by generation I believe it has been with the great Tree of Life, which fills with its dead and broken branches the crust of the earth, and covers the surface with its ever branching and beautiful ramifications.

样。在大树幼小时，现在的主枝曾是生芽的小枝，后来主枝分出大枝、大枝分出小枝。这种由分枝相连的旧芽和新芽的关系，可以代表所有已绝灭的和现存的物种在互相隶属的类群中的分类关系。当大树还十分矮小时，它有许多繁茂的小枝条，其中只有两三枝长成主枝干，它们一直支撑着其他的树枝并存在至今。物种的情况也是如此，那些生活在远古地质时期的物种中，能够遗传下来生存着的变异后代的，确实寥寥无几。从大树有生以来，许多主枝、大枝都枯萎、脱落了，这些脱落的大小枝干，可以代表那些今天已没有遗留下来存活的后代，而仅有化石可作考证的整个的目、科和属。有时，也许看到一条细而散乱的小枝条，从大树根基部蔓生出来，由于某种有利的条件，至今枝端还在生长。这就像我们偶然看到如鸭嘴兽或肺鱼那样的动物一样，它们可以通过亲缘关系，哪怕是以微弱的程度，去连接两条生物分类的大枝。显然，这些低等生物是因为生活在有庇护的场所，才得以从激烈的生存斗争中存活下来的。这棵大树不断地生长发育，从旧芽上发出新芽，使强壮的新芽能生长出枝条向四处伸展，遮盖住许多柔弱的枝条。我想，代代相传的巨大生命之树也是如此，它用枯枝落叶去填充地壳，用不断滋生的美丽枝条去覆盖大地。

《DNA：生命的秘密》导引

回顾人类历史，似乎总有某些时刻特别被善与美的火光所眷顾，中国的百家争鸣、盛唐气象，西方的古希腊哲学、文艺复兴等莫不如此，而对当代思想和科学影响最深远的，当属20世纪上半叶开始的科学革命，量子力学、生命科学以及信息科学的开创与发展不仅影响了我们对世界的认识，重塑了我们对自然的理解，也深切地参与到每个人的生活方式当中。

让我们聚焦于当代生命科学，分子生物学无疑是20世纪生命科学皇冠上最璀璨的明珠，这门学科以DNA双螺旋结构的发现为起点，短短几十年间就极大地推进了人类对生命世界的深入理解，并将影响力渗透到多个学科领域。

《DNA：生命的秘密》的作者詹姆斯·沃森正是DNA双螺旋结构破解者之一，年仅25岁就与同伴共同完成了DNA双螺旋结构的构建，34岁获得诺贝尔生理学或医学奖，参与了分子生物学领域最重要科学机理发现的全过程。沃森从不畏惧选择明显超前的研究目标，并总能在这些前人未曾踏足的领域做出卓越成绩。他自1968年起就职美国冷泉港实验室主任，将冷泉港实验室发展为全美乃至全球的分子生物学前沿研究重镇，并将研究范围拓宽至肿瘤生物学、神经生物学等重要领域。1988年至1993年沃森与美国国立卫生院合作，筹资启动和实施人类基因组计划并担任首位负责人，为破译人类基因组信息作出了卓越贡献。

《DNA：生命的秘密》一书出版于DNA双螺旋结构发现50周年之际，回溯历史，这一发现标志着生物学从科学家的实验室走向公众生活的各个层面，成为当代的核心科技。作者基于自己多年的科研经历，对历史进行回顾，并对这一过程中引发的一系列问题和造成的影响进行了反思和追问。全书共13章，第1~4章回顾了前分子生物学年代现代遗传学的发展历史以及分子生物学发端之初里程碑式发现的科研历程；第5~6章展示了分子生物学在制药业、农业等方面发挥的作用；第7~9章详述了人类基因组计划的原理、内容以及对人类进化历史研究的影响；第10~12章关注现代生物学与临床医学的发展；第13章则反思背离科学研究的方法和规律造成的对生命科学的误读，对社会生活的各个方面可能造成的影响。在尾声部分，通过回顾和追问，对生命科学的未来进行了展望，并发出强有力的呼声"如果我们能无惧地接受DNA所揭露的真相，就不会再为我们子孙的未来而绝望悲叹了"！

我们将和大家一起阅读该书的第2章《双螺旋：生命之所在》，与科学家一起发现线索，大胆推测，进行科学实证并得出结论，再现DNA双螺旋结构的发现过程，从而深刻理解这一发现在人类科学史上的重要意义。

20世纪三四十年代，物理学和化学取得了举世瞩目的科学成就，无数青年学子投身其中，该书的作者沃森虽然出于兴趣没有选择物理学专业，但同样抱有"化学与物理学才是'真正的'科学，而理论物理学更是科学翘楚"的观点。与此同时，一位成就卓越富于远见的理论物理学家数次发表演讲，表达对生命科学的展望和启示，并将这些思想结集成册，这就是薛定谔与他影响深远的著作《生命是什么?》在书中薛定谔天才般地预言：可以从储存与传递生物信息的观点来思索生命，这些信息必须压缩成所谓的遗传密码脚本，要了解生命，就必须辨识这些分子，破解它们的密码。

在DNA双螺旋结构破译之前，随着由孟德尔开创、摩尔根发扬光大的现代遗传学的发展，人们已经意识到，达尔文所提出的"伴随变异的继承"

(descent with modification)是通过某种实际存在的物质完成，找到这种物质并破解其结构和运作原理，就可以深入理解生命的秘密。

1947年，年仅19岁的沃森在印第安纳大学开始攻读博士学位，在研究DNA生物学功能的过程中他意识到要回答这个问题“必须彻底解开它的分子结构和所有化学细节才行”。回顾自己的学术生涯，沃森认为选择自己感兴趣且明显超前的目标才能解决重大问题，因此他选择自己最感兴趣的DNA化学结构作为研究目标。尽管当时化学家和生物学家普遍认为要解开DNA结构之谜至少还需要10年时间，沃森仍然选择在1951年远赴英国开始他的DNA结构探索之旅，并在这里遇到了将与他共同解开生命秘密的几位合作伙伴：对错综复杂问题非常着迷的克里克，参与过曼哈顿计划的前物理学家威尔金斯，以及讲求逻辑和精确、对自己的专业执著得不得了的富兰克林。有意思的是，这几位科学家包括沃森自己，几乎都是受到《生命是什么?》的影响而走上了生命科学研究之路。

解决科学问题不能仅依靠勇气，而是要具有科学判断能力，否则很可能用自己一辈子的学术生涯去为渺茫的终点线冒险，因此选择几年内可能出现重大突破的合适的研究目标和方法非常关键。利用衍射方法研究分子结构是当时最具挑战性的工作，而这对于四人小组简直再合适不过：DNA可以获得纯度很高的结晶，并具有相对简单的分子结构，非常适合利用X光衍射进行结构分析，威尔金斯和富兰克林将之作为研究分子结构的对象，沃森和克里克则根据衍射实验获得的数据进行分析并构建模型。

在追求这么重要的学术目标的过程中，必定会与其他研究者发生竞争，有价值的竞争者可以确保研究目标的确值得投入，与X光衍射团队进行竞争的是当时科学界的超级巨星泡令，他以自己“身为结构化学家的丰富经验，大胆推论哪种类型的螺旋结构最符合多肽链的化学特性”，从而破解了蛋白质的α螺旋结构，之后他又挑战DNA的结构并率先提出了三链螺旋模型。遗憾的是，这次幸运之神没有眷顾泡令——他竟然搞错了脱氧核糖核酸的基本化学性质。

与此同时，四人小组中也产生了一定分歧：完全根据实验数据进行推演计算还是引入化学家构建模型的思路？科学研究的道路上，方法选择的岔路口往往意味着不同的结果走向，与学术同行的交流在这种时刻显得尤为必要，哪些科学家的研究给沃森和同伴以重要启示？通过阅读文本，相信你一定会得到答案。

在1953年2月的最后一天，一切都水到渠成完美地整合在一起，衍射数据、被纠正过来的碱基配对概念以及生物学家的直觉结合在一起，DNA双螺旋结构破解了！“这么简单美丽的构造，绝对错不了。”

双螺旋结构的提出揭示了遗传物质的分子结构，不仅如此，碱基在双链排列上具有互补特性，“这显然是细胞分裂前染色体进行复制时，基因的遗传信息能精准重现的原因。DNA分子会像拉链一样‘拉开’，形成独立的两股，每一股都可以作为新股合成时的模板，于是一条双螺旋就变成了两条”。

对于这一“像拉链一样拉开”的复制假设，两个聪明的年轻学者梅索森和史塔尔设计了“生物学上最完美的实验”，简单明了而巧妙地证实了DNA的半保留复制机制。聪明的读者，你也可以自己尝试根据文本提供的知识，做出假设并设计实验，然后再和文本对照，看看“最完美的实验”究竟有何精妙之处？

双螺旋结构的拼图完整了，沃森、克里克和威尔金斯实至名归获得了诺贝尔奖，当然我们同样不能忘记杰出的科学家富兰克林在DNA发现过程中的重要贡献。

双螺旋结构的发现者之一克里克曾经戏称自己的研究成果是揭开了“生命的秘密”，这也正是该书标题的由来，现在看来，这个说法毫无夸耀之嫌，正如作者所言：“双螺旋的发现敲响了生机论的丧钟。认真的科学家，甚至有宗教信仰的科学家都已发现，要对生命有完整的了解，不需要寻找新的自然定律。生命不过就是物理与化学——尽管是极为精密复杂的物理与化学。”

该书写作风格平易近人，史料丰富，细节准确，逻辑清晰，层层递进，带给读者美妙的阅读体验。亲爱的读者，在阅读时，你是否既感觉像亲历了这项伟大的发现一样激动，又会不时停下来思考自己如果身在其中会做出什么样的科学探索和判断呢？希望这些思考会使你从全新的角度理解科学的严谨之美，感受生命的复杂之美，体验自然的和谐之美，从而对自然科学少一些畏难之怯，多一份亲近之情。

（杜润蕾）

DNA: The Secret of Life

James D. Watson

Chapter 2 The Double Helix: This is life

1 I got hooked on the gene during my third year at the University of Chicago. Until then, I had planned to be a naturalist and looked forward to a career far removed from the urban bustle of Chicago's South Side, where I grew up. My change of heart was inspired not by an unforgettable teacher but a little book that appeared in 1944, *What Is Life?* by the Austrian-born father of wave mechanics, Erwin Schrödinger. It grew out of several lectures he had given the year before at the Institute for Advanced Study in Dublin. That a great physicist had taken the time to write about biology caught my fancy. In those days, like most people, I considered chemistry and physics to be the "real" sciences, and theoretical physicists were science's top dogs.

2 Schrödinger argued that life could be thought of in terms of storing and passing on biological information. Chromosomes were thus simply information bearers. Because so much information had to be packed into every cell, it must be compressed into what Schrödinger called a "hereditary code-script" embedded in the molecular fabric of chromosomes. To understand life, then, we would have to identify these molecules, and crack their code. He even speculated that understanding life—which would involve finding the gene—might take us beyond the laws of physics as we then understood them. Schrödinger's book was tremendously influential. Many of those who would become major players in Act 1 of molecular biology's great drama, including Francis Crick (a former physicist himself), had, like me, read *What Is Life?* and been impressed.

DNA：生命的秘密

詹姆斯·沃森

第二章　双螺旋：生命之所在

1　我在芝加哥大学三年级时，迷上了基因。我原本想当博物学家，向往日后能离开自幼成长的芝加哥南区，到没有都市尘嚣的地方发展职业生涯。让我改变心意的，并不是某位难忘的老师，而是1944年出版的一本薄薄的小书《生命是什么?》(*What Is Life*?)，作者是奥地利籍的波动力学之父薛定谔(Erwin Schrodinger)。这本书辑录了他前一年在都柏林的高等研究院(Institute for Advanced Study)发表的数场演讲。这么伟大的物理学家竟会花时间写生物学的书，引起了我的兴趣。当时我和大多数的人都认为，化学与物理学才是"真正的"科学，而理论物理学更是科学翘楚。

2　薛定谔认为，我们可以从储存与传递生物信息的观点来思索生命。因此，染色体只是信息的携带者。由于每个细胞都要容纳这么多的信息，因此这些信息必须压缩成薛定谔所谓的遗传密码脚本(hereditary code-script)，植入染色体的分子结构内。要了解生命，就必须辨识这些分子，破解它们的密码。他甚至臆测，了解生命(包括找到基因)说不定能让我们超越当时所知的物理定律。薛定谔这本著作的影响甚巨，日后许多在分子生物学这出大戏的序幕中成为要角的人物，包括我和克里克(他先前是物理学者)在内，都拜读过《生命是什么?》，而且深受感动。

本文选自[美]詹姆斯·沃森：《DNA：生命的秘密》，陈雅云译，上海人民出版社2007年版，第30-51页。中译本由时报文化出版企业股份有限公司授权。

3 In my own case, Schrödinger struck a chord because I too was intrigued by the essence of life. A small minority of scientists still thought life depended upon a vital force emanating from an all-powerful god. But like most of my teachers, I disdained the very idea of vitalism. If such a "vital" force were calling the shots in nature's game, there was little hope life would ever be understood through the methods of science. On the other hand, the notion that life might be perpetuated by means of an instruction book inscribed in a secret code appealed to me. What sort of molecular code could be so elaborate as to convey all the multitudinous wonder of the living world? And what sort of molecular trick could ensure that the code is exactly copied every time a chromosome duplicates?

4 At the time of Schrödinger's Dublin lectures, most biologists supposed that proteins would eventually be identified as the primary bearers of genetic instruction. Proteins are molecular chains built up from twenty different building blocks, the amino acids. Because permutations in the order of amino acids along the chain are virtually infinite, proteins could, in principle, readily encode the information underpinning life's extraordinary diversity. DNA then was not considered a serious candidate for the bearer of code-scripts, even though it was exclusively located on chromosomes and had been known about for some seventy-five years. In 1869, Friedrich Miescher, a Swiss biochemist working in Germany, had isolated from pus-soaked bandages supplied by a local hospital a substance he called "nuclein." Because pus consists largely of white blood cells, which, unlike red blood cells, have nuclei and therefore DNA-containing chromosomes, Miescher had stumbled on a good source of DNA. When he later discovered that "nuclein" was to be found in chromosomes alone, Miescher understood that his discovery was indeed a big one. In 1893, he wrote: "Inheritance insures a continuity in form from generation to generation that lies even deeper than

3 以我为例，薛定谔触动了我的心弦，因为我对生命的本质也很感兴趣。当时仍有少数科学家认为，生命仰赖全能的上帝赋予生命力。不过如同我大多数的老师，我也鄙视生机论的观念。如果这种“生命力”是大自然运作的主宰，我们势必很难经由科学方法来了解生命。反之，一想到生命可能借由一本以密码写成的指令书而永续长存，就令我神往不已。什么样的分子密码能够复杂到足以传递众多的生命奇迹？又有什么样的分子秘密，让染色体在复制时都能复制出一模一样的密码？

4 在薛定谔于都柏林演讲的年代，大多数的生物学家都认为，最终科学界会证明蛋白质是遗传指令的主要携带者。蛋白质是由 20 种不同的建构单元（氨基酸）所组成的分子链。由于氨基酸沿分子链排列的顺序可以说有无限多种，因此原则上蛋白质是有可能隐含造成生命如此多样的密码信息的。虽然 DNA 就位于染色体上，为世人所知也有 75 年之久，但当时 DNA 并未被视为密码脚本的可能携带者。1869 年，在德国工作的瑞士生化学家弗雷德里希·米歇尔（Friedrich Miescher）从当地医院沾满脓的绷带上分离出一种物质，并称之为“核素”（nuclein），脓大多由具有细胞核的白血球构成（红血球没有细胞核），因此也具有包含 DNA 的染色体，弗雷德里希·米歇尔等于在无意间发现了 DNA 的良好来源。稍后当他发现唯有在染色体里才找得到“核素”时，就知道自己有了重大发现。1893 年，他写道：“遗传确保形态能世代延续，而这一切就隐藏在比化学分子还深的层次。它隐藏在结构化的原子群组内。因此，我支持化学遗传论。”

the chemical molecule. It lies in the structuring atomic groups. In this sense, I am a supporter of the chemical heredity theory."

5 Nevertheless, for decades afterward, chemistry would remain unequal to the task of analyzing the immense size and complexity of the DNA molecule. Only in the 1930s was DNA shown to be a long molecule containing four different chemical bases: adenine (A), guanine (G), thymine (T), and cytosine (C). But at the time of Schrödinger's lectures, it was still unclear just how the subunits (called deoxynucleotides) of the molecule were chemically linked. Nor was it known whether DNA molecules might vary in their sequences of the four different bases. If DNA were indeed Schrödinger's code-script, then the molecule would have to be capable of existing in an immense number of different forms. But back then it was still considered a possibility that one simple sequence like AGTC might be repeated over and over along the entire length of DNA chains.

6 DNA did not move into the genetic limelight until 1944, when Oswald Avery's lab at the Rockefeller Institute in New York City reported that the composition of the surface coats of pneumonia bacteria could be changed. This was not the result he and his junior colleagues, Colin MacLeod and Maclyn McCarty, expected.

7 For more than a decade Avery's group had been following up on another most unexpected observation made in 1928 by Fred Griffith, a scientist in the British Ministry of Health. Griffith was interested in pneumonia and studied its bacterial agent, Pneumococcus. It was known that there were two strains, designated "smooth" (S) and "rough" (R) according to their appearance under the microscope. These strains differed not only visually but also in their virulence. Inject S bacteria into a mouse, and within a few days

5 然而，数十年后，化学仍无法分析庞大复杂的 DNA 分子。一直到 20 世纪 30 年代，科学家才证明 DNA 是由四种不同的化学碱基所构成的长分子，即腺嘌呤（A）、鸟嘌呤（G）、胸腺嘧啶（T）与胞嘧啶（C）。不过在薛定谔发表演说的年代，科学界尚不明白 DNA 分子上的这些次单位（称为脱氧核糖核苷酸［deoxynucleotides］）在化学上如何链接，也不知道 DNA 分子的四种化学碱基序列是否有差异。如果 DNA 真的是薛定谔所谓的密码脚本，那么这种分子应该有极多种不同的形式。不过在当时，一般仍认为整条 DNA 链有可能是由一个简单序列（例如 AGTC）一再重复出现而构成的。

6 一直到 1944 年，DNA 才成为遗传界的焦点，当时艾弗里（Oswald Avery）在纽约洛克菲勒研究所（Rockefeller Institute）的实验室发表报告说，肺炎病菌的外膜组成可以改变。这个结果出乎他和麦克劳德（Colin Macleod）与麦卡提（Maclyn McCarty）这两位资历较浅的同事意料。

7 艾弗里的研究小组花了 10 余年的时间，持续追踪英国卫生部科学家格里菲斯（Fred Griffith）在 1928 年视察到的奇特现象。格里菲斯对肺炎很感兴趣，潜心研究肺炎的致病菌——肺炎双球菌（*Pneumococcus*）。当时已知肺炎双球菌有两种形态，依照它们在显微镜下的外观而分为“平滑”（smooth）的 S 型与“粗糙”（rough）的 R 型。两者不仅外观不同，毒性也不同。将 S 型菌注入老鼠体内，几天内老鼠就会死去，但是注入 R 型菌的老鼠则依旧健康。后来发现 S 型菌的细胞有荚膜，可以防止老鼠的免疫系统认出它是入侵者。R 型菌的细胞没有荚膜，因此会受到老鼠免疫系统的攻击。

the mouse dies; inject R bacteria and the mouse remains healthy. It turns out that S bacterial cells have a coating that prevents the mouse's immune system from recognizing the invader. The R cells have no such coating and are therefore readily attacked by the mouse's immune defenses.

8 Through his involvement with public health, Griffith knew that multiple strains had sometimes been isolated from a single patient, and so he was curious about how different strains might interact in his unfortunate mice. With one combination, he made a remarkable discovery: when he injected heat-killed S bacteria (harmless) and normal R bacteria (also harmless), the mouse died. How could two harmless forms of bacteria conspire to become lethal? The clue came when he isolated the Pneumococcus bacteria retrieved from the dead mice and discovered living S bacteria. It appeared the living innocuous R bacteria had acquired something from the dead S variant; whatever it was, that something had allowed the R in the presence of the heat-killed S bacteria to transform itself into a living killer S strain. Griffith confirmed that this change was for real by culturing the S bacteria from the dead mouse over several generations: the bacteria bred true for the S type, just as any regular S strain would. A genetic change had indeed occurred to the R bacteria injected into the mouse.

9 Though this transformation phenomenon seemed to defy all understanding, Griffith's observations at first created little stir in the scientific world. This was partly because Griffith was intensely private and so averse to large gatherings that he seldom attended scientific conferences. Once, he had to be virtually forced to give a lecture. Bundled into a taxi and escorted to the hall by colleagues, he discoursed in a mumbled monotone, emphasizing an obscure corner of his microbiological work but making no mention of bacterial transformation. Luckily, however, not everyone overlooked Griffith's breakthrough.

8 格里菲斯从参与公共卫生的经验中得知，单一病人身上有时能分离出多种类型的菌株，因此他很好奇，不同类型的菌株在实验鼠身上会如何交互作用。后来他在一种组合上有了重大发现：当他将加热杀死的 S 型菌株(已变得无害)及正常的 R 型菌株(原本就无害)同时注入老鼠体内时，老鼠会死亡。两种无害的菌株在混合后，怎么可能变得致命呢？后来他分离出死老鼠身上的肺炎双球菌，发现里面有活的 S 型菌，线索于是出现。无害的 R 型菌似乎会从已死的 S 型菌取得不明物质；无论此物质为何，它显然会使 R 型菌在有加热致死的 S 型菌存在时，转型为活的杀手型 S 菌。格里菲斯从死老鼠身上培育出数代的 S 型菌，证实这种变化的确存在：这些细菌繁殖成 S 型菌，如同任何正常的 S 型菌株。注入老鼠体内的 R 型菌真的发生了“遗传”变化。

9 尽管这种转型现象似乎与当时的了解相悖，但格里菲斯的观察结果起初并未在科学界激起太大的涟漪。部分原因在于格里菲斯非常注重隐私，厌恶大型聚会，鲜少参加科学会议，有一次还是在别人强迫下才发表演讲。那次他被同事架到出租车上，护送至演讲厅，然后以含糊单调的声调发表演说，谈的是他在微生物领域所作的艰涩研究，完全没有提及细菌转型。幸好，并非所有人都忽视格里菲斯的突破。

10 Oswald Avery was also interested in the sugarlike coats of the Pneumococcus. He set out to duplicate Griffith's experiment in order to isolate and characterize whatever it was that had caused those R cells to change to the S type. In 1944 Avery, MacLeod, and McCarty published their results: an exquisite set of experiments showing unequivocally that DNA was the transforming principle. Culturing the bacteria in the test tube rather than in mice made it much easier to search for the chemical identity of the transforming factor in the heat-killed S cells. Methodically destroying one by one the biochemical components of the heat-treated S cells, Avery and his group looked to see whether transformation was prevented. First they degraded the sugarlike coat of the S bacteria. Transformation still occurred: the coat was not the transforming principle. Next they used a mixture of two protein-destroying enzymes, trypsin and chymotrypsin, to degrade virtually all the proteins in the S cells. To their surprise, transformation was again unaffected. Next they tried an enzyme (RNase) that breaks down RNA (ribonucleic acid), a second class of nucleic acids similar to DNA and possibly involved in protein synthesis. Again transformation occurred. Finally, they came to DNA, exposing the S bacterial extracts to the DNA-destroying enzyme, DNase. This time they hit a home run. All S-inducing activity ceased completely. The transforming factor was DNA.

11 In part because of its bombshell implications, the resulting February 1944 paper by Avery, MacLeod, and McCarty met with a mixed response. Many geneticists accepted their conclusions. After all, DNA was found on every chromosome; why shouldn't it be the genetic material? By contrast, however, most biochemists expressed doubt that DNA was a complex enough molecule to act as the repository of such a vast quantity of biological information. They continued to believe that proteins, the other component of chromosomes, would prove to be the hereditary substance. In principle, as

10 艾弗里对肺炎双球菌糖衣般的荚膜也很感兴趣。他复制格里菲斯的实验，试图分离出使R型菌变成S型菌的物质，找出它的特征。1944年，艾弗里、麦克劳德与麦卡提公布了他们的研究结果：他们以精心设计的一组实验明确地证实，DNA就是造成这种变化的“转化因子”(transforming factor)。艾弗里及其研究小组用试管培养细菌，而不是用老鼠，因此更容易在加热致死的S菌细胞上，找出哪个化学物质是转型因子。他们有系统地一一破坏经过加热处理的S型菌的生化成分，看要摧毁哪一种成分，才能阻止转型发生。首先他们使S型菌糖衣般的荚膜水解，但转型仍旧发生，这证明荚膜并非转化因子。接着他们使用两种酶(胰蛋白酶[trypsin]与胰凝乳蛋白酶[chymotrypsin])的混合制剂，这两种酶都会破坏蛋白质，结果S型菌的蛋白质几乎全遭破坏。出乎他们意料，转化仍然继续发生。他们又尝试一种会分解核糖核酸(ribonucleic acid，RNA)的核糖核酸酶(ribonuclease，RNase)，但是转化再次发生。RNA也是一种核酸，与DNA类似，而且可能与蛋白质的合成有关。最后他们把目标锁定DNA，让从S型菌取出的萃取物接触会破坏DNA的脱氧核糖核酸酶(deoxyribonuclease，DNase)，这次他们总算命中目标，R型菌不再转型为S型菌，转型因子就是DNA。

11 艾弗里、麦克劳德与麦卡提在1944年2月提交研究报告后，科学界反应不一，部分原因在于这项发现太过惊人。有许多遗传学家接受他们的结论，毕竟DNA在每个染色体上都找得到，它为什么不能是遗传物质？但是，也有许多生化学家对DNA分子是否复杂到能储存庞大的生物信息表示存疑。他们仍旧认为，最终会证实同为染色体构成要素的蛋白质才是遗传物质。其实也难怪生化学家会这么想，因为基本上蛋白质有20个氨基酸字母可以编码庞大的信息，这要比只有4个核酸字母的DNA容易得多。跟艾弗里同在洛克菲勒研究所任职的蛋白质化学家墨斯基

the biochemists rightly noted, it would be much easier to encode a vast body of complex information using the twenty-letter amino-acid alphabet of proteins than the four-letter nucleotide alphabet of DNA. Particularly vitriolic in his rejection of DNA as the genetic substance was Avery's own colleague at the Rockefeller Institute, the protein chemist Alfred Mirsky. By then, however, Avery was no longer scientifically active. The Rockefeller Institute had mandatorily retired him at age sixty-five.

12 Avery missed out on more than the opportunity to defend his work against the attacks of his colleagues: He was never awarded the Nobel Prize, which was certainly his due, for identifying DNA as the transforming principle. Because the Nobel committee makes its records public fifty years following each award, we now know that Avery's candidacy was blocked by the Swedish physical chemist Einar Hammarsten. Though Hammarsten's reputation was based largely on his having produced DNA samples of unprecedented high quality, he still believed genes to be an undiscovered class of proteins. In fact, even after the double helix was found, Hammarsten continued to insist that Avery should not receive the prize until after the mechanism of DNA transformation had been completely worked out. Avery died in 1955; had he lived only a few more years, he would almost certainly have gotten the prize.

13 When I arrived at Indiana University in the fall of 1947 with plans to pursue the gene for my Ph. D thesis, Avery's paper came up over and over in conversations. By then, no one doubted the reproducibility of his results, and more recent work coming out of the Rockefeller Institute made it all the less likely that proteins would prove to be the genetic actors in bacterial transformation. DNA had at last become an important objective for chemists setting their sights on the next breakthrough. In Cambridge, England, the canny Scottish chemist Alexander Todd rose to the challenge of identifying

（Alfred Mirsky）更是激烈反对 DNA 是遗传物质的说法，不过那时艾弗里已不再活跃于科学界，洛克菲勒研究所强制他在 65 岁时退休。

12 艾弗里错失的，不仅是反击同事、为自己的研究成果辩护的机会，他也错失了获得诺贝尔奖的机会，身为 DNA 是转化因子的发现者，他其实很有获奖资格。由于诺贝尔奖委员会在各个奖项颁发 50 年后会公布记录，现在我们已经知道当时阻挡艾弗里获得候选资格的人，是瑞典籍的物理化学家哈马斯滕（Einar Hammarsten）。虽然哈马斯滕的声望主要奠基于他能萃取出质量绝佳的 DNA 样本，但他仍然相信基因是某类尚未辨识出来的蛋白质。事实上，即使在发现双螺旋后，一直到 DNA 转型的机制完全公布前，哈马斯滕仍坚持艾弗里不应获得诺贝尔奖。艾弗里于 1955 年过世，若是他再多活几年，肯定可以拿到诺贝尔奖。

13 1947 年秋，我到印第安纳大学，计划以研究基因作为博士论文的题目，那时我们经常讨论艾弗里的论文。当时已没有人怀疑他实验结果的正确性，而洛克菲勒研究所也有更多的研究结果出炉，显示蛋白质不太可能是细菌转型过程中的遗传因子。至此，化学家终于把下一次重大突破的目标放在 DNA 上。英国剑桥精明干练的化学家托德（Alexander Todd）迎接挑战，开始鉴定连接 DNA 核苷酸的化学键。到了 1951 年初，他的实验室证实这些连接总是相同的，亦即 DNA 分子的骨干非常规则。同一时期，在奥地利出生，后来逃到美国，任职于哥伦比亚大学的内外科学

the chemical bonds that linked together nucleotides in DNA. By early 1951, his lab had proved that these links were always the same, such that the backbone of the DNA molecule was very regular. During the same period, the Austrian-born refugee Erwin Chargaff, at the College of Physicians and Surgeons of Columbia University, used the new technique of paper chromatography to measure the relative amounts of the four DNA bases in DNA samples extracted from a variety of vertebrates and bacteria. While some species had DNA in which adenine and thymine predominated, others had DNA with more guanine and cytosine. The possibility thus presented itself that no two DNA molecules had the same composition.

14 At Indiana I joined a small group of visionary scientists, mostly physicists and chemists, studying the reproductive process of the viruses that attack bacteria (bacteriophages—"phages" for short). The Phage Group was born when my Ph.D. supervisor, the Italian-trained medic Salvador Luria and his close friend, the German-born theoretical physicist Max Delbrück, teamed up with the American physical chemist Alfred Hershey. During World War II both Luria and Delbrück were considered enemy aliens, and thus ineligible to serve in the war effort of American science, even though Luria, a Jew, had been forced to leave France for New York City and Delbrück had fled Germany as an objector to Nazism. Thus excluded, they continued to work in their respective university labs—Luria at Indiana and Delbrück at Vanderbilt—and collaborated on phage experiments during successive summers at Cold Spring Harbor. In 1943, they joined forces with the brilliant but taciturn Hershey, then doing phage research of his own at Washington University in St. Louis.

15 The Phage Group's program was based on its belief that phages, like all viruses, were in effect naked genes. This concept had first been proposed in

院的查加夫(Erwin Chargaff)使用滤纸层析法(paper chromatography)这种新技术，萃取出多种脊椎动物与细菌的DNA样本，测量DNA内四种碱基的相对含量。他发现，有些物种的DNA以腺嘌呤与胸腺嘧啶居多，有些物种则是鸟嘌呤与胞嘧啶较多。因此，有可能任何两个DNA分子的组成都不同。

14 在印第安纳大学，我加入一小群以物理学家和化学家为主、很有远见的科学家之中，共同研究感染细菌的病毒“噬菌体”(bacteriophage，简称phage)的繁殖过程。后来我的博士论文指导老师，也就是在意大利受训的医师卢里亚(Salvador Luria)，以及他在德国出生的理论物理学家好友德尔布吕克(Max Delbrück)，和美国理化学家赫尔希(Alfred Hershey)携手合作，成立了噬菌体研究小组(the Phage Group)。在第二次世界大战期间，尽管犹太裔的卢里亚是被迫离开法国投奔纽约，德尔布吕克因为反对纳粹主义而逃离德国，但他们仍被视为敌侨，没有资格和美国科学界共同为战争效力。虽然遭到排挤，但他们还是在各自的大学实验室里努力工作(卢里亚在印第安纳大学，德尔布吕克则在范德比尔特大学)，并且接连几年夏季都到冷泉港实验室合作进行噬菌体实验。1943年，他们与才华横溢但沉默寡言的赫尔希合作，当时赫尔希正在圣路易的华盛顿大学研究噬菌体。

15 噬菌体小组认为，如同所有的病毒，噬菌体其实就是赤裸裸的基因(病毒就只是一个蛋白质外鞘包着核酸)，并依据这个想法来规划研究计划。这个概念是想象力丰富的美国遗传学家穆勒(Herman J. Muller)于1922

1922 by the imaginative American geneticist Herman J. Muller, who three years later demonstrated that X rays cause mutations. His belated Nobel Prize came in 1946, just after he joined the faculty of Indiana University. It was his presence, in fact, that led me to Indiana. Having started his career under T. H. Morgan, Muller knew better than anyone else how genetics had evolved during the first half of the twentieth century, and I was enthralled by his lectures during my first term. His work on fruit flies (Drosophila), however, seemed to me to belong more to the past than to the future, and I only briefly considered doing thesis research under his supervision. I opted instead for Luria's phages, an even speedier experimental subject than Drosophila: genetic crosses of phages done one day could be analyzed the next.

16 For my Ph. D thesis research, Luria had me follow in his footsteps by studying how X rays killed phage particles. Initially I had hoped to show that viral death was caused by damage to phage DNA. Reluctantly, however, I eventually had to concede that my experimental approach could never give unambiguous answers at the chemical level. I could draw only biological conclusions. Even though phages were indeed effectively naked genes, I realized that the deep answers the Phage Group was seeking could be arrived at only through advanced chemistry. DNA somehow had to transcend its status as an acronym; it had to be understood as a molecular structure in all its chemical detail.

17 Upon finishing my thesis, I saw no alternative but to move to a lab where I could study DNA chemistry. Unfortunately, however, knowing almost no pure chemistry, I would have been out of my depth in any lab attempting difficult experiments in organic or physical chemistry. I therefore took a postdoctoral fellowship in the Copenhagen lab of the biochemist Herman Kalckar in the fall of 1950. He was studying the synthesis of the small

年首先提出的，三年后，他证实 X 光会引起突变。但是直到 1946 年，穆勒到印第安纳大学任教不久之后，才获得了迟来的诺贝尔奖。事实上，正是因为他在印第安纳大学，我才会到这里就读。穆勒是在摩根的手下展开他的职业生涯的，遗传学在 20 世纪上半叶的发展过程，没人比他更清楚，我在第一个学期时就对他的讲学非常着迷。然而，他在果蝇方面的研究，对我而言似乎属于过去，而非未来，因此我只短暂考虑过请他指导我的论文。后来我选择卢里亚的噬菌体实验，这个实验做起来比果蝇快得多：噬菌体的遗传杂交（genetic cross）子代在隔天就可以进行分析。

16 为了完成博士论文研究，我在卢里亚的要求下跟随他的步履，研究 X 光如何杀死噬菌体粒子。刚开始时，我希望能证明病毒死亡是因为噬菌体的 DNA 遭到破坏。但最后我不得不承认，我的实验方法在化学上永远无法获得确切的答案，只能得到生物学上的结论。虽然噬菌体的确是裸露的基因，但我知道噬菌体研究小组所要的深奥答案，唯有通过高深的化学才找得到。DNA 不再只是一个笼统的缩写名词，我们必须彻底解开它的分子结构和所有的化学细节才行。

17 完成论文后，我发现自己别无选择，只能到可以让我研究 DNA 化学组成的实验室。然而，不幸的是，由于我几乎毫无理论化学的基础，实在不够格转战任何以有机化学或物理化学来进行艰难实验的实验室。后来在 1950 年秋，我拿到博士后研究奖学金，到生化学家开尔卡（Herman Kalckar）在哥本哈根的实验室作研究。当时他正在研究构成 DNA 的小分子的合成作用，但我很快就发现，他的生化方法永远无法解开基因的本质。在他的实验室多待一天，就会晚一天了解 DNA 如何携带遗传信息。

molecules that make up DNA, but I figured out quickly that his biochemical approach would never lead to an understanding of the essence of the gene. Every day spent in his lab would be one more day's delay in learning how DNA carried genetic information.

18 My Copenhagen year nonetheless ended productively. To escape the cold Danish spring, I went to the Zoological Station at Naples during April and May. During my last week there, I attended a small conference on X-ray diffraction methods for determining the 3D structure of molecules. X-ray diffraction is a way of studying the atomic structure of any molecule that can be crystallized. The crystal is bombarded with X rays, which bounce off its atoms and are scattered. The scatter pattern gives information about the structure of the molecule but, taken alone, is not enough to solve the structure. The additional information needed is the "phase assignment," which deals with the wave properties of the molecule. Solving the phase problem was not easy, and at that time only the most audacious scientists were willing to take it on. Most of the successes of the diffraction method had been achieved with relatively simple molecules.

19 My expectations for the conference were low. I believed that a three-dimensional understanding of protein structure, or for that matter of DNA, was more than a decade away. Disappointing earlier X-ray photos suggested that DNA was particularly unlikely to yield up its secrets via the X-ray approach. These results were not surprising since the exact sequences of DNA were expected to differ from one individual molecule to another. The resulting irregularity of surface configurations would understandably prevent the long thin DNA chains from lying neatly side by side in the regular repeating patterns required for X-ray analysis to be successful.

18 不过，我在哥本哈根的那一年仍然获益良多。为了避开丹麦寒冷的春天，我在四五月间前往意大利那不勒斯动物研究所。在那里的最后一周，我参加了一场小型研讨会，主题是以 X 光衍射法(X-ray diffraction)决定分子的三维结构。X 光衍射法可以研究任何能够形成晶体的分子的原子结构。X 光在轰击晶体后，会在撞到原子时弹开而散射。从 X 光的散射图形可以获得有关分子结构的信息。但是只靠 X 光，尚不足以解决结构的问题，还需要“相分配”(phase assignment)的额外信息，来处理分子的波性质(wave properties)。要解决“相”的问题并不容易，当时只有胆量最大的科学家愿意面对这种挑战。以衍射法成功研究的对象，大多是比较简单的分子。

19 原本我对这场研讨会的期望不高，因为我认为要解开蛋白质或 DNA 的三维结构，起码还要 10 年光景。从早期令人失望的 X 光照片看来，要通过 X 光来解开 DNA 的秘密，尤其不可能。会有这种结果也是很自然的，因为当时大家都预期，每个分子的 DNA 序列应该不尽相同。在表面构造不规则的情况下，DNA 细长的分子链势必不可能按照规律重复的模式整齐地排列，X 光分析自然也无法成功。

20 It was therefore a surprise and a delight to hear the last-minute talk on DNA by a thirty-four-year-old Englishman named Maurice Wilkins from the Biophysics Lab of King's College, London. Wilkins was a physicist who during the war had worked on the Manhattan Project. For him, as for many of the other scientists involved, the actual deployment of the bomb on Hiroshima and Nagasaki, supposedly the culmination of all their work, was profoundly disillusioning. He considered forsaking science altogether to become a painter in Paris, but biology intervened. He too had read Schrödinger's book, and was now tackling DNA with X-ray diffraction.

21 He displayed a photograph of an X-ray diffraction pattern he had recently obtained, and its many precise reflections indicated a highly regular crystalline packing. DNA, one had to conclude, must have a regular structure, the elucidation of which might well reveal the nature of the gene. Instantly I saw myself moving to London to help Wilkins find the structure. My attempts to converse with him after his talk, however, went nowhere. All I got for my efforts was a declaration of his conviction that much hard work lay ahead.

22 While I was hitting consecutive dead ends, back in America the world's preeminent chemist, Caltech's Linus Pauling, announced a major triumph: he had found the exact arrangement in which chains of amino acids (called polypeptides) fold up in proteins, and called his structure the a-helix (alpha helix). That it was Pauling who made this breakthrough was no surprise: he was a scientific superstar. His book *The Nature of the Chemical Bond* essentially laid the foundation of modern chemistry, and, for chemists of the day, it was the Bible. Pauling had been a precocious child. When he was nine, his father, a druggist in Oregon, wrote to the Oregonian newspaper requesting suggestions of reading matter for his bookish son, adding that he

20 因此，当我听到来自伦敦国王学院(King's College)生物物理实验室的英国人威尔金斯(Maurice Wilkins)在最后发表有关 DNA 的演讲时，不禁惊喜交加。34 岁的威尔金斯是物理学家，战时曾参与制造原子弹的曼哈顿计划(Manhattan Project)。对他和许多参与这个计划的科学家而言，原子弹投到广岛和长崎，应该是他们研究工作的最高成就，结果这却造成他们的理想破灭。威尔金斯曾考虑完全放弃科学，到巴黎去当画家，但生物学引起了他的兴趣——他也读过薛定谔的书。当时他正设法用 X 光衍射法解开 DNA 的秘密。

21 威尔金斯在演讲中展示了一张最近拍到的 X 光衍射图，上面有许多明确的反射影像，显示它是高度规则的结晶体。由此可以得出推论，DNA 必定具有规则的结构，只要能解开此结构，就可以揭露基因的本质。我立即开始幻想自己搬到伦敦去，协助威尔金斯找出这个结构。演讲过后，我去找他谈话，但却一无所获。他只对我表示，未来还有更多艰辛的工作要做。

22 就在我连续碰壁时，加州理工学院享誉国际的化学家泡令(Linus Carl Pauling)宣布获得了重大的成就：他发现蛋白质里氨基酸链(称为多肽[polypeptide])的排列结构，并且将这个结构取名为 α 螺旋(alpha helix)。这个突破会由泡令获得其实并不意外，他是科学界的超级巨星。他所著的《化学键的本质》(*The Nature of the Chemical Bond*)奠立了现代化学的基础，被当时的化学家奉为圣经。泡令非常早熟，他在俄勒冈州长大，父亲是位药剂师。他 9 岁时，父亲曾写信给《俄勒冈人》报，希望对方能提供他那好学不倦的孩子可以阅读的书籍，还说他儿子已经读完《圣经》和达尔文的《物种起源》。泡令的父亲不幸早逝，家中经济陷入困境，但这位前途无量的年轻人仍然完成了学业，相当难能可贵。

had already read the Bible and Darwin's *Origin of Species*. But the early death of Pauling's father, which brought the family to financial ruin, makes it remarkable that the promising young man managed to get an education at all.

23 As soon as I returned to Copenhagen I read about Pauling's a-helix. To my surprise, his model was not based on a deductive leap from experimental X-ray diffraction data. Instead, it was Pauling's long experience as a structural chemist that had emboldened him to infer which type of helical fold would be most compatible with the underlying chemical features of the polypeptide chain. Pauling made scale models of the different parts of the protein molecule, working out plausible schemes in three dimensions. He had reduced the problem to a kind of three-dimensional jigsaw puzzle in a way that was simple yet brilliant.

24 Whether the a-helix was correct—in addition to being pretty—was now the question. Only a week later, I got the answer. Sir Lawrence Bragg, the English inventor of X-ray crystallography and 1915 Nobel laureate in Physics, came to Copenhagen and excitedly reported that his junior colleague, the Austrian-born chemist Max Perutz, had ingeniously used synthetic polypeptides to confirm the correctness of Pauling's a-helix. It was a bittersweet triumph for Bragg's Cavendish Laboratory. The year before, they had completely missed the boat in their paper outlining possible helical folds for polypeptide chains.

25 By then Salvador Luria had tentatively arranged for me to take up a research position at the Cavendish. Located at Cambridge University, this was the most famous laboratory in all of science. Here Ernest Rutherford first described the structure of the atom. Now it was Bragg's own domain, and I

23 我一回到哥本哈根，立即拜读了泡令有关 α 螺旋的研究。令人惊讶的是，他并非根据 X 光衍射的实验数据推论出模型，而是根据身为结构化学家的丰富经验，大胆推论哪种类型的螺旋结构最符合多肽链的化学特性。泡令制作蛋白质分子不同部分的比例模型，找出可能的三维结构。他将问题简化成一种三维拼图游戏，既简单又聪明。

24 α 螺旋相当美丽，但现在的问题在于它是否正确。短短一星期后，我得到了答案。发明 X 光晶体学（X-ray chrystailography）的 1915 年诺贝尔物理学奖得主布拉格爵士（Sir Lawrence Bragg）来到哥本哈根，兴奋地宣布，比他资浅的同事、奥地利籍化学家佩鲁茨（Max Perutz）巧妙地用合成多肽证实了泡令的 α 螺旋是正确的。对布拉格的卡文迪什实验室而言，这是个苦乐参半的胜利，因为前一年他们在论文中列举多肽链可能具有的螺旋形态时，完全不得要领。

25 那时，卢里亚尝试替我安排卡文迪什实验室的研究职位。卡文迪什实验室位于剑桥大学，是科学界最著名的实验室，卢瑟福（Ernest Rutherford）就是在这里首先描述出原子的结构。当时那是布拉格的研究领域，而我则被安排跟英国化学家肯德鲁（John Kendrew）实习，他的兴

was to work as apprentice to the English chemist John Kendrew, who was interested in determining the 3D structure of the protein myoglobin. Luria advised me to visit the Cavendish as soon as possible. With Kendrew in the States, Max Perutz would check me out. Together, Kendrew and Perutz had earlier established the Medical Research Council (MRC) Unit for the Study of the Structure of Biological Systems.

26 A month later in Cambridge, Perutz assured me that I could quickly master the necessary X-ray diffraction theory and should have no difficulty fitting in with the others in their tiny MRC Unit. To my relief, he was not put off by my biology background. Nor was Lawrence Bragg, who briefly came down from his office to look me over.

27 I was twenty-three when I arrived back at the MRC Unit in Cambridge in early October. I found myself sharing space in the biochemistry room with a thirty-five-year-old ex-physicist, Francis Crick, who had spent the war working on magnetic mines for the Admiralty. When the war ended, Crick had planned to stay on in military research, but, on reading Schrödinger's *What Is Life*? he had moved toward biology. Now he was at the Cavendish to pursue the 3D structure of proteins for his Ph.D.

28 Crick was always fascinated by the intricacies of important problems. His endless questions as a child compelled his weary parents to buy him a children's encyclopedia, hoping that it would satisfy his curiosity. But it only made him insecure: he confided to his mother his fear that everything would have been discovered by the time he grew up, leaving him nothing to do. His mother reassured him (correctly, as it happened) that there would still be a thing or two for him to figure out.

趣是找出肌红素(myoglobin)这种蛋白质的三维结构。卢里亚建议我尽快前往卡文迪什实验室，因为肯德鲁那时在美国，佩鲁茨会审核我的资格。肯德鲁与佩鲁茨早先曾一起建立医学研究委员会(Medical Research Council，MRC)作为生物系统结构研究的单位。

26 一个月后在剑桥，佩鲁茨向我保证，我很快就能精通必要的 X 光衍射理论，应该也很快就能融入他们人数不多的研究单位。我松了一口气，因为他并没有因为我的生物学背景而拒绝我。布拉格也没有，他还从办公室下来看了我一下。

27 我在 10 月初抵达剑桥医学研究委员会的研究单位，那年我 23 岁。我和 35 岁的前物理学家克里克共用生化研究室，他在战时曾替英国海军研究磁性水雷。战争结束后，克里克原本计划留在军方的研究机构，但在拜读了薛定谔的《生命是什么?》后，决定朝生物学发展。当时他在卡文迪什实验室以研究蛋白质的三维结构为博士论文主题。

28 克里克对重要问题的错综复杂总是非常着迷。小时候，他老爱问问题，被问腻的双亲只好买一套儿童百科全书给他，希望能满足他的好奇心，结果这反而让他没有安全感，他告诉母亲，他怕长大时，所有的事物都已被人发现，而他将无事可做。母亲向他保证，日后一定还会有一两件事等着他发现。事后证明她说得很准。

29 A great talker, Crick was invariably the center of attention in any gathering. His booming laugh was forever echoing down the hallways of the Cavendish. As the MRC Unit's resident theoretician, he used to come up with a novel insight at least once a month, and he would explain his latest idea at great length to anyone willing to listen.

30 The morning we met he lit up when he learned that my objective in coming to Cambridge was to learn enough crystallography to have a go at the DNA structure. Soon I was asking Crick's opinion about using Pauling's model-building approach to go directly for the structure. Would we need many more years of diffraction experimentation before modeling would be practicable? To bring us up to speed on the status of DNA structural studies, Crick invited Maurice Wilkins, a friend since the end of the war, up from London for Sunday lunch. Then we could learn what progress Wilkins had made since his talk in Naples.

31 Wilkins expressed his belief that DNA's structure was a helix, formed by several chains of linked nucleotides twisted around each other. All that remained to be settled was the number of chains. At the time, Wilkins favored three on the basis of his density measurements of DNA fibers. He was keen to start model-building, but he had run into a roadblock in the form of a new addition to the King's College Biophysics Unit, Rosalind Franklin.

32 A thirty-one-year-old Cambridge-trained physical chemist, Franklin was an obsessively professional scientist; for her twenty-ninth birthday all she requested was her own subscription to her field's technical journal, Acta Crystallographica. Logical and precise, she was impatient with those who acted otherwise. And she was given to strong opinions, once describing her Ph.D thesis adviser, Ronald Norrish, a future Nobel Laureate, as "stupid,

29 克里克很擅长说话，无论是哪种聚会，他总是众人注意的焦点。在卡文迪什实验室的走廊上，总是可以听到他爽朗的笑声。他是医学研究委员会研究单位的专任理论家，每个月至少会提出一个新构想，而且只要有人愿意听，他总是很乐意花许多时间仔细解释。

30 我和克里克相遇的那个早晨，他听说我来剑桥是为了学习大量有关晶体学的知识以便破解 DNA 的结构，大为高兴。不久之后，我就请克里克谈谈他对使用泡令的模型建构法来破解 DNA 结构的看法。我们是不是还得做许多年的衍射实验，才能实际下手去建构模型？为了加快我们研究 DNA 结构的速度，克里克邀请自战后就认识的朋友威尔金斯，在星期天从伦敦过来共进午餐，这样我们就能得知自从那不勒斯的演讲后，威尔金斯还有哪些进展。

31 威尔金斯表示，他认为 DNA 的结构是螺旋状，由数条链接的核苷酸互相缠绕而成，而唯一尚待解决的问题在于链的数目。当时威尔金斯根据他测量的 DNA 纤维密度，认为应该有三条核苷酸链。他迫切地想着手建造模型，却碰上了一个障碍：刚加入国王学院生物物理学研究单位的富兰克林(Rosalind Franklin)。

32 31 岁的富兰克林是出身剑桥的物理化学家，她是个对自己的专业执著得不得了的科学家，在 29 岁生日时，她只要求订阅自己所属领域的技术期刊《晶体学报》(*Acta Crystallogrsphica*)作为生日礼物。她讲求逻辑和精确，对于欠缺这些特质的人没有什么耐性。她习惯发表措辞强烈的看法，一度将她的博士论文指导教授，未来的诺贝尔奖得主诺里什(Ronald Norrish)描述为“愚蠢、固执、奸诈、态度恶劣、专制”。在实验室外，

bigoted, deceitful, ill-mannered and tyrannical." Outside the laboratory, she was a determined and gutsy mountaineer, and, coming from the upper echelons of London society, she belonged to a more rarefied social world than most scientists. At the end of a hard day at the bench, she would occasionally change out of her lab coat into an elegant evening gown and disappear into the night.

33 Just back from a four-year X-ray crystallographic investigation of graphite in Paris, Franklin had been assigned to the DNA project while Wilkins was away from King's. Unfortunately, the pair soon proved incompatible. Franklin, direct and data-focused, and Wilkins, retiring and speculative, were destined never to collaborate. Shortly before Wilkins accepted our lunch invitation, the two had had a big blowup in which Franklin had insisted that no model-building could commence before she collected much more extensive diffraction data. Now they effectively didn't communicate, and Wilkins would have no chance to learn of her progress until Franklin presented her lab seminar scheduled for the beginning of November. If we wanted to listen, Crick and I were welcome to go as Wilkins's guests. Crick was unable to make the seminar, so I attended alone and briefed him later on what I believed to be its key take-home messages on crystalline DNA. In particular, I described from memory Franklin's measurements of the crystallographic repeats and the water content. This prompted Crick to begin sketching helical grids on a sheet of paper, explaining that the new helical X-ray theory he had devised with Bill Cochran and Vladimir Vand would permit even me, a former bird-watcher, to predict correctly the diffraction patterns expected from the molecular models we would soon be building at the Cavendish.

34 As soon as we got back to Cambridge, I arranged for the Cavendish machine shop to construct the phosphorous atom models needed for short sections of

她是果决勇敢的登山家，来自伦敦的上流社会，相较于大多数的科学家，她属于高尚的社交界。在工作台辛苦一整天后，她偶尔会脱下实验室的外套，换上优雅的晚礼服，消失在夜色中。

33 富兰克林才从法国回来，她在巴黎用 X 光晶体学的技术研究了 4 年的石墨。她接受聘任，加入国王学院的 DNA 计划时，威尔金斯刚好不在。不幸的是，后来证明这两人根本合不来。富兰克林的个性率直，重视数据，威尔金斯则拘谨而勇于猜想，他们注定无法合作。在威尔金斯接受我们的午餐邀约前不久，他们两人才大吵一次，富兰克林坚持在她搜集更多衍射数据前，不能着手建立模型。他们俩显然无法沟通，而在富兰克林于 11 月初举办实验室研讨会之前，威尔金斯无从得知她的进展。不过若我们想参加这次研讨会，威尔金斯很乐意邀请我和克里克前往。后来克里克因故未能参加研讨会，由我独自前往，稍后再把我认为与 DNA 晶体有关的重要信息告诉他。我特别根据记忆，描述了富兰克林关于晶体重复与含水量的测量值。克里克听了之后，开始在纸上绘制螺旋网格，并说就连我这种先前以赏鸟为业的人也可以应用他和科克兰(Bill Cochran)及凡德(Vladimir Vand)提出的新螺旋 X 光理论，准确预测我们即将建造的分子模型有哪些衍射图。

34 我们一回到剑桥，我就安排卡文迪什的机械部门建造磷的原子模型，以便用于建造 DNA 里磷酸糖骨干的片段。等这些模型做好后，我们开始

the sugar phosphate backbone found in DNA. Once these became available, we tested different ways the backbones might twist around each other in the center of the DNA molecule. Their regular repeating atomic structure should allow the atoms to come together in a consistent, repeated conformation. Following Wilkins's hunch, we focused on three-chain models. When one of these appeared to be almost plausible, Crick made a phone call to Wilkins to announce we had a model we thought might be DNA.

35 The next day both Wilkins and Franklin came up to see what we had done. The threat of unanticipated competition briefly united them in common purpose. Franklin wasted no time in faulting our basic concept. My memory was that she had reported almost no water present in crystalline DNA. In fact, the opposite was true. Being a crystallographic novice, I had confused the terms "unit cell" and "asymmetric unit." Crystalline DNA was in fact water-rich. Consequently, Franklin pointed out, the backbone had to be on the outside and not, as we had it, in the center, if only to accommodate all the water molecules she had observed in her crystals.

36 That unfortunate November day cast a very long shadow. Franklin's opposition to model-building was reinforced. Doing experiments, not playing with Tinkertoy representations of atoms, was the way she intended to proceed. Even worse, Sir Lawrence Bragg passed down the word that Crick and I should desist from all further attempts at building a DNA model. It was further decreed that DNA research should be left to the King's lab, with Cambridge continuing to focus solely on proteins. There was no sense in two MRC-funded labs competing against each other. With no more bright ideas up our sleeves, Crick and I were reluctantly forced to back off, at least for the time being.

测试骨干在DNA分子中央彼此缠绕的不同方法。它们规则重复的原子结构，应该会让原子形成一致且重复的构造。我们听从威尔金斯的直觉，把重点放在三链模型上。当其中一个模型看似很有可能是答案时，克里克打电话给威尔金斯，宣称我们可能找到了DNA的模型。

35 第二天，威尔金斯与富兰克林一起来访，查看我们的成果。在这意外出现的竞争威胁下，他们俩为了共同目标而难得地暂时合作。富兰克林立即挑出我们在基本概念上的错误。我记得她在报告时指出：DNA晶体几乎不含水。其实是我自己弄错了。由于才刚开始学晶体学，我把晶胞（unit cell）与不对称单位（asymmetric unit）这两个术语搞混了。其实，DNA晶体富含水分。因此，富兰克林指出，光是要容纳她在晶体内观察到的水分子，骨干就得在分子外面，而不是像我们所做的在分子中央。

36 11月的那一天实在不幸，让未来蒙上浓浓阴影。富兰克林更加坚定地反对建造模型，她打算继续做实验，不玩看似小孩玩具的原子模型。更惨的是，布拉格也说话了，叫我和克里克不要再尝试建造DNA模型，后来还进一步决定，DNA研究应交由国王学院实验室来做，剑桥只需继续研究蛋白质即可。两家同样由医学研究委员会赞助的实验室居然彼此竞争，实在没有道理。在无计可施下，我和克里克不情愿地暂时罢手。

37 It was not a good moment to be condemned to the DNA sidelines. Linus Pauling had written Wilkins to request a copy of the crystalline DNA diffraction pattern. Though Wilkins had declined, saying he wanted more time to interpret it himself, Pauling was hardly obliged to depend upon data from King's. If he wished, he could easily start serious X-ray diffraction studies at Caltech.

38 The following spring, I duly turned away from DNA and set about extending prewar studies on the pencil-shaped tobacco mosaic virus using the Cavendish's powerful new X-ray beam. This light experimental workload gave me plenty of time to wander through various Cambridge libraries. In the zoology building, I read Erwin Chargaff's paper describing his finding that the DNA bases adenine and thymine occurred in roughly equal amounts, as did the bases guanine and cytosine. Hearing of these one-to-one ratios Crick wondered whether, during DNA duplication, adenine residues might be attracted to thymine and vice versa, and whether a corresponding attraction might exist between guanine and cytosine. If so, base sequences on the "parental" chains (e.g., ATGC) would have to be complementary to those on "daughter" strands (yielding in this case TACG).

39 These remained idle thoughts until Erwin Chargaff came through Cambridge in the summer of 1952 on his way to the International Biochemical Congress in Paris. Chargaff expressed annoyance that neither Crick nor I saw the need to know the chemical structures of the four bases. He was even more upset when we told him that we could simply look up the structures in textbooks as the need arose. I was left hoping that Chargaff's data would prove irrelevant. Crick, however, was energized to do several experiments looking for molecular "sandwiches" that might form when adenine and thymine (or alternatively, guanine and cytosine) were mixed together in solution. But his experiments went nowhere.

37 在此时退出DNA研究，实在不是时候。泡令已写信给威尔金斯，请他提供一份DNA晶体的衍射图。虽然威尔金斯拒绝，表示自己需要更多的时间来解读，但泡令其实也不见得要依赖国王学院的资料。如果他愿意的话，大可在加州理工学院自行研究X光衍射。

38 来年春天，我不再研究DNA，反而继续做那些战前的研究，用卡文迪什实验室强大的新X光束，研究铅笔状的烟草花叶病毒(tobacco mosaic virus)。这个实验的工作量很少，我有许多时间游走剑桥众多的图书馆。我在动物学系看到查加夫的论文，他发现在DNA中，腺嘌呤和胸腺嘧啶的数量大致相同，而鸟嘌呤则与胞嘧啶的数量差不多。在听到这一比一的比例后，克里克想到，在DNA复制时，腺嘌呤和胸腺嘧啶是否互相吸引，而胞嘧啶与鸟嘌呤之间是否也存有类似的吸引力。若是如此，DNA“亲代”链上的碱基序列(如ATGC)应该会与“子代”链上的互补(亦即TACG)。

39 这些原本都只是空想，直到1952年夏天，查加夫在前往巴黎参加国际生化会议时途经剑桥为止。我和克里克认为不需要了解4种碱基的化学结构，但查加夫对这看法颇不赞同。又听到我们说，如有必要的话，可以到教科书里查它们的结构时，他更是不悦。我只希望能证实查加夫的数据与DNA结构并不相关。不过克里克却兴致勃勃地要做一些实验，寻找腺嘌呤与胸腺嘧啶(或鸟嘌呤与胞嘧啶)在溶液中混合时，可能会形成的分子“三明治”。但是，他的实验没获得任何结果。

40 Like Chargaff, Linus Pauling also attended the International Biochemical Congress, where the big news was the latest result from the Phage Group. Alfred Hershey and Martha Chase at Cold Spring Harbor had just confirmed Avery's transforming principle: DNA was the hereditary material! Hershey and Chase proved that only the DNA of the phage virus enters bacterial cells; its protein coat remains on the outside. It was more obvious than ever that DNA must be understood at the molecular level if we were to uncover the essence of the gene. With Hershey and Chase's result the talk of the town, I was sure that Pauling would now bring his formidable intellect and chemical wisdom to bear on the problem of DNA.

41 Early in 1953, Pauling did indeed publish a paper outlining the structure of DNA. Reading it anxiously I saw that he was proposing a three-chain model with sugar phosphate backbones forming a dense central core. Superficially it was similar to our botched model of fifteen months earlier. But instead of using positively charged atoms (e. g., Mg^{2+}) to stabilize the negatively charged backbones, Pauling made the unorthodox suggestion that the phosphates were held together by hydrogen bonds. But it seemed to me, the biologist, that such hydrogen bonds required extremely acidic conditions never found in cells. With a mad dash to Alexander Todd's nearby organic chemistry lab my belief was confirmed: The impossible had happened. The world's best-known, if not best, chemist had gotten his chemistry wrong. In effect, Pauling had knocked the A off of DNA. Our quarry was deoxyribonucleic acid, but the structure he was proposing was not even acidic.

42 Hurriedly I took the manuscript to London to inform Wilkins and Franklin they were still in the game. Convinced that DNA was not a helix, Franklin had no wish even to read the article and deal with the distraction of Pauling's

40 泡令跟查加夫一样，也参加了国际生化会议，那时的大新闻是噬菌体研究小组的最新结果。冷泉港的赫尔希(Hershey)与蔡斯(Martha Chase)才刚证实了艾弗里的转化因子：DNA 就是遗传物质！赫尔希和蔡斯证明，进入细菌细胞的只有噬菌体病毒的 DNA，它的蛋白质鞘(protein coat)留在外面。看来如果我们想揭开基因的本质，势必得了解 DNA 的分子。在赫尔希与蔡斯成为大家的话题之后，我确定泡令也会将他的才华与化学知识，全力投注在解决 DNA 的问题上。

41 早在 1953 年，泡令就发表过描述 DNA 结构的论文。我急切地拜读了大作，发现他提出的是三链模型，以磷酸糖的骨干形成稠密的中央核心。乍看之下，它跟我们在 15 个月前所做的拙劣模型类似。但是，泡令没有采用带正电的原子(例如 Mg^{2+})来稳定带负电的骨干，而是采取非正统的做法，以氢链来连接磷酸盐。不过，看在身为生物学家的我眼中，这种氢键所需的极酸状态从不曾见于细胞内。我发疯似地冲到托德在附近的有机化学实验室，立刻就证实了我的看法：不可能的事居然发生了！全世界最优秀或至少最著名的化学家竟然弄错了化学基本原理。实际上，泡令等于把 DNA 里代表酸的缩写 A 除掉了。我们研究的对象是脱氧核糖核酸，但是他所提出的结构甚至不属于酸类。

42 我立刻把论文带到伦敦，告诉威尔金斯与富兰克林，他们仍有成功机会。但深信 DNA 并非螺旋的富兰克林甚至不想看这篇文章，以免受到泡令的螺旋观念影响，连在我提出克里克的螺旋论点后也没改变想法。

helical ideas, even when I offered Crick's arguments for helices. Wilkins, however, was very interested indeed in the news I brought; he was now more certain than ever that DNA was helical. To prove the point, he showed me a photograph obtained more than six months earlier by Franklin's graduate student Raymond Gosling, who had X-rayed the so-called B form of DNA. Until that moment, I didn't know a B form even existed. Franklin had put this picture aside, preferring to concentrate on the A form, which she thought would more likely yield useful data. The X-ray pattern of this B form was a distinct cross. Since Crick and others had already deduced that such a pattern of reflections would be created by a helix, this evidence made it clear that DNA had to be a helix! In fact, despite Franklin's reservations, this was no surprise. Geometry itself suggested that a helix was the most logical arrangement for a long string of repeating units such as the nucleotides of DNA. But we still did not know what that helix looked like, nor how many chains it contained.

43 The time had come to resume building helical models of DNA. Pauling was bound to realize soon enough that his brainchild was wrong. I urged Wilkins to waste no time. But he wanted to wait until Franklin had completed her scheduled departure for another lab later that spring. She had decided to move on to avoid the unpleasantness at King's. Before leaving, she had been ordered to stop further work with DNA and had already passed on many of her diffraction images to Wilkins.

44 When I returned to Cambridge and broke the news of the DNA B form, Bragg no longer saw any reason for Crick and me to avoid DNA. He very much wanted the DNA structure to be found on his side of the Atlantic. So we went back to model-building, looking for a way the known basic components of DNA—the backbone of the molecule and the four different

倒是威尔金斯对我带来的消息很感兴趣；他现在更确定DNA是螺旋状。为了证明这一点，他拿出一张6个月前，由富兰克林的研究生葛斯林(Raymond Gosling)用X光拍下的照片，即所谓的B型DNA。在那之前，我甚至不知道有B型的存在。富兰克林不理会这张照片，把注意力集中在A型DNA上，因为她认为研究A型比较可能获得有用的资料。B型DNA的X光图是一个清晰的"十"字。既然克里克和其他人早已推论出，这类的反射图案是由螺旋所造成，这项证据清楚说明DNA必定是螺旋状！事实上，尽管富兰克林持保留态度，但这个发现并不出人意外。几何学本身就显示，螺旋结构是一长串重复的单元(例如DNA的核苷酸)最合理的排列方式。不过，我们仍不知道这个螺旋的外观，也不知道它含有多少链。

43 现在终于到了我们继续建构DNA螺旋模型的时候。泡令肯定不久就会发现，他的精心杰作出现了谬误。我敦促威尔金斯不要再浪费时间，但是他想等到富兰克林在该年春天稍后到另一家实验室工作后才开始。她选择离去，避开在国王学院的不愉快。在离开前，她奉命停止对DNA作更进一步的研究，并将许多衍射照片交给威尔金斯。

44 当我回到剑桥，向布拉格报告B型DNA的消息之后，他认为不应再禁止我和克里克研究DNA，而且很希望DNA的结构能由大西洋这一岸破解。于是我们再度着手建构模型，设法把已知的DNA基本成分凑成螺旋结构。这些基本成分就是分子骨干以及4个不同的碱基(腺嘌呤、胸腺嘧啶、鸟嘌呤与胞嘧啶)。我委托卡文迪什的工厂替我们做一套锡制的碱基模型，但是他们的制造速度对我来说不够快，最后我只得拿硬纸板来剪出粗略的模型。

bases, adenine, thymine, guanine, and cytosine—could fit together to make a helix. I commissioned the shop at the Cavendish to make us a set of tin bases, but they couldn't produce them fast enough for me: I ended up cutting out rough approximations from stiff cardboard.

45 By this time I realized the DNA density-measurement evidence actually slightly favored a two-chain, rather than three-chain, model. So I decided to search out plausible double helices. As a biologist, I preferred the idea of a genetic molecule made of two, rather than three, components. After all, chromosomes, like cells, increase in number by duplicating, not triplicating.

46 I knew that our previous model with the backbone on the inside and the bases hanging out was wrong. Chemical evidence from the University of Nottingham, which I had too long ignored, indicated that the bases must be hydrogen-bonded to each other. They could only form bonds like this in the regular manner implied by the X-ray diffraction data if they were in the center of the molecule. But how could they come together in pairs? For two weeks I got nowhere, misled by an error in my nucleic acid chemistry textbook. Happily, on February 27, Jerry Donahue, a theoretical chemist visiting the Cavendish from Caltech, pointed out that the textbook was wrong. So I changed the locations of the hydrogen atoms on my cardboard cutouts of the molecules.

47 The next morning, February 28, 1953, the key features of the DNA model all fell into place. The two chains were held together by strong hydrogen bonds between adenine-thymine and guanine-cytosine base pairs. The inferences Crick had drawn the year before based on Chargaff's research had indeed been correct. Adenine does bond to thymine and guanine does bond to cytosine, but not through flat surfaces to form molecular sandwiches. When Crick arrived, he took it all in rapidly, and gave my base-pairing scheme his

45 此时我已经发现，DNA 的密度测量证据比较倾向于双链，而非三链的模型结构。因此，我决定寻找可能的双螺旋体。身为生物学家，我偏好遗传分子是由两个，而非三个成分组成的概念。毕竟，染色体就像细胞一样，数量是以复制成两倍而非三倍的方式增加。

46 我知道先前将骨干置于里面，而碱基悬挂在外的模型是错误的。诺丁汉大学提出的化学证据显示，碱基必须由氢键彼此连接，但这项证据一直被我忽略。如果碱基位于分子中央的话，就只能按照 X 光衍射数据所显示的规律方式形成这种键。但是它们怎么会成双作对呢？在错误的核酸化学教科书误导下，我连续两个星期毫无进展。幸好 2 月 27 日加州理工学院的理论化学家多纳休（Jerry Donahue）到卡文迪什访问，他指出教科书的错误。于是我改变了氢原子在硬纸板分子模型上的位置。

47 隔天早晨，1953 年 2 月 28 日，DNA 模型的重要特征全都各就各位。它的两条链由腺嘌呤–胸腺嘧啶，以及鸟嘌呤–胞嘧啶这两对碱基对之间的强氢键连在一起。克里克一年前根据查加夫的研究所得到的推论，真的是正确的。腺嘌呤的确与胸腺嘧啶连接，而鸟嘌呤也与胞嘧啶连接，但是它们并非通过平坦的表面形成分子三明治。当克里克抵达时，他很快就了解了状况，并且认同我的碱基配对（base-pairing）方式。而且他当下就发现，这会造成双螺旋的双股以相反方向连接。

blessing. He realized right away that it would result in the two strands of the double helix running in opposite directions.

48 It was quite a moment. We felt sure that this was it. Anything that simple, that elegant just had to be right. What got us most excited was the complementarity of the base sequences along the two chains. If you knew the sequence—the order of bases—along one chain, you automatically knew the sequence along the other. It was immediately apparent that this must be how the genetic messages of genes are copied so exactly when chromosomes duplicate prior to cell division. The molecule would "unzip" to form two separate strands. Each separate strand then could serve as the template for the synthesis of a new strand, one double helix becoming two.

49 In *What Is Life?* Schrödinger had suggested that the language of life might be like Morse code, a series of dots and dashes. He wasn't far off. The language of DNA is a linear series of As, Ts, Gs, and Cs. And just as transcribing a page out of a book can result in the odd typo, the rare mistake creeps in when all these As, Ts, Gs, and Cs are being copied along a chromosome. These errors are the mutations geneticists had talked about for almost fifty years. Change an "i" to an "a" and "Jim" becomes "Jam" in English; change a T to a C and "ATG" becomes "ACG" in DNA.

50 The double helix made sense chemically and it made sense biologically. Now there was no need to be concerned about Schrödinger's suggestion that new laws of physics might be necessary for an understanding of how the hereditary code-script is duplicated: genes in fact were no different from the rest of chemistry. Later that day, during lunch at the Eagle, the pub virtually adjacent to the Cavendish Lab, Crick, ever the talker, could not help but tell everyone we had just found the "secret of life." I myself, though no less electrified by the thought, would have waited until we had a pretty three-dimensional model to show off.

48 这真是令人难忘的时刻，我们觉得这次肯定对了。这么简单美丽的构造，绝对错不了。最令我们兴奋的是碱基序列沿着双链排列的互补特性，只要知道一条链上的序列(碱基的顺序)，自然就能推知另一条链上的序列。这显然是细胞分裂前染色体在进行复制时，基因的遗传信息能精准重现的原因。DNA 分子会像拉链一样"拉开"，形成独立的两股。每一股都可以作为新股合成时的模板，于是一条双螺旋就变成了两条。

49 在《生命是什么?》中，薛定谔提议，生命的语言也许就像摩斯密码，是一系列的点与线。这个讲法倒是蛮接近事实的。DNA 的语言是由 A，T，G 与 C 构成的线性序列。就像我们在誊写书籍时，偶尔也会写错字一样，所有的 A，T，G，C 在沿染色体复制时，也会出现极少量的错误。这些错误就是遗传学家近 50 年来所一直探讨的突变。在英文中，将 i 变成 a，Jim 就会变成 Jam，而在 DNA 中，将 T 变为 C，ATG 就变成了 ACG。

50 无论从化学或生物学的观点来看，双螺旋都很合理。我们现在无须担忧薛定谔所说，要了解遗传密码如何复制有可能需要新的物理定律，事实上，基因的组成与其他的化学作用并没有两样。那天稍后，在紧邻卡文迪什实验室的鹰吧吃午餐时，向来爱说话的克里克忍不住告诉大家我们刚发现了"生命的奥秘"。我虽同样激动，但宁可等到做出漂亮的三维模型时才炫耀。

51 Among the first to see our demonstration model was the chemist Alexander Todd. That the nature of the gene was so simple both surprised and pleased him. Later, however, he must have asked himself why his own lab, having established the general chemical structure of DNA chains, had not moved on to asking how the chains folded up in three dimensions. Instead the essence of the molecule was left to be discovered by a two-man team, a biologist and a physicist, neither of whom possessed a detailed command even of undergraduate chemistry. But paradoxically, this was, at least in part, the key to our success: Crick and I arrived at the double helix first precisely because most chemists at that time thought DNA too big a molecule to understand by chemical analysis.

52 At the same time, the only two chemists with the vision to seek DNA's 3D structure made major tactical mistakes: Rosalind Franklin's was her resistance to model-building; Linus Pauling's was a matter of simply neglecting to read the existing literature on DNA, particularly the data on its base composition published by Chargaff. Ironically, Pauling and Chargaff sailed across the Atlantic on the same ship following the Paris Biochemical Congress in 1952, but failed to hit it off. Pauling was long accustomed to being right. And he believed there was no chemical problem he could not work out from first principles by himself. Usually this confidence was not misplaced. During the Cold War, as a prominent critic of the American nuclear weapons development program, he was questioned by the FBI after giving a talk. How did he know how much plutonium there is in an atomic bomb? Pauling's response was "Nobody told me. I figured it out". Over the next several months Crick and (to a lesser extent) I relished showing off our model to an endless stream of curious scientists. However, the Cambridge biochemists did not invite us to give a formal talk in the biochemistry building. They started to refer to it as the "WC," punning our initials with those used in Britain for

51 化学家托德是最早看到我们这个模型的人之一。基因的本质如此简单，让他非常惊喜。然而，稍后他必然曾扪心自问，为什么自己的实验室在建立 DNA 链的一般化学结构后，未能进一步研究这些链在三维空间的组成方式，反而让由生物学家和物理学家组成的双人组找出这种分子的本质，这两人对大学程度的化学甚至都不是那么了解。然而话说回来，这正是我们成功的关键，至少是部分关键：我和克里克之所以能率先获得双螺旋的结论，正是因为当时大多数的化学家认为，DNA 的分子太大，无法用化学分析来了解。

52 同时，唯一两位具有远见、知道要寻找 DNA 三维结构的化学家，又犯了策略上的错误：富兰克林不愿建构模型，泡令则忽略了有关 DNA 的现有文献，特别是查加夫所发表的 DNA 碱基组成数据。1952 年巴黎的生化会议后，泡令和查加夫还曾同船越过大西洋，两人却不投缘。泡令习惯于自己总是对的，而且相信自己可以用基本原理来解开任何化学问题。他的自信通常很有道理。冷战期间，他是批评美国核武器发展计划的主要人士，有一次在发表演说后，FBI 警员质问他怎么知道原子弹含有多少钸？泡令傲然答道："没人告诉我，我自己想出来的。"在其后的几个月，克里克和我(虽然我没那么热衷)乐得把我们的模型拿给川流不息、充满好奇的科学家们看。然而，剑桥的生化学家并未邀请我们到生化大楼发表正式演说。他们戏称这是"WC"，拿我们的名字缩写开双关语的玩笑。我们没做实验就找到双螺旋，令他们恼怒。

the toilet or water closet. That we had found the double helix without doing experiments irked them.

53 The manuscript that we submitted to *Nature* in early April was published just over three weeks later, on April 25, 1953. Accompanying it were two longer papers by Franklin and Wilkins, both supporting the general correctness of our model. In June, I gave the first presentation of our model at the Cold Spring Harbor symposium on viruses. Max Delbriick saw to it that I was offered, at the last minute, an invitation to speak. To this intellectually high-powered meeting I brought a three-dimensional model built in the Cavendish, the adenine-thymine base pairs in red and the guanine-cytosine base pairs in green.

54 In the audience was Seymour Benzer, yet another ex-physicist who had heeded the clarion call of Schrödinger's book. He immediately understood what our breakthrough meant for his studies of mutations in viruses. He realized that he could now do for a short stretch of bacteriophage DNA what Morgan's boys had done forty years earlier for fruit fly chromosomes: he would map mutations—determine their order—along a gene, just as the fruit fly pioneers had mapped genes along a chromosome. Like Morgan, Benzer would have to depend on recombination to generate new genetic combinations, but, whereas Morgan had the advantage of a ready mechanism of recombination—the production of sex cells in a fruit fly—Benzer had to induce recombination by simultaneously infecting a single bacterial host cell with two different strains of bacteriophage, which differed by one or more mutations in the region of interest. Within the bacterial cell, recombination—the exchange of segments of molecules—would occasionally occur between the different viral DNA molecules, producing new permutations of mutations—so-called "recombinants." Within a single astonishingly produc-

53 我们在 4 月初将发现双螺旋的报告交给《自然》(*Nature*)杂志，并于 3 个星期后，也就是 1953 年 4 月 25 日刊出。同期还有两篇由富兰克林与威尔金斯执笔、篇幅较长的论文，两篇文章都认为我们的模型大致正确。6 月，我在冷泉港实验室的病毒研讨会上，首次就我们的模型提出报告。德尔布吕克亲自出马确保了我在最后一刻受邀发表演说。我带了在卡文迪什制造的三维模型，参加这场精英荟萃的会议，我使用的腺嘌呤-胸腺嘧啶碱基对模型是红色，而鸟嘌呤-胞嘧啶碱基对则是绿色。

54 同样受到薛定谔那本著作感召的前物理学家本泽(Seymour Benzer)也出席了这场盛会。他立刻了解到我们的突破性发现，对于他的病毒突变研究具有重要意义。他发现现在他能够以“摩根的孩子们”在 40 年前研究果蝇染色体的方式，来研究一小段的噬菌体 DNA：他可以在基因图上标出突变位置，就好像当年研究果蝇的先驱们在染色体上标出基因的位置。本泽跟摩根一样，也得靠重组作用来产生新的遗传组合。不过摩根可以利用现成的重组机制，即果蝇性细胞的产物。本泽则得用两种不同的噬菌体同时感染一个的细菌宿主细胞，借此引发重组机制。这两种不同的噬菌体在重要区段有一个或多个突变的差异。在细菌细胞里，重组作用(分子片段的交换)有时会发生在不同的病毒 DNA 分子之间，产生新的突变置换，即所谓的重组体(recombinant)。在普渡大学(Purdue University)的实验室里，本泽短短一年内就有惊人的丰富成果，他制作了噬菌体 rⅡ基因的图谱，显示出一连串的突变(遗传脚本上的所有错误)在病毒 DNA 上的线状排列方式。这种“生命语言”既简单又呈直线形状，就像书页上的一行文字。对于我在冷泉港所发表有关双螺旋的演讲，匈牙利物理学家齐拉特(Leo Szilard)的反应跟学术比较无关。他问

tive year in his Purdue University lab, Benzer produced a map of a single bacteriophage gene, rII, showing how a series of mutations—all errors in the genetic script—were laid out linearly along the virus DNA. The language was simple and linear, just like a line of text on the written page. The response of the Hungarian physicist Leo Szilard to my Cold Spring Harbor talk on the double helix was less academic. His question was, "Can you patent it?" At one time Szilard's main source of income had been a patent that he held with Einstein, and he had later tried unsuccessfully to patent with Enrico Fermi the nuclear reactor they built at the University of Chicago in 1942. But then as now patents were given only for useful inventions and at the time no one could conceive of a practical use for DNA. Perhaps then, Szilard suggested, we should copyright it.

55 There remained, however, a single missing piece in the double helical jigsaw puzzle: our unzipping idea for DNA replication had yet to be experimentally verified. Max Delbrück, for example, was unconvinced. Though he liked the double helix as a model, he worried that unzipping it might generate horrible knots. Five years later, a former student of Pauling's, Matt Meselson, and the equally bright young phage worker Frank Stahl put to rest such fears when they published the results of a single elegant experiment.

56 They had met in the summer of 1954 at the Marine Biological Laboratory at Woods Hole, Massachusetts, where I was then lecturing, and agreed—over a good many gin martinis—that they should get together to do some science. The result of their collaboration has been described as "the most beautiful experiment in biology".

57 They used a centrifugation technique that allowed them to sort molecules according to slight differences in weight; following a centrifugal spin,

我："你能申请专利吗？"有一阵子，齐拉特的主要收入来源是他和爱因斯坦共享的一项专利，后来他还试图和费米（Enrico Fermi）一起申请1942年他们在芝加哥大学所建造的核子反应炉专利，但并未成功。不过当时和现在一样，专利只授予有用的发明，当时没有人想到DNA有什么实际用途。于是，齐拉特建议，我们应该申请版权。

55 然而，在双螺旋的拼图中，还是少了一块：对于DNA"像拉链一样拉开"的复制方式，我们还需要以实验来证实。德尔布吕克就不相信我们的概念，他喜欢双螺旋的模型，但是担心把双螺旋拉开，会产生可怕的打结情况。5年后，泡令以前的学生梅索森（Matt Meselson）及同样聪明的研究噬菌体的年轻学者史塔尔（Frank Stahl），发表了一个简单明了的实验结果，一举扫除了这类疑虑。

56 1954年夏天，这两人在马萨诸塞州伍兹霍尔（Woods Hole）的海洋生物实验室结识，当时我在那里讲学。喝了不少马丁尼之后，两人决定他们应该搭挡研究科学。他们的合作结果赢得了"生物学上最完美的实验"之称。

57 他们使用离心技术，按照重量的些微差异来分离分子。在离心旋转后，较重的分子会比较轻的分子落到更接近试管底部之处。氮原子（N）是

heavier molecules end up nearer the bottom of the test tube than lighter ones. Because nitrogen atoms (N) are a component of DNA, and because they exist in two distinct forms, one light and one heavy, Meselson and Stahl were able to tag segments of DNA and thereby track the process of its replication in bacteria. Initially all the bacteria were raised in a medium containing heavy N, which was thus incorporated in both strands of the DNA. From this culture they took a sample, transferring it to a medium containing only light N, ensuring that the next round of DNA replication would have to make use of light N. If, as Crick and I had predicted, DNA replication involves unzipping the double helix and copying each strand, the resultant two "daughter" DNA molecules in the experiment would be hybrids, each consisting of one heavy N strand (the template strand derived from the "parent" molecule) and one light N strand (the one newly fabricated from the new medium). Meselson and Stahl's centrifugation procedure bore out these expectations precisely. They found three discrete bands in their centrifuge tubes, with the heavy-then-light sample halfway between the heavy-heavy and light-light samples. DNA replication works just as our model supposed it would.

58 The biochemical nuts and bolts of DNA replication were being analyzed at around the same time in Arthur Romberg's laboratory at Washington University in St. Louis. By developing a new, "cell-free" system for DNA synthesis, Kornberg discovered an enzyme (DNA polymerase) that links the DNA components and makes the chemical bonds of the DNA backbone. Romberg's enzymatic synthesis of DNA was such an unanticipated and important event that he was awarded the 1959 Nobel Prize in Physiology or Medicine, less than two years after the key experiments. After his prize was announced, Romberg was photographed holding a copy of the double helix model I had taken to Cold Spring Harbor in 1953.

DNA 的成分之一，而且原本就有两种不同的形态，一种较轻，另一种较重，因此梅索森和史塔尔能借由标注 DNA 片段，追踪 DNA 在细菌里的复制过程。起初所有的细菌都在含有重氮的培养基里培养，让重氮进入 DNA 的双股上。然后他们从这个培养菌中取出样本，转移至仅含轻氮的培养基，确保下一次 DNA 在复制时，只会用到轻氮。如果 DNA 的复制如同我和克里克的预测，是将双螺旋拉开，然后各复制一股，那么实验所制造出的两个“子代”DNA 分子将会是混种，每个分子都包括一个重氮股(来自“亲代”分子的模板股)，以及一个轻氮股(用培养基制造出来的新股)。梅索森与史塔尔的离心处理程序完全证实了这种预测。他们发现离心试管分离出三个明显的区段，分别是重—重、轻—轻，以及介于两者之间、先重后轻的样本。DNA 的复制方式跟我们的模型所预测的完全相符。

58 大约同一时间，酶学专家科恩伯格(Arthur Kornberg)位于圣路易市华盛顿大学的实验室也在分析 DNA 复制过程的生化细节。科恩伯格发展出一种新的合成 DNA 的无细胞系统，从而发现了一种称做 DNA 聚合肽(polymerase)的酶，这种酶连接构成 DNA 的不同小单元，形成 DNA 骨干的化学链。科恩伯格发现 DNA 酶合成机制，是惊人的重要大事，因此在这些实验完成后不到两年，他就于 1959 年获得了诺贝尔生理医学奖的殊荣。在公布得奖后，科恩伯格拿着我在 1953 年带到冷泉港实验室的双螺旋模型复制品摄影留念。

59 It was not until 1962 that Francis Crick, Maurice Wilkins, and I were to receive our own Nobel Prize in Physiology or Medicine. Four years earlier, Rosalind Franklin had died of ovarian cancer at the tragically young age of thirty-seven. Before then Crick had become a close colleague and a real friend of Franklin's. Following the two operations that would fail to stem the advance of her cancer, Franklin convalesced with Crick and his wife, Odile, in Cambridge.

60 It was and remains a long-standing rule of the Nobel Committee never to split a single prize more than three ways. Had Franklin lived, the problem would have arisen whether to bestow the award upon her or Maurice Wilkins. The Swedes might have resolved the dilemma by awarding them both the Nobel Prize in Chemistry that year. Instead, it went to Max Perutz and John Kendrew, who had elucidated the three-dimensional structures of hemoglobin and myoglobin respectively.

61 The discovery of the double helix sounded the death knell for vitalism. Serious scientists, even those religiously inclined, realized that a complete understanding of life would not require the revelation of new laws of nature. Life was just a matter of physics and chemistry, albeit exquisitely organized physics and chemistry. The immediate task ahead would be to figure out how the DNA-encoded script of life went about its work. How does the molecular machinery of cells read the messages of DNA molecules? As the next chapter will reveal, the unexpected complexity of the reading mechanism led to profound insights into how life first came about.

59 直到1962年，克里克、威尔金斯和我才获得诺贝尔生理学或医学奖。在此4年前，富兰克林已经因卵巢癌而不幸早逝，享年37岁。那时克里克已和她成为亲密的同事与真正的好友。富兰克林动了两次手术，但都未能遏阻癌细胞蔓延，之后她曾回到剑桥，在克里克和他太太欧蒂莉(Odile)的照顾下疗养。

60 诺贝尔委员会向来不曾将单一奖项颁给超过三个人，倘若当时富兰克林还在人世，他们势必得面对要将这个奖颁给她或威尔金斯的问题。瑞典人可能会授予他们诺贝尔化学奖以解决这个难题。最后，化学奖颁给了佩鲁茨和肯德鲁，他们分别发现了血红素与肌红素的三维结构。

61 双螺旋的发现敲响了生机论的丧钟。认真的科学家，甚至有宗教信仰的科学家都已发现，要对生命有完整的了解，不需要寻找新的自然定律。生命不过就是物理与化学——尽管是极为精密复杂的物理与化学。接下来的工作是要找出，生命如何上演出隐藏在DNA中的密码脚本。细胞的分子机器如何读取DNA分子携带的信息？下一章①将介绍，极度复杂的读取机制如何引导我们深入了解生命的形成。

① 此处的“下一章”指的是[美]詹姆斯·沃森著，陈雅云译，上海人民出版社2007年出版的《DNA：生命的秘密》的第3章，而非本导引的下一章。

《惊人的假说》导引

正如美国著名哲学家 Nagel 所言："意识使身心问题变得棘手，如果没有意识，心身问题就会索然无味，可一旦有了意识，要想搞清楚它又渺无希望。"千百年来，有太多的哲人围绕意识这一谜题开展了系统的思考。早在古希腊时期，柏拉图就提出，人是囚禁在必有一死的肉体之中的不死灵魂，而意识正来源于不死灵魂。到了近代，笛卡儿秉承这种二元论的学说，将广延实体(res extensa)和思维实体(res cogitans)区分开，认为广延实体是空间上的物理实体，而思维实体是人类独有，是意识的来源。但貌似逻辑严密的二元论却难以得到现实的证实，很多学者认为这种"机器中的幽灵"不属于科学探讨的范畴。

然而，这一观点随着《惊人的假说》的问世受到了巨大的动摇。《惊人的假说》是由英国著名的生物学家、诺贝尔生理学或医学奖得主弗朗西斯·克里克(Francis Crick)所著。克里克于 1933 年考入伦敦大学学院学习物理，辅修数学，1937 年获理学学士学位，并进入伦敦大学攻读博士，其博士研究的主题为"负责测量水的黏度"。但是第二次世界大战爆发，克里克中止了他的博士学习，并在战争结束后转向生物学的研究。据克里克所言："我的间接收获是物理学家的自豪感，感到物理学作为一种学科来说非常成功。为什么其他学科不能也这样呢？这促使我战后终于转到生物物理。这种感觉也有助于矫正我和生物学家交往时遇到的沉闷而谨慎的态度。”当时，有两个重大问题科学无法回答，一是生物如何与非生物世界相区别；二是意识的生物学本质究竟是什么。克里克首先对第一个问题进行了深入探索，他于 1951 年结识了 23 岁的詹姆斯·沃森(James Watson)，两人在剑桥大学卡文迪什实验室并肩探索 DNA 的结构，并于 1953 年 4 月在《自然》杂志发表了关于 DNA 双螺旋结构模型的论文，也由此共同获得 1962 年诺贝尔生理学奖。然而克里克并未止步于此，1976 年(时年 60 岁)开始了第二个重大问题的研究，探索意识的生物学本质，直至他生命的终结。

克里克对于意识的探索成果以《惊人的假说》命名出版。全书共分三篇十八章。第一篇论述了意识研究的前科学思想、意识的一般性质及当代视觉心理学的基本理论。第二篇主要针对大脑及视觉系统的神经生物学常识进行了介绍。前两篇为进入第三篇提供了必要的背景知识，第三篇详细论述了各种可能的研究视觉意识的实验方法及初步的结论。《自然科学经典导引》节选了《惊人的假说》第二章、第十二章、第十四章内容，使读者能够通过阅读万字左右的摘录对该书以及该书所提出的意识的神经生物学本质问题有着初步、直接但较为全面的认识。

究竟“惊人的假说”如何惊人的呢？或者说，克里克对于意识的见解较之于前人有多大的不同呢？这两个问题是理解《惊人的假说》以及克里克在意识研究上学术贡献的关键。在此只需罗列出克里克的假说内容便能略窥一二。克里克认为：“你的喜悦、悲伤、记忆和抱负，你的本体感觉和自由意志，实际上都只不过是一大群神经细胞及其相关分子的集体行为。”简而言之，人不过是一对神经元的集合，没有臆想的灵魂，更没有所谓的不朽。抹掉了意识所具有的浪漫主义色彩，带来的是一种严肃的既视感。确实如克里克所说“它可以真正被认为是惊人的。”

任何的科学假说都需要通过科学的方法来进行验证，要么证实、要么证伪。克里克在抛出了“惊人的假说”后，也确实通过一系列的科学方法对该假说进行了初步的论证。简而言之，克里克进行意识探秘的方法主要有二：第一，通过还原论的方法，探讨在人们说自己有意识时，其大脑做了什么，即寻找意识的“神经关联物”。克里克认为：“在任意时刻意识将会与瞬间的神经元集合的特定活动类型相对应”。但与传统的机械还原论不同，克里克并不简单认为复杂系统的行为分解为部分，并来源于部分的行为，他认为“某一特定部分的行为可能需要用它的各个组成部分及其相互作用的特性加以解释”。第二，大力倡导和使用实验的研究方法。克里克认为，哲学家们倾向于从外部观察世界，而心理学家倾向于通过内省来分析意识。这些都无助于意识神经生物学本质的讨论。因此，必须通过实验“从神经元的角度考虑问题，考察它们的内部成分以及它们之间复杂的、出人意料的相互作用的方式，这才是问题的实质”。

基于科学的研究方法，克里克从视觉切入，研究并试图解决两大问题：第一是意识物质载体的确定问题；第二是意识的统一性问题，即不同感知所引起的神经元活动如何协调整合为一个意识。当克里克着手解决这两个问题之时，科学技术的发展也为他提供了有力的支持。脑电图、X 射线断层照相以及磁共振成像技术可以“打开”大脑，探测脑内电流活动，从而侦查到活动的神经元。据此，克里克首先通过不同的实验确定了大脑皮层各区域的神经元对不同视觉刺激的反应情况。同时，提出了“捆绑”理论来解释意识的统一性问题，即 40Hz 的同步神经元激发的作用可以解释不同的知觉系统捆绑不同的刺激成单一的意识。当然，“捆绑”理论的提出也似乎是克里克“脑洞”大开的结果。他在“捆绑”理论之后，加了一句注释“现在还不能完全肯定捆绑问题如我所说的那样存在，还是脑通过某种未知的技巧绕了过去”。但我们能够感受到，克里克在提倡一种脚踏实地的科学研究方法的同时，也在发挥其天马行空的想象力，尝试为困扰人类多年的科学难题提供可能的解决思路。在“惊人的假说”中，克里克也尝试着论证意识统一性的物质载体为大脑中的屏状核。据 2000 年诺贝尔生理学或医学奖得主埃里克 · 坎德尔（Eric Richard Kandel）在回忆录《追寻记忆的痕迹》中记载“在克里克离世前三个星期，我去他家中探望。他癌症晚期，正在进行化疗，很疼。然而，他仍然没停止他最后的研究课题。就在去世之前被送往医院的几个小时里，他还在修改一篇论文，该论文认为位于皮层下脑组织的屏状核是整合意识统一性的区域。”我们相信如果上天能够再给克里克多一些时间，他一定能够揭开意识更多的秘密。

当然，《惊人的假说》并没有完全解决意识的神经生物学本质问题，甚至从某种意义上讲《惊人的假说》仅仅提供了一个可从神经生物学角度研究意识的思路和基本范式，大量悬而未决的问题及答案尚隐匿在神经生物学的大门之内。例如，其他形式感觉的神经作用机制、无意识的神经作用机制等问题。回答诸如此类的问题颇为不易，甚至在克里克看来“意识的许多方面，如可感知的特性，完全有可能是科学所不能解释的”。然而，这并不意味着饱有热情的科学家们会停止他们的探索步伐，人类“已经学会了生活在认知的局限之中(例如量子力学的局限)，这种局限将常伴人类的生活”。正如柏拉图《理想国》中“洞喻”之景，洞中之人可能一直都不知道自己看到的究竟是事物的真相还是只是真理之光投射的阴影，但当他们坚信能够以某种方式(科学研究的方法)看到事实真相。正如克里克所言“现在，唯一的问题是如何着手去研究它以及何时开始。我极力主张应该现在立刻开始研究！”我确实相信，只要我们保持这种探索，我们迟早会达到这种理解。这一天或许很远，但我们越早开始我们就越早地得到对自然本质的清晰的认识！

（王传毅）

The Astonishing Hypothesis

Francis Crick

Chapter 2 The General Nature of Consciousness

"In any field find the strangest thing and then explore it."

—John Archibald Wheeler

1 To come to grips with the problem of consciousness, we first need to know what we have to explain. Of course, in a general way we all know what consciousness is like. Unfortunately that is not enough. Psychologists have frequently shown that our commonsense ideas about the workings of the mind can be misleading. The obvious first step, then, is to find out what psychologists over the years have considered to be the essential features of consciousness. They may not have got it quite right, but at least their ideas on the subject will provide us with a starting point.

2 Since the problem of consciousness is such a central one, and since consciousness appears so mysterious, one might have expected that psychologists and neuroscientists would now direct major efforts toward understanding it. This, however, is far from being the case. The majority of modern psychologists omit any mention of the problem, although much of what they study enters into consciousness. Most modern neuroscientists ignore it.

本文选自 Francis Crick 著 *The Astonishing Hypothesis*, 西蒙 · 休斯特出版公司 1995 年版。

惊人的假说

弗朗西斯·克里克

第二章 意识的本质

在任何一个领域内发现最神奇的东西，然后去研究它。

——惠勒(John Archibald Wheeler)

1 要研究意识问题，首先就要知道哪些东西需要我们去解释。当然，我们大体上知道什么是意识。但遗憾的是，仅仅如此是不够的。心理学家常向我们表明，有关心理活动的常识可能把我们引入歧途。显然，第一步就是要弄清楚多年来心理学家所认定的意识的本质特征。当然，他们的观点未必完全正确，但至少他们对此问题的某些想法将为我们提供一个出发点。

2 既然意识问题是如此的重要和神秘，人们自然会期望，心理学家和神经科学家应该把主要精力花在研究意识上。但事实远非如此。大多数现代心理学家都回避这一问题，尽管他们的许多研究都涉及意识。大多数现代神经科学家则完全忽略了这一问题。

本文选自[英]弗朗西斯·克里克：《惊人的假说》，汪云九等译，湖南科学技术出版社 2018 年版，第 15~26 页，第 180~197 页，第 228~242 页。

3 This was not always so. When psychology began as an experimental science, largely in the latter part of the nineteenth century, there was much interest in consciousness, even though it was admitted that the exact meaning of the word was unclear. The major method of studying it, especially in Germany, was by detailed and systematic introspection. It was hoped that psychology might become more scientific by refining introspection until it became a reliable technique.

4 The American psychologist William James (the brother of novelist Henry James) discussed consciousness at some length. In his monumental work *The Principles of Psychology*, first published in 1890, he described five properties of what he called "thought." Every thought, he wrote, tends to be part of personal consciousness. Thought is always changing, is sensibly continuous, and appears to deal with objects independent of itself. In addition, thought focuses on some objects to the exclusion of others. In other words, it involves attention. Of attention he wrote, in an oft-quoted passage: "Everyone knows what attention is. It is the taking possession by the mind, in clear and vivid form, of one out of what seem several simultaneously possible objects or trains of thought. ... It implies withdrawal from some things in order to deal effectively with others."

5 In the nineteenth century we can also find the idea that consciousness is closely associated with memory. James quotes the Frenchman Charles Richet, who wrote in 1884 that "to suffer for only a hundredth of a second is not to suffer at all; and for my part I would readily agree to undergo a pain, however acute and intense it might be, provided it should last only a hundredth of a second, and leave after it neither reverberation nor recall."

3 情况也并非总是这样。大约在19世纪后期，当心理学开始成为一门实验科学的时候，就有许多人对意识问题怀有极大的兴趣。尽管这个词的确切含义当时还不太清楚。那时研究意识的主要方法就是进行详细的、系统的内省，尤其是在德国。人们希望，在内省成为一项可靠的技术之前，通过对它的精心改进而使心理学变得更加科学。

4 美国心理学家威廉·詹姆斯(William James)(与小说家亨利·詹姆斯是兄弟)较详尽地讨论了意识问题，在他1890年首次出版的巨著《心理学原理》中，他描述了被他称为“思想”(thought)的五种特性。他写道，每一个思想都是个人意识的一部分。思想总是在变化之中，在感觉上是连续的，并且似乎可以处理与自身无关的问题。另外，思想可以集中到某些物体而舍弃其他物体。换句话说，它涉及注意。关于注意，他写下了这样一段经常被人引用的话：“每个人都知道注意是什么，它以清晰和鲜明的方式，利用意向从若干个同时可能出现的物体或一系列思想中选取其中的一个……这意味着舍掉某些东西以便更有效地处理另外一些。”

5 在19世纪，我们还可以发现意识与记忆紧密联系的想法。詹姆斯曾引用法国人查尔斯·理迟特(Charles Richet)1884年发表的一段话：“片刻的苦痛微不足道。对我而言，我宁愿忍受疼痛，哪怕它是剧烈的，只要它持续的时间很短，而且，在疼痛过去之后，永远不再出现并永远从记忆中消失。”

6 Not all operations of the brain were thought to be conscious. Many psychologists believed that some processes are subliminal or subconscious. For example, Hermann von Helmholtz, the nineteenth-century German physicist and physiologist, often spoke of perception in terms of "unconscious inference." By this he meant that perception was similar in its logical structure to what we normally mean by inference, but that it was largely unconscious.

7 In the early twentieth century the concepts of the preconscious and the unconscious were made widely popular, especially in literary circles, by Freud, Jung, and their associates, mainly because of the sexual flavor they gave to them. By modern standards, Freud can hardly be regarded as a scientist but rather as a physician who had many novel ideas and who wrote persuasively and unusually well. He became the main founder of the new cult of psychoanalysis.

8 Thus, as long as one hundred years ago we see that three basic ideas were already current:

(1) Not all the operations of the brain correspond to consciousness.

(2) Consciousness involves some form of memory, probably a very short term one.

(3) Consciousness is closely associated with attention.

9 Unfortunately, a movement arose in academic psychology that denied the usefulness of consciousness as a psychological concept. This was partly because experiments involving introspection did not appear to be leading anywhere and partly because it was hoped that psychology could become more scientific by studying behavior (in particular, animal behavior) that could be observed unambiguously by the experimenter. This was the

6 并非脑的全部操作都是有意识的。许多心理学家相信，存在某些下意识或潜意识的过程。例如，19世纪德国物理学家和生理学家赫尔曼·冯·亥姆霍兹(Hermann von Helmholtz)在谈到知觉时就经常使用“无意识推论”这种术语。他想借此说明，在逻辑结构上，知觉与通常推论所表达的含义类似，但基本上又是无意识的。

7 20世纪初期，潜意识和无意识的概念变得非常流行，特别是在文学界。这主要是因为弗洛伊德(Freud)、荣格(Jung)及其合作者给文学赋予了某种性的情趣。按现代的标准看，弗洛伊德不能算作科学家，而应该被视为既有许多新思想、又有许多优秀著作的医生。正因为如此，他成为精神分析学派的奠基人。

8 早在100年前，3个基本的观点就已经盛行：

(1)并非大脑的全部操作都与意识有关；

(2)意识涉及某种形式的记忆，可能是极短时的记忆；

(3)意识与注意有密切的关系。

9 但不幸的是，心理学研究中兴起了一场运动，它否定意识的应用价值，把它看成是一个纯心理学概念。产生这场运动的一方面原因是由于涉及内省的实验不再是研究的主流；另一方面，人们希望通过研究行为，特别是动物的行为，使心理学研究更具科学性。因为，对实验者而言，行为实验具有确定的观察结果。这就是行为主义运动，它回避谈论精神事件。一切行为都必须用刺激和反应去解释。

Behaviorist movement. It became taboo to talk about mental events. All behavior had to be explained in terms of the stimulus and the response.

10 Behaviorism was especially strong in the United States, where it was started by John B. Watson and others before World War I. It flourished in the 1930s and 1940s when B. F.Skinner was its most celebrated exponent. Other schools of psychology existed in Europe, such as the Gestalt school, but it was not until the rise of cognitive psychology in the late 1950s and 1960s that it became intellectually respectable, at least in the United States, for a psychologist to talk about mental events. It then became possible to study visual imagery for example, and to postulate psychological models for various mental processes, usually based on concepts used to describe the behavior of digital computers. Even so, consciousness was seldom mentioned and there was little attempt to distinguish between conscious and unconscious activity in the brain.

11 Much the same was true of neuroscientists studying the brains of experimental animals. Neuroanatomists worked almost entirely on dead animals (including human beings) while neurophysiologists largely studied anesthetized animals who are not conscious. For example, they do not feel any pain during such experiments. This was especially true after the epoch-making discovery by the neurobiologists David Hubel and Torsten Wiesel in the late fifties. They found that nerve cells in the visual cortex of the brain of an anesthetized cat showed a whole series of interesting responses when light was shone on the cat's opened eyes, even though its brain waves showed it to be more asleep than awake. For this and subsequent work they were awarded a Nobel Prize in 1981.

10　约翰·沃森(John B. Watson)等人在“第一次世界大战”前发起的这场行为主义运动，在美国盛行一时。并且，由于以斯金纳(B. F. Skinner)为代表的许多著名鼓吹者的影响，该运动在20世纪三四十年代达到顶峰。尽管在欧洲还存在以格式塔(Gestalt)为代表的心理学派，但至少在美国，在20世纪50年代后期和20世纪60年代认知心理学成为受科学界尊重的学科之前，心理学家从不谈论精神事件。在此之后，才有可能去研究视觉意象，并且在原来用于描述数字计算机行为的概念基础之上，提出各种精神过程的心理学模型。即便如此，意识还是很少被人提及，也很少有人去尝试区分脑内的有意识和无意识活动。

11　神经科学家在研究实验动物的大脑时也是如此。神经解剖学几乎都是研究死亡后的动物(包括人类)，而神经生理学家大多只研究麻醉后丧失意识的动物。此时受试对象已不可能具有任何痛苦的感觉了。特别是20世纪50年代后期，戴维·休伯(David Hubel)和托斯滕·威塞尔(Torsten Wiesel)有了划时代的发现以后，情况更是如此。他们发现，麻醉后的猫大脑视皮质上的神经细胞，对入射到其眼内的光照模式呈现一系列有趣的反应特性。尽管脑电波显示，此时猫处于睡眠而非清醒的状态。由于这一发现及其后续的工作，他们获得了1981年诺贝尔奖。

12 It is far more difficult to study the response of these brain cells in an alert animal (not only must its head be restrained but eye movements must either be prevented or carefully observed). For this reason, very few experiments have been done to compare the responses of the same brain cell to the same visual signals under two conditions: when the animal is asleep and again when it is awake. Not only have neuroscientists traditionally avoided the problem of consciousness because of these experimental difficulties but also because they considered the problem both too subjective and too "philosophical," and thus not easily amenable to experimental study. It would not have been easy for a neuroscientist to get a grant for the purpose of studying consciousness.

13 Physiologists are still reluctant to worry about consciousness, but in the last few years a number of psychologists have begun to address the matter. I will sketch briefly the ideas of three of them. What they have in common is a neglect or, at best, a distant interest in nerve cells. Instead, they hope to contribute to an understanding of consciousness mainly by using the standard methods of psychology. They treat the brain as an impenetrable "black box" of which we know only the outputs—the behaviors it produces—caused by various inputs, such as those signalled by the senses. And they construct models that use general ideas based on our commonsense understanding of the mind, which they express in engineering or computing terms. All three authors would probably describe themselves as cognitive scientists.

14 Philip Johnson-Laird, now a professor of psychology at Princeton University, is a distinguished British cognitive scientist with a major interest in language, especially in the meaning of words, sentences, and narratives. These are issues unique to humans. It is not surprising that Johnson-Laird pays little attention to the brain, since much of our detailed information about the

12 要研究清醒状态下动物脑神经反应的特性，是一件更加困难的事情(此时不仅需要约束头部运动，还要禁止眼动或详细记录眼动)。因此，很少有人做比较同一个大脑细胞在清醒和睡眠两种状态下，对同一视觉信号的反应特性的实验。传统的神经科学家回避意识问题，这不仅仅是因为有实验上的困难，还因为他们认为这一问题太具哲学意味，很难通过实验加以观测。一个神经科学家要想专门去研究意识问题，很难获得资助。

13 生理学家们至今还不大关心意识问题，但在近几年，某些心理学家开始涉及这一问题。我将简述一下他们中的3个人的观点。他们的共同点，就是忽视神经细胞或者说对它们缺少兴趣。相反，他们主要想用标准的心理学方法对理解意识作出贡献。他们把大脑视为一个不透明的“黑箱”，我们只知道它的各种输入(如感觉输入)所产生的输出(它产生的行为)。他们根据对精神的常识性了解和某些一般性概念建立模型。该模型使用工程和计算术语表达精神。上述3个作者也许会标榜自己是认知科学家。

14 现任普林斯顿大学心理系教授的菲力普·约翰逊-莱尔德(Philip Johnson-Laind)是一位杰出的英国认知心理学家。他主要的兴趣是研究语言，特别是字、语句和段落的意义。这是人类才有的问题。莱尔德不大注意大脑，这是不足为奇的。因为我们有关灵长类大脑的主要信息是从猴子身上获得的，而它们并没有真正的语言。他的两部著作《心理模型》(*Mental Models*)和《计算机与思维》(*The Computer and the Mind*)的着眼

primate brain is derived from monkeys and they have no true language. His two books, *Mental Models* and *The Computer and the Mind*, are concerned with the problem of how to describe the mind—the activities of the brain—and the relevance of modern computers to that description. He stresses that the brain is, as we shall see, a highly parallel mechanism (meaning that millions of processes are going on at the same time) and that we are unconscious of much of what it does.

15 Johnson-Laird believes that any computer, and especially a highly parallel computer, must have an operating system that controls the rest of its functions, even though it may not have complete control over them. He proposes that it is the workings of this operating system that correspond most closely to consciousness and that it is located at a high level in the hierarchies of the brain.

16 Ray Jackendoff, professor of linguistics and cognitive science at Brandeis University, is a well-known American cognitive scientist with a special interest in language and music. Like most cognitive scientists, he believes that the mind is best thought of as a biological information-processing system. However, he differs from many of them in that he regards, as one of the most fundamental issues of psychology, the question "What makes our conscious experience the way it is?"

17 His Intermediate-Level Theory of Consciousness states that awareness is derived neither from the raw elements of perception nor from high-level thought but from a level of representation intermediate between the most peripheral (sensationlike) and the most central (thoughtlike). He rightly emphasizes that this is a quite novel point of view.

点放在怎样描述精神的问题(大脑的活动)以及现代计算机与这一描述的关系。他强调，大脑具有高度并行的机制(即数以万计的过程可以同时进行)，但它做的多数工作我们是意识不到的。①

15 约翰逊-莱尔德确信，任何一台计算机，特别是高度并行的计算机，必须有一个操作系统用以控制(即使不是彻底地控制)其余部分的工作。他认为，操作系统的工作与位于脑的高级部位的意识存在着紧密的联系。

16 布兰迪斯大学(Brandeis University)语言学和认知学教授雷·杰肯道夫(Ray Jackendoff)是一位著名的美国认知科学家。他对语言和音乐有特殊的兴趣。与大多数认知科学家类似，他认为最好把脑视为一个信息加工系统。但与大多数科学家不同的是，他把"意识是怎样产生的"看作心理学的一个最基本的问题。

17 他的意识的中间层次理论(intermediate-level theory of consciousness)认为，意识既不是来自未经加工的知觉单元，也不是来自高层的思想，而是来自介于最低的周边(类似于感觉)和最高的中枢(类似于思想)之间的一种表达层次。他恰当地突出了这个十分新颖的观点。

① 约翰逊-莱尔德尤其对自我反应和自我意识感兴趣。出于策略上的考虑，这些问题先放在一边。

18 Like Johnson-Laird, Jackendoff has been strongly influenced by the analogy of the brain to a modern computer. He points out that this analogy provides some immediate dividends. For example, in a computer, a great deal of the information is stored but only a small part is active at any one time. The same is true of the brain.

19 However, not all of the activity in the brain is conscious. He thus makes a distinction not merely between the brain and the mind but between the brain, the computational mind, and what he calls the "phenomenological mind," meaning (roughly) what we are conscious of. He agrees with Johnson-Laird that what we are conscious of is the *result* of computations rather than the computations themselves.

20 He also believes there is an intimate connection between awareness and short-term memory. He expresses this by saying that "awareness is supported by the contents of short-term memory," going on to add that short-term memory involves "fast" processes and that slow processes have no direct phenomenological effect.

21 As to attention, he suggests that the computational effect of attention is that the material being attended to is undergoing especially intensive and detailed processing. He believes that this is what accounts for attention's limited capacity.

22 Jackendoff and Johnson-Laird are both functionalists. Just as it is not necessary to know about the actual wiring of a computer when writing programs for it, so a functionalist investigates the information processed by the brain, and the computational processes the brain per forms on this information, without considering the neurological implementation of these

18 与约翰逊-莱尔德类似，杰肯道夫在很大程度上也受到脑和现代计算机之间类比的影响。他指出，这种类比可以带来某些直接的好处。比如，计算机中存储了大量信息，但在某一时刻，只有一小部分信息处于活动状态。大脑中亦是如此。

19 然而，并非大脑的全部活动都是有意识的。因此，他不仅在脑和思维之间，而且在脑(计算思维)与所谓的现象学思维(大体指我们所能意识到的)之间作了严格的区分。他同意莱尔德的观点——我们意识到的只是计算的结果，而非计算本身。①

20 他还认为，意识与短时记忆存在紧密的联系。他所说的“意识需要短时记忆的内容来支持”就表达了这样一种观点。但还应补充的是，短时记忆涉及快速过程，而慢变化过程没有直接的现象学效应。

21 谈到注意时他认为，注意的计算效果就是使被注意的材料经历更加深入和细致的加工。他认为这样就可以解释为何注意容量如此有限。

22 杰肯道夫与约翰逊-莱尔德都是功能主义者。正如在编写计算机程序时并不需要了解计算机的实际布线情况一样，功能主义者在研究大脑的信息加工和大脑对这些信息执行的计算过程时，并没有考虑到这些过程的神经生物学实现机制。他们认为，这种考虑是无关紧要的，至少目前考

① 杰肯道夫用自己的行话表达这一点。他把我称为“结果”的东西叫作“信息结构”。

processes. He usually regards such considerations as totally irrelevant or, at best, premature.

23 This attitude does not help when one wants to *discover* the workings of an immensely complicated apparatus like the brain. Why not look inside the black box and observe how its components behave? It is not sensible to tackle a very difficult problem with one hand tied behind one's back. When, eventually, we know in some detail how the brain works, then a high-level description (which is what functionalism is) may be a useful way to think about its overall behavior. Such ideas can always be checked for accuracy by using detailed information from lower levels, such as the cellular or molecular levels. Our provisional high-level descriptions should be thought of as rough guides to help us unravel the intricate operations of the brain.

24 Bernard J. Baars, an institute faculty professor at the Wright Institute in Berkeley, California, has written a book entitled *A Cognitive Theory of Consciousness*. Although Baars is a cognitive scientist, he is rather more interested in the human brain than either Jackendoff or Johnson-Laird.

25 He calls his basic idea the Global Workspace. He identifies the information that exists at any one time in this workspace as the content of consciousness. The workspace, which acts as a central informational exchange, is connected to many unconscious receiving processors. These specialists are highly efficient in their own domains but not outside them. In addition, they can cooperate and compete for access to the workspace. Baars elaborates this basic model in several ways. For example, the receiving processors can reduce uncertainty by interacting until they reach agreement on only one active interpretation.

虑它们为时过早。①

23 然而，在试图揭示像大脑这样一个极端复杂的装置的工作方式时，这种态度并没有什么好处。为什么不打开黑箱去观察其中各单元的行为呢？处理一个复杂问题时，把一只手捆在背后是不明智的。一旦我们了解了大脑工作的某些细节，功能主义者关心的高层次描述就会成为考虑大脑整体行为的有用方法。这种想法的正确性可以用由低水平的细胞和分子所获得的详细资料精确地加以检验。高水平的尝试性描述应当被看作帮助我们阐明大脑的复杂操作的初步向导。

24 加利福尼亚州伯克利的赖特研究所的伯纳德·巴尔斯(Bemard J. Baars)教授写了《意识的认知理论》一书。虽然巴尔斯也是一位认知科学家，但与杰肯道夫或约翰逊-莱尔德相比，他更关心人的大脑。

25 他把自己的基本思想称为全局工作空间(global workspace)。他认为，在任一时刻存在于这一工作空间内的信息都是意识的内容。作为中央信息交换的工作空间，它与许多无意识的接收处理器相联系。这些专门的处理器只在自己的领域之内具有高效率。此外，它们还可以通过协作和竞争获得工作空间。巴尔斯以若干种方式改进了这一模型。例如，接收处理器可以通过相互作用减小不确定性，直到它们符合一个唯一有效的解释。②

① 遗传学也关心各代之间和个体内部的信息传递。但真正的突破是在DNA结构把该习语所表达的信息显示得一清二楚之后。

② 我不想赘述巴尔斯模型的所有复杂性。为了解释意识问题的各个方面，如自我意识、自我监控以及其他一些心理活动，如无意识的断章取义、意志、催眠等，他的模型附加了许多复杂性。

26 In broader terms, he considers consciousness to be profoundly active and that there are attentional control mechanisms for access to consciousness. He believes we are conscious of some items in shortterm memory but not all.

27 These three cognitive theoreticians share a loose consensus on three points about the nature of consciousness. They all agree that not all the activities of the brain correspond directly with consciousness, and that consciousness is an active process. They all believe that attention and some form of short-term memory is involved in consciousness. And they would probably agree that the information in consciousness can be fed into both long-term episodic memory as well as into the higher, planning levels of the motor system that controls voluntary movements. Beyond that their ideas diverge somewhat.

28 Let us keep all three sets of ideas in mind while exploring an approach that tries to see what we can learn by combining them with our growing knowledge of the structure and activity of the nerve cells in the brain.

29 Most of my own ideas on consciousness have been developed in collaboration with a younger colleague, Christof Koch, now a professor of computation and neural systems at the California Institute of Technology (Caltech). Christof and I have known each other from the time in the early eighties when he was a graduate student of Tomaso Poggio in Tübingen. Our approach is essentially a scientific one. We believe that it is hopeless to try to solve the problems of consciousness by general philosophical arguments; what is needed are suggestions for new experiments that might throw light on these problems. To do this we need a tentative set of theoretical ideas that may have to be modified or discarded as we proceed. It is characteristic of a scientific approach that one does not try to construct some all-embracing theory that purports to explain *all* aspects of consciousness. Nor does such an

26 广义上讲，他认为意识是极为活跃的，而且注意到控制机制可进入意识。我们意识到的是短时记忆的某些项目而非全部。

27 这三位认知理论家对意识的属性大致达成了三点共识。他们都同意并非大脑的全部活动都直接与意识有关，而且意识是一个主动的过程；他们都认为意识过程有注意和某种形式的短时记忆参与；他们大概也同意，意识中的信息既能够进入长时情景记忆(long-term episodic memory)中，也能进入运动神经系统(motor system)的高层计划水平，以便控制随意运动。除此之外，他们的想法存在着这样或那样的分歧。

28 让我们把这三点想法铭记在心，并将它们与我们日益增长的脑内神经细胞的结构和活动的知识结合起来，看看这样的研究方法能够得到什么结果。

29 我自己的大多数想法是在与我的年轻同事——加州理工学院计算与神经系统副教授克里斯托弗·科赫(Christof Koch)的合作研究中形成的。科赫与我相识于20世纪80年代初，当时他还是托马索·波吉奥(Tomaso Poggio)在蒂宾根(Tübingen，德国城市)的研究生。我们的探索在本质上是科学的。① 我们认为，泛泛的哲学争论无助于解决意识问题。真正需要的是提出有希望解决这些问题的新的实验方法。为了做到这一点，我们还需要一个尝试性的思想体系，它随着我们工作的进展不断加以改进和扬弃。一个科学方法的特点应是不试图建立包罗万象的理论，从而一下子解释意识问题的所有方面。也不能把研究的重点放在语言上，因为只有人类才有语言。应选择在当时看来对研究意识最有利的系统，并

① 下文我将广泛引述科赫和我于1990年在《神经科学研讨》(*Seminars in the Neurosciences*, SIN)杂志[b]上发表的一篇关于该问题的文章中的思想。

approach concentrate on studying language simply because it is uniquely human. Rather, one tries to select the most favorable system for the study of consciousness, as it appears at this time, and study it from as many aspects as possible. In a battle, you do not usually attack on all fronts. You probe for the weakest place and then concentrate your efforts there.

30 We made two basic assumptions. The first is that there is something that requires a scientific explanation. There is general agreement that people are not conscious of all the processes going on in their heads, although exactly which might be a matter of dispute. While you are aware of many of the results of perceptual and memory processes, you have only limited access to the processes that produce this awareness (e.g., "How did I come up with the first name of my grandfather?"). In fact, some psychologists have suggested that you have only very limited introspective access to the origins of even higher order cognitive processes. It seems probable, however, that at any one moment some active neuronal processes in your head correlate with consciousness, while others do not. *What are the differences between them*?

31 Our second assumption was tentative: that all the different aspects of consciousness, for example pain and visual awareness, employ a basic common mechanism or perhaps a few such mechanisms. If we could understand the mechanisms for one aspect, then we hope we will have gone most of the way to understanding them all. Paradoxically, consciousness appears to be so odd and, at first sight, so difficult to understand that only a rather special explanation is likely to work. The general nature of consciousness may be easier to discover than more mundane operations, such as how the brain processes information so that you see in three dimensions, which can, in principle, be explained in many different ways. Whether this is really true remains to be seen.

从尽可能多的方面加以研究。正如在战争中，通常并不采取全面进攻，而是往往找出最薄弱的一点，集中力量加以突破。

30 我们作出了两条基本假设。第一条就是我们需要对某件事情作出科学解释。尽管对哪些过程能够意识到还可能有争议，但大家基本认同的是，人们不能意识到头脑中发生的全部过程。当你意识到许多知觉和记忆过程的结果时，你对产生该意识的过程可能了解很有限。(比如，“我如何想起了我祖父的名字呢?”)实际上，某些心理学家已经暗示，即使对较高级的认知过程的起源，你也只有很有限的内省能力。在任一时刻，可能都有某些活跃的神经过程与意识有关，而另一些过程与意识无关。它们之间的差别是什么呢?

31 我们的第二条假设是尝试性的：意识的所有不同方面，如痛觉和视觉意识(visual awareness)，都使用一个基本的共同机制或者也许几个这样的机制。如果我们能够了解其中某一方面的机制，我们就有希望借此了解其他所有方面的机制。自相矛盾的是，意识似乎如此古怪，初看起来又是如此费解，只有某种相当特殊的解释才有可能行得通。意识的一般本质也许比一些较常见的操作更容易被发现。像脑如何处理三维信息，在原则上可以用很多不同的方法去解释。这一点是否正确还有待于进一步观察。

32 Christof and I suggested that several topics should be set aside or merely stated outright without further discussion, for experience has shown that otherwise much valuable time can be wasted arguing about them.

33 Everyone has a rough idea of what is meant by consciousness. It is better to avoid a *precise* definition of consciousness because of the dangers of premature definition. Until the problem is understood much better, any attempt at a formal definition is likely to be either misleading or overly restrictive, or both.

34 Detailed arguments about what consciousness are probably premature, although such an approach may give useful hints about its nature. It is, after all, a bit surprising that one should worry too much about the function of something when we are rather vague about what it is. It is known that without consciousness you can deal only with familiar, rather routine, situations or respond to very limited information in new situations.

35 It is plausible that some species of animals—and in particular the higher mammals—possess some of the essential features of consciousness, but not necessarily all. For this reason, appropriate experiments on such animals may be relevant to finding the mechanisms underlying consciousness. It follows that a language system (of the type found in humans) is not essential for consciousness—that is, one can have the key features of consciousness without language. This is not to say that language may not considerably enrich consciousness.

36 It is not profitable at this stage to argue about whether "lower" animals, such as octopus, fruit flies, or nematodes, are conscious. It is probable, however, that consciousness correlates to some extent with the degree of complexity of

32 克里斯托弗和我认为，某些问题可以暂且放在一边或者只是无保留地陈述一遍，根本无需进一步讨论。因为，经验告诉我们，如果不是这样的话，很多宝贵的时间就会耗费在无休止的争论上。

33 关于什么是意识，每个人都有一个粗略的想法。因此，最好先不要给它下精确的定义，因为过早下定义是危险的。在对这一问题有较深入的了解之前，任何正式的定义都有可能引起误解或过分的限制。①

34 详细争论什么是意识还为时过早，尽管这种探讨可能有助于理解意识的属性。当我们对某种事物的定义还含糊不清时，过多地考虑该事物的功能毕竟是令人奇怪的。众所周知，没有意识你就只能处理一些熟悉的日常情况，或者只能对新环境下非常有限的信息作出反应。

35 某些种类的动物，特别是高等哺乳动物可能具有意识的某些(而不需要全部)重要特征。因此，用这些动物进行的适当的实验有助于揭示意识的内在机制。因此，语言系统(人类具有的那种类型)对意识来说不是本质的东西，也就是说，没有语言仍然可以具有意识的关键特征。当然，这并不是说语言对丰富意识没有重要作用。

36 在现阶段，争论某些低等动物如章鱼、果蝇或线虫等是否具有意识是无益的。因为意识可能与神经系统的复杂程度有关。当我们在原理上和细节上都清楚地了解了人类的意识时，这才是我们考虑非常低等动物的意识问题的时候。

---

① 如果这看来像是唬人的话，你不妨给我定义一下“基因”(gene)这个词。尽管我们对基因已经了解许多，但任何一个简单的定义很可能都是不充分的。可想而知，当我们对某一问题知之甚少时，去定义一个生物学术语是多么困难。

any nervous system. When we clearly understand, both in detail and in principle, what consciousness involves in humans, then will be the time to consider the problem of consciousness in much lower animals.

37 For the same reason I won't ask whether some parts of our own nervous system have a special, isolated, consciousness of their own. If you say, "Of course my spinal cord is conscious but it's not telling me," I am not, at this stage, going to spend time arguing with you about it.

38 There are many forms of consciousness, such as those associated with seeing, thinking, emotion, pain, and so on. Self-consciousness—that is, the self-referential aspect of consciousness—is probably a special case of consciousness. In our view, it is better left to one side for the moment. Various rather unusual states, such as the hypnotic state, lucid dreaming, and sleep walking, will not be considered here since they do not seem to have special features that would make them experimentally advantageous.

39 How can we approach consciousness in a scientific manner? Consciousness takes many forms, but as I have already explained, for an initial scientific attack it usually pays to concentrate on the form that appears easiest to study. Christof Koch and I chose visual awareness rather than other forms of consciousness, such as pain or selfawareness, because humans are very visual animals and our visual awareness is especially vivid and rich in information. In addition, its input is often highly structured yet easy to control. For these reasons much experimental work has already been done on it.

40 The visual system has another advantage. There are many experiments that, for ethical reasons, cannot be done on humans but can be done on animals.

37 出于同样原因，我们也不会提出，我们自身的神经系统的某些部分是否具有它们特殊的、孤立的意识这样的问题。如果你偏要说："我的脊髓当然有意识，只不过是它没有告诉我而已。"那么，在现阶段，我不会花时间与你争论这一问题。

38 意识具有多种形式，比如与看、思考、情绪、疼痛等相联系的意识形式。自我意识，即与自身有关的意识，可能是意识的一种特殊情况。按照我们的观点，姑且先将它放在一边为好。某些相当异常的状态，如催眠、白日梦、梦游等，由于它们没有能给实验带来好处的特殊特征，我们在此也不予考虑。

39 我们怎样才能科学地研究意识呢？意识具有多种形式。正如我们已经解释过的，初始的科学探索通常把精力集中到看起来最容易研究的形式。科赫和我之所以选择视觉意识而不是痛觉意识或自我感受等其他形式，就是因为人类很大程度上依赖于视觉。而且，视觉意识具有特别生动和丰富的信息。此外，它的输入高度结构化，也易于控制。正是由于这些原因，许多实验工作已围绕它展开。

40 视觉系统还有另外的优点。由于伦理学上的原因，很多实验不能在人身上进行，但是可以在动物身上进行。幸运的是，高等灵长类动物的视觉系统似乎与人类有某些相似之处。许多视觉实验已经在诸如恒河猴等灵长类动物身上完成了。倘若我们选择语言系统去研究，我们就不会有合适的实验动物。

Fortunately, the visual system of higher primates appears somewhat similar to our own, and many experiments on vision have already been done on animals such as the macaque monkey. If we had chosen to study the language system there would have been no suitable experimental animal to work on.

41 Because of our datailed knowledge of the visual system in the primate brain, we can see how the visual parts of the brain take the picture (the visual field) apart, but we do not yet know how the brain puts it all together to provide our highly organized view of the world—that is, what we see. It seems as if the brain needs to impose some global unity on certain activities in its different parts so that the attributes of a single object—its shape, color, movement, location, and so on—are in some way brought together without at the same time confusing them with the attributes of other objects in the visual field.

42 This global process requires mechanisms that could well be described as "attention" and involves some form of very short term memory. It has been suggested that this global unity might be expressed by the *correlated* firing of the neurons involved. Loosely speaking, this means that the neurons that respond to the properties of that particular object tend to fire at the same moment, whereas other active neurons, corresponding to other objects, do not fire in synchrony with the correlated set. To approach the problem we must first understand something of the psychology of vision.

Chapter 12 Brain Damage

"Babylon in all its desolation is a sight not so awful as that of the human mind in ruins."

—Scrope Davies

41 由于我们对灵长类大脑的视觉系统具有的详尽认知，因而我们知道大脑的各个视觉部分是如何分解视野的图像的。但我们还不清楚，大脑是怎样把它们整合在一起，以形成像我们看到的那样的高度组织化的外部世界的景观。看来，大脑就如同把某种整体的统一性叠加到了各视觉部分的神经活动之中。这样，某一物体的各个属性(形状、颜色、运动、位置等)就可以组装在一起，不至于与视野中的其他物体发生混淆。

42 这一全局过程所需要的机制，可以用“注意”很好地去描述，并且还涉及某种形式的短时记忆。有人已提出建议，这种全局的统一性，可以用有关神经元的相关发放进行表达。粗略地讲，这意味着对某个物体特性进行响应的神经元趋于同步发放，对其他物体响应的神经元的发放则与这一相关发放集并不同步。为了探索这一问题，我们需要先对视觉心理学有一些了解。

第十二章　脑损伤

巴比伦所有的废墟看上去远不如人类的思想的毁灭那样可怕。

——斯克罗普·戴维斯(Scrope Davies)

43 Over the years, neurologists have examined people whose brains have been damaged in various ways—by strokes, blows to the head, gunshot wounds, infections. Quite a number of these injuries alter some aspects of visual awareness while leaving other abilities, like speech or motor activities, more or less intact. The evidence suggests that there is a remarkable degree of functional specialization in the cortex, often in rather surprising ways.

44 In many cases, the damage to the brain is not very "clean" or specific. A speeding bullet is no respecter of cortical areas. (The living cortex has the texture of a rather soft jelly. Bits of it can easily be removed by sucking on a pipette.) Typically, the damage is likely to involve several cortical areas. The most dramatic effects are produced by damage to corresponding places on the two sides of the head, although such cases are rather rare.

45 Many neurologists have time only for a short examination of an injured patient—just long enough to be able to make an educated guess as to where the damage is likely to be. Lately, even this form of detective work has been largely superseded by brain scans. In the recent past it was the usual practice to report on a dozen or so similar cases together, since it was felt that to describe a single, isolated injury was unscientific. This unfortunately tended to lump together what were really somewhat distinct types of damage.

46 Recent trends have corrected these practices to some extent. Particular attention is now often given to the small number of cases in which one particular aspect of perception or behavior is altered while most other aspects are spared. These patients are likely to have suffered more limited and therefore more specific damage. An effort is also made to localize the damage via brain scans. The living patient, if he is cooperative, can be given a whole battery of psychological and other tests in order to discover just what he can

43 近些年来，神经病学家对脑部受到损伤的病人进行了研究。可能造成这些损伤的方式有多种，如中风、头部受到打击、枪伤、感染等。许多损伤改变了病人的视觉意识的某些方面，但病人的其他一些机能(如语言或运动行为)则基本未受影响。这些证据表明皮质具有显著的功能分化，而这种分化的方式通常是相当令人吃惊的。

44 在许多情况下，脑受到的损伤并不是单一的、专门化的。一粒高速射入的子弹对各皮质区域一视同仁(活的皮质组织是相当柔软的胶体，用移液管吸吮能很容易地移去其中一小部分)。通常情况下，损伤可能包括几个皮质区域。对头部两侧对应区域同时造成伤害的后果最为严重，不过这种情况非常罕见。

45 许多神经病学家对病人做简短的检查——仅够作出一个关于损伤的可能部位的合理猜测。后来，甚至连这种形式的检查工作也大部分被脑扫描取代了。近来，描述一个单独的、隔离的脑损伤被认为是不科学的，因此习惯上同时报告许多相似的病症。遗憾的是，这导致了将一些实际不同的损伤形式混为一谈。

46 当前的趋势在某种程度上纠正了这种做法。有少数病例中病人的感觉或行为的某个特定方面发生了改变，而其他大部分方面未受伤害。现在往往特别注意这些病例。这些病人受到的伤害很可能比较有限，因而更加专门化。人们还努力通过脑扫描来定位这些损伤。① 如果病人合作的话，他将在清醒状态下进行完整的一组心理学及其他一些测试，用来发现哪些是他所能或不能看到或做到的。在某些情况下，这种测试会进行

① 原则上，如果这种损伤不是进行性的，则可以在病人死后立即详细检查他的脑而确定损伤的位置，但这通常是不可能的(进行性的例子，如癌症及阿尔茨海默症)。

or cannot see or do. In some cases such tests have extended over a number of years. As ideas about visual processing have become more sophisticated, experiments to test these ideas have become more subtle and extensive. They can now be combined with brain scans that record the *activity* in the brain during these different tasks. These results on several patients can be used to compare and contrast patients with either similar damage, or similar symptoms, or both.

47 The obvious example to begin with would be damage to Vl, the striate cortex. If V1 is totally destroyed on one side of the brain, the patient appears to be blind in the opposite hemifield. Toward the end of this chapter a strange phenomenon known as "blindsight" will be discussed in detail. At this point let us look instead at the effects of damage to one of the highest parts of the visual hierarchy, and to the right-hand side of the head only. This is known as unilateral neglect or hemineglect. The damaged area corresponds broadly to area 7a in the macaque's brain. It is usually caused by vascular disorders in one of the cerebral arteries, commonly known as strokes.

48 In its early stages, the symptoms can be extreme—the patient's eyes and head may be rotated to the right. In really severe cases, where the damage may be so extensive that control and feeling have been lost on the patient's left side, he may deny that his own left leg belongs to him. One man was so incensed at having somebody else's leg in bed with him that he threw it out of the bed and was surprised to find himself lying on the floor.

49 Most cases are not as severe as this. After some days the extreme symptoms usually lessen or disappear. The patient may, for example, fail to pick up food from the lett side ot his plate. If asked to draw a clock, or a face, he will typically only draw the right side of it. After some weeks, as the brain

好几年。由于关于视觉处理的理论变得越来越深奥，检验这些观点的实验也变得更加广泛和精细。现在，它们可以和脑扫描技术相结合。该技术可以记录脑在完成这些不同任务时的行为。这些结果可以在具有相似损伤或相似病症(或者二者皆有的)病人之间进行比较和对照。

47 对 V1 区(条纹皮质)的损伤是一个明显的例子，现就以此作为开始进行讨论。如果脑一侧的 V1 区被完全破坏，病人的表现是看不见对侧的半个视野。在本章的结尾我将详细讨论一个被称作"盲视"的奇怪现象。在这里让我们先看一下对视觉等级最高层部分损伤的结果，并将损伤局限在头的右手侧。这是人们所知的单侧忽略。损伤区域大致对应于猕猴的 7a 区。这通常由大脑动脉血管疾病(如中风)引起的。

48 在早期阶段，症状可能非常严重——病人的眼睛和头会转向右侧。在最严重的病例中，损伤的范围可能很大，以致病人失去了左侧的控制和感觉。他会否认他自己的左腿是属于他的。有一个人对于"别人的腿"出现在他的床上感到极度愤怒，于是他把它扔到了床外。结果他惊讶地发现他自己躺在了地板上。

49 大多数情况并没有这么严重。通常几天以后严重的病症就会减轻或消失。例如，这时病人可能无法拿起盘中左侧的食物。如果让他画一个钟，或者一张脸，他通常只画其中的右侧。在几周以后，随着脑得到部分恢复，他对半边的忽略程度进一步下降，但他对左侧的注意仍显得比右侧弱。如果让他平分一条直线，他会将中点画到右边。不过他对左侧并不完全是盲的。如果那里有一个孤立的物体，他会看见它。但如果在

partially recovers, the severity of this hemineglect diminishes further, but the patient still appears to attend less to the left side than to the right. If asked to bisect a line he will draw the midway point over to the right. He is not, however, truly blind on the left. He can see an object there if it is in isolation, but he may fail to notice it if there is also something significant on his right-hand side. Moreover, he often denies that anything is amiss and does not report seeing an empty space on the left side of his visual field.

50 Hemineglect is not limited to visual perception. It can also apply to visual imagination. The classical example was reported by the Italian Edoardo Bisiach and his colleagues. Their patients were asked to imagine that they were standing at one end of the main square of Milan, facing the cathedral, and to describe what they could remember. They reported details mainly of buildings that, from this point of view, were on their right-hand side. They were then asked to do the same, but this time as if they were standing at the opposite end of the square, with the cathedral behind them. They then reported mainly the details on the side that they had previously neglected to mention, again visualized on their right.

51 Another remarkable form of brain damage produces a partial or comeplete loss of color vision. The patient sees everything in shades of gray. This is known as "achromatopsia"—a case was reported by Robert Boyle (the "father of chemistry") as early as 1688. In 1987, Oliver Sacks and Robert Wasserman described such a case in *The New York Review of Books*. The patient was a New York abstract painter, Jonathan I., who had been profoundly interested in color, so much so that listening to music had produced "a rich tumult of inner colors." This synesthesia, as it is called, disappeared after his accident, so that music lost much of its appeal to him.

右侧也有某个明显的物体，他就无法注意到左侧的物体。此外，他经常否认有什么东西是斜的，而且不承认看到了视野左侧的没有物体的空间。

50 单侧忽略并不限于视觉感知。它也会出现在视觉想象中。意大利的埃德瓦尔多·比西阿奇(Edoardo Bisiach)和同事们报告了一个典型的例子。他们要求病人想象自己站在米兰市的一个主要广场的一端，面对教堂，叙述他们所回忆起的景象。他们描述的主要是从该视点看到的右侧建筑的细节。随后病人被要求想象站在广场的对侧，教堂则在他们身后，再重复上述过程。此刻他们讲述的主要是先前他们叙述时忽略的那一侧的细节，此时仍是视野的右侧。

51 另一种显著的脑损伤形式造成了颜色视觉部分或全部丧失。患者看到的所有物体仅具有不同浓淡的灰色。这是众所周知的“全色盲”——早在1688年，被称为“化学之父”的罗伯特·波义耳(Robert Boyle)就报告过。1987年，奥立佛·萨克斯(Oliver Sacks)和罗伯特·瓦赛曼(Robert Wasserman)在《纽约书评》中讲述了这样一个病例，病人是纽约的抽象派画家乔纳森(Jonathan I.)。他对颜色有特殊的兴趣，以致他听音乐时会产生了“丰富的内部颜色的一阵激发”。这被称作联觉。在一次事故后他的这种联觉消失了，因而音乐对他的感染力也大大消失了。

52 The damage was a result of a rather minor car accident. Jonathan I. probably suffered from concussion but otherwise appeared to be unhurt. He was able to give a clear account of the accident to the police but later developed a bad headache and also amnesia for the accident. The next morning, after stuporous sleep, he found he could not read, although this inability disappeared five days later. He had difficulty distinguishing colors but no subjective sense that colors had altered.

53 This developed on the next day. While he knew it was a bright sunny morning, the world looked to him, as he drove to his studio, as if it were in some sort of fog. The full reality of his deficit only struck him when he arrived there and saw his own brilliantly colored paintings. They were now "utterly gray and void of color."

54 Sacks and Wasserman's account of the psychological effects of this cruel deficit is vivid and detailed. Although his problem could be judged no worse than looking at an old black-and-white movie, this was not how Mr. I. felt. Most foods appeared disgusting to him—tomatoes appeared black, for example. His wife's skin seemed to him to be rat-colored, and he could not bear to make love to her. It did not help him if he shut his eyes. His highly developed visual imagination had also become blind to color. Even his dreams had lost their previously vivid coloration.

55 Mr. I.'s scale of grays was compressed, especially in strong sunlight, so that he could not see delicate tonal graduations. His overall response to the wavelengths ot light was normal, except that he had an additional peak of sensitivity in the short wavelength ("blue") region of the spectrum. This may explain why he could not see white clouds against a background of blue sky. He also had trouble identifying faces unless they were close. His vision

52 损伤是一次相当轻微的车祸造成的。乔纳森·艾可能受到了撞击，但除此以外他好像并未受伤。他能够向警察清楚地叙述事故的原因。但后来他感到头疼得很厉害，并经常忘记这次事故。昏睡之后，次日清晨他发现自己不能阅读了。不过这种障碍在五天后就消失了。虽然他对颜色的主观感觉并未改变，但他很难区别颜色了。

53 这种情况在第二天又进一步发展。尽管他知道那是一个阳光灿烂的早晨，在他驱车前往工作室时，整个世界看上去像是在雾中。只有当他到达那里并看见自己的那些色彩绚丽的绘画现在变得“完全是灰色而缺乏色彩”时，他才被自己有这样缺陷惊呆了。

54 这种缺陷是残酷的。萨克斯和瓦赛曼形象而具体地解释了这种心理效应。虽然可以判断他的问题并不比看老式的黑白电影更糟，但是艾先生并不这样认为。大多数食物让他感到厌恶——例如，西红柿看上去是黑的。在他看来他妻子的皮肤就像白鼠的颜色，他无法忍受同她做爱。即使他闭上眼睛也无济于事。他那高度发达的视觉想象力也变得色盲了。连他的梦也失去了往日的色彩。

55 艾先生所感受的灰度尺度被压缩了，特别在强光下更严重。因此他不能辨别细微的色调等级。他对所有波长的光的反应是一样的，只在光谱的短波区(“蓝色”)有一个额外的敏感峰。这可以解释他为什么看不见蓝天上的白云。他在识别面孔时也遇到了困难，除非他们离得很近他才能认出来。但由于突出来的物体具有显著的对比，十分清晰，几乎像剪影

seemed sharper to him because objects stood out with considerable contrast and clarity, almost like silhouettes. He was abnormally sensitive to movement, and reported, "I can see a worm wriggling a block away." At night he claimed he could see so well that he could read license plates from four blocks away. Because of this he became, in his own words, "a night-person." While wandering about at night his vision was no worse than other people's.

56 Mr. I.'s loss of awareness of color has interfered rather little with other aspects of his vision, except that the loss has altered his sensitivity to shades of gray and may have produced extra sensitivity to movement. The damage was clearly bilateral, since both halves ot the visual field were affected (some cases of achromatopsia affect only one half) and delayed, since the full loss of color awareness took about two days to develop. Were it not for the increased response to the shorter wave-lengths of light (blue) it might appear to be a defect in the P system (the one more interested in shape and color) that had left most of the work of seeing to an undamaged M system (the one more interested in movement).

57 Mr. I.'s brain was given both an MRI scan and a CAT scan (although the latter was on rather a coarse scale), but no damage could be seen, so it is still not yet clear that the damage was truly cortical. However, previous cases have shown that achromatopsia usually involves cortical damage at a fairly high level of the human visual system (the ventromedial sector of the occipital lobe).

58 Another very striking deficit produced by brain damage is the inability to recognize faces, known as "prosopagnosia." One of England's prime ministers in the last century had this difficulty. He even failed to recognize

一样，因此他的视觉显得更敏锐了。他对运动异常敏感。他报告说："我可以看到一条街区外的一条虫在蠕动。"在夜间他声称自己能看得非常清楚，能读出四条街区外的车牌。因此，用他自己的话说，他成了一个"夜行者"。在夜间徘徊时，他的视觉并不比别人差。

56 艾先生失去的颜色意识对视觉的其他方面影响极小，这种丧失只改变了他对灰度浓淡的敏感性并使他对运动更敏锐。这种损伤显然是双侧的，因为两侧视野都受到了影响(有些情况下全色盲仅对一侧有影响)。这种损伤还是一种延迟过程，因为对颜色意识的完全丧失是在两天内发展起来的。如果不是他对短波长的光(蓝光)有增强反应的话，这很像是P系统有缺陷(P系统对形状和颜色更敏感)，而大部分视觉任务由未受损伤的M系统(对运动更敏感)来完成。

57 艾先生的脑也进行了MRI扫描和CAT扫描(尽管后者尺度较粗糙)，但未发现任何损伤，因而尚不清楚损伤的部位是否在皮质上。不管怎样，上述情况表明全色盲通常包括了人视觉系统中相当高层次皮质的损伤(枕叶的腹侧正中部分)。

58 另一种损伤造成的缺陷非常惊人，这就是面容失认症(pmsopagnosia)。19世纪的一位英国首相就遇到了这种困难。他甚至认不出自己的长子的脸。面容失认症有多种不同的形式，这可能是因为不同病人的脑损伤的

the face of his eldest son. There are many varieties of prosopagnosia, probably because the exact nature of the brain damage varies from patient to patient. The problem is usually not that of recognizing that a face is a face but of recognizing whose face it is, whether that of a wife, a child, or close friends. The affected person often cannot recognize his own face in a photograph, or even in a mirror, although he knows that it must be his face because it winks when he winks. He can usually recognize his wife from her voice, or the way she walks, but not just from seeing her face.

59 Unless the damage is extreme, he can describe the features of a tace—the eyes, nose, mouth and so on—and how they are related to each other. Moreover, his visual scanning mechanism is normal. In some cases a patient can discriminate between faces when asked to match differently lit photographs of unfamiliar faces, yet he cannot say whose faces they are photographs of, even if they were previously well known to him.

60 Prosopagnosia often accompanies achromatopsia when the latter is bilateral, but it should be remembered that there is no reason why the damage (often due to a'stroke) should affect only a single cortical area. Indeed, prosopagnosia can occur in association with other specific defects.

61 The neurologist Antonio Damasio has made several important contributions to the study of prosopagnosia. The condition is not necessarily limited to difficulties in face recognition. A case was already known of a farmer who could no longer recognize his cows, each of which he previously knew by name. But Damasio went further than this. He and his colleagues showed that in many cases patients could not recognize the individual members of a class of rather similar objects. For example, one patient could easily recognize a car as a car, yet could no longer tell whether it was a Ford or a Rolls-Royce,

实质各有不同。问题通常不是他们认不出那是一张脸，而是识别不出那是谁的脸，不知那是他的妻子的、孩子的还是一个老朋友的脸。病人常常认不出照片上他自己的脸。他甚至不能认出镜子中的自己，尽管他知道那肯定是他的脸，因为当他眨眼时镜中的像也在眨眼。他常常能从妻子的声音或走路的样子中认出她来，但只看她的脸时却不能。

59 除非损伤很严重，否则他能描述一张脸的特性(如眼睛、鼻子、嘴，等等)以及它们的相对位置。此外，他的目视扫描机制也正常。在一些情况下，让他辨认某些在不同光照下拍摄的不熟悉的照片时，他能区分这些不同的面孔。但即便他和他们早就很熟悉，他也不能说出哪张照片是谁的脸。

60 双侧全色盲患者常常同时患有面容失认症。但应当记住，没理由认为损伤(通常由中风引起)只影响单个皮质区。事实上，面容失认症可以和其他几种缺陷一同出现。

61 神经病学家安东尼奥·达马西欧(Antonio Damasio)对面容失认症的研究作出了不少重要的贡献。情况并不局限于面孔识别困难。在一个病例中，一个农夫再也不能识别他的牛，虽然原先他能叫出其中每一头牛的名字。但达马西欧的研究更深入一步。他和同事们表明，许多病例中病人不能在一组相类似的物体中识别出单个成员，例如，病人可能很容易认出一辆小汽车，但无法说出它是福特牌轿车还是劳斯莱斯轿车；不过

although he could identify an ambulance or a fire engine, presumably because they were sufficiently distinct from a typical car. A shirt could be recognized as a shirt but not as a "dress" shirt.

62 Damasio and his colleagues have also discovered that there are some patients who, while they cannot recognize the identity of a face, can recognize the meaning of facial expressions and can also estimate age and gender. Other prosopagnosia patients lack these abilities. These results suggest that different aspects of facial recognition are handled in different parts of the brain.

63 There is some controversy as to exactly how prosopagnosia and the underlying mechanisms should be described. Damasio stresses that it is not a general disorder of memory, since such a memory can be triggered through other sensory channels (auditory, for instance). Exactly what the mechanisms are in each type of case remains to be discovered.

64 A remarkable case of a patient lacking awareness of most types of movement has been reported by the psychologist Joseph Zihl and his colleagues. The damage was bilateral, located in several regions on each side of the cortex. When first examined, the patient was in a very frightened condition. This is not surprising since objects or persons she saw in one place suddenly appeared in another without her being aware they were moving. This was particularly distressing if she wanted to cross a road, since a car that at first seemed far away would suddenly be very close. When she tried to pour tea into a cup she saw only a glistening, frozen arc of liquid. Since she did not notice the tea rising in the cup, it often overflowed. She experienced the world rather as some of us might see the dance floor in the strobe lighting of a discotheque.

他能识别救护车或消防车，可能是因为它们与典型的汽车有显著差异。他能认出一件衬衫，但不知道那是不是礼服衬衫。

62 达马西欧和同事们还发现，尽管有些病人不能分辨面孔，他们能识别面部表情的含义并能估计年龄和性别。其他面容失认症患者则没有这种能力。这些结果表明面孔不同方面特征的识别是在脑的不同部位完成的。

63 目前对如何准确描述面容失认症及其内在机制尚有争议。达马西欧强调这不是一种普通的记忆疾病，因为这种记忆可以通过其他感觉通道(如听觉)激发出来。每种情况下的准确机理尚有待发现。

64 心理学家约瑟夫·齐尔(Joseph Zihl)和同事们报告了一个令人吃惊的病例：病人对大多数形式的运动没有意识。病人所受的损伤是双侧的，位于皮质的多个区域。第一次接受检查时，病人处于非常惊恐的状态。这并不令人奇怪，因为她看见在一个地方的人和物体突然出现在另一个地方，而她并未感觉到他们的运动。当她想过马路时就感到特别沮丧，因为原先在很远处的汽车会突然离她很近。当她试图把茶倒入杯子时，她只看到了一道凝固的液体弧的反光。她注意不到杯子中茶的上升，因而茶经常溢出来。她所体验的世界与我们某些人在夜总会中看到的频闪灯光下的舞池的地板很相似。

65 We all have this trouble on a vastly slower time scale. The hour hand of a clock does not appear to move, yet if we glance at it some time later it is in another place. We are thus familiar with the idea that things can be moving even if we are not directly aware of the movement, but we normally do not have this difficulty within the usual time scales of everyday lite. Clearly we must have a special system to detect movement, in and of itself, without having to infer it logically from two distinct observations separated by time.

66 Detailed tests showed that the patient could detect certain forms of motion, probably corresponding to a severely impaired short-range mechanism, but the mechanism that made more global associations of movement had been disrupted. Her vision had some other defects, mostly connected with motion, but she could see color and recognize faces, for example, and showed no signs of the type of neglect described carlier in this chapter.

67 There are many other kinds of visual defects produced by brain damage. Two cases have been reported in which the sufferer had lost perception of depth and saw the world and the people and other objects in it as perfectly flat, so that "the most corpulent individual might be a moving cardboard figure, for his body is represented by an outline only." Other patients can recognize an object if seen from a normal straight-on point of view but not from an unusual one, such as looking at a saucepan from directly above.

68 Two British psychologists, Glyn Humphreys and Jane Riddoch, studied a patient over a period of five years who had a number of visual defects—for example, he had lost his color vision and also could not recognize faces. They showed that his major visual problem was that while he could see the *local* features of an object, he could not bind them together. Therefore, he could not recognize what the object was although he could copy a drawing

65 在极慢的时间尺度上我们也遇到过这个问题。钟的时针看上去并不动，但是过一段时间后我们再看时，它已在另一个位置上。我们对这样一种观念很熟悉：一个物体可能是动的，即便我们并不能直接感受到它的运动。但在日常生活的一般时间尺度上我们通常没有这种困难。显然我们必定有一个特殊的系统自行来检测运动，而不必由时间分隔的两次不同的观察中从逻辑上推断它。

66 仔细的测试表明病人可以检测某些形式的运动，可能是一种严重受损后残存的短时机制作用的结果，而形成关于运动的更为全局的联系机制则已被破坏。她的视觉还有其他一些缺陷，大多数都与运动有关。但她能看见颜色并能识别面孔，也未表现出有本章前面描述的各种类型的忽视的征兆。

67 还有许多其他种类的脑损伤引起的视觉缺陷。报道中有两个病例，患者失去了深度感知，看到世间万物和人都完全是平的，因而“由于人的身体仅由轮廓线表示，最胖的人看上去也只是运动的纸板人形而已”。其他病人仅从通常的正对方向看物体时才能识别出它来，而从非常规角度观看，如从正上方看一个平底锅，无法识别。

68 英国的两位心理学家格林·汉弗莱斯(Glyn Humphreys)和简·里多克(Jane Riddoch)用了五年时间研究一个病人。他有多种视觉缺陷，例如，他失去了颜色视觉，也不能识别面孔。他们表明他的主要的视觉问题在于，当他看见一个物体的局部特征时，他不能把它们组合在一起。因

fairly well, was articulate, and could produce fluent verbal descriptions of things he had known about before his stroke. Such cases are important because they show that a person who has lost some of his high-level vision can still be visually aware at a lower level. This supports the claim that there is no single cortical area that registers everything we can see.

69 There is one kind of visual defect that is so surprising that people have been known to doubt that it could possibly exist. This is known as Anton's syndrome or "blindness denial." The patient is clearly unable to see but is unaware of that fact Asked to describe the doctor's tie the patient may say that it is a blue tie with red spots when in fact the doctor is not wearing a tie at all. When pressed the patient may volunteer the information that the light in the room seems a little dim.

70 At first, such a condition seems impossible. Alternatively the medical diagnosis might be hysteria, which is not very helpful. But consider the following possibility. I have often found that when talking to someone over the telephone whom I have not met I spontaneously form a crude visual image of his or her appearance. I held several long telephone conversations with one man whom I must have pictured in his fifties, rather thin, with rimless glasses. When, eventually, he came to see me I found he was in his thirties and was decidedly fat. It was only my surprise at his appearance that made me realize that I had previously imagined him otherwise.

71 I suspect that a person suffering from blindness denial produces such images, probably because the brain damage is such that these images do not have to compete with the normal visual input from the eyes. In addition, due to damage elsewhere, they may have lost the critical faculty that would normally alert them that something was wrong. Whether this explanation is

此，尽管他能很好地复制一幅地图，能清晰地发音，并流利地用语言描述他中风前所知道的事情，却不能认出物体是什么。这些病例很重要，它表明一个人失去了部分高层视觉后仍会有低层次上的视觉意识。它支持这样一种主张：没有一个单独的皮质区标记了我们能看到的所有事物。

69 有一种视觉缺陷是那么令人惊异，以致知道此事的人都怀疑它是否可能存在。这就是安通综合征(Anton's syndrome)，或称"失明否认症"。病人显然看不见东西，但并不知道这个事实。当让他描述医生的领带时，病人会说那是一条有红色斑点的蓝色领带，而事实上医生根本没戴领带。进一步追问病人，他会主动告诉你房间的灯显得有些暗。

70 最初，这种情况显得不可能是真的。医学诊断其是歇斯底里症，但这并没有多大帮助。不过考虑如下的可能性：我经常发现，当我与从未见过面的人通过电话交谈时，我会在脑海里自然而然地形成他(或她)的外貌的粗略影像。我和一个男子进行过多次电话长谈，我想象他有五十来岁，相当瘦，戴着度数很高的眼镜。当他终于来看我时我发现他只有三十多岁，明显发胖。我对他的外貌感到很惊讶，这才使我意识到我原来把他想象成别的样子了。

71 我猜想那些失明否认症患者产生了这种影像。或许是由于脑损伤导致这些影像不必与来自眼睛的正常视觉输入竞争。此外，在正常人脑中可能有某些重要机能可以提醒它们某些影像是错的，而这些患者由于其他部

correct remains to be seen, but at least it makes the condition appear not totally incomprehensible.

72 Are there any trends in the way the different cortical areas react to damage? Damasio has pointed out that in the human temporal region (at the side of the head) brain damage toward the back of the head produces effects of a different character from those produced by damage that occurs farther forward. Damage located toward the rear of the temporal region concerns rather general ("categorical") matters. As one proceeds forward the effects of damage become rather less general till, near the hippocampus, the loss concerns very individual ("episodic") events. Thus the distinction between categorical and episodic memory may be too sharp a distinction. There may be a gradual transition from areas that deal with general objects and events to those that deal with unique ones.

73 This suggestion of Damasio's is very much in line with my description (on page 158) of how a cortical area functions. Each area constructs new features from the combination of features already extracted by other areas (usually lower in the hierarchy) that feed into its middle layers.

74 As one ascends the visual hierarchy, for example, one goes from cortical area VI, which deals with rather simple visual features (such as oriented lines) that occur all the time, to areas that deal with objects like faces, which occur much less frequently, until one reaches the cortex associated with the hippocampus, where the combination of signals it responds to (both visual and others) corresponds largely to unique events.

75 What we've covered is enough to establish two general points: The visual system works in strange and mysterious ways, and its behavior is not

位的损伤而丧失了这些机能。这种解释是否正确尚有待研究，但它至少使得这种情况显得并不完全难以理解。

72 在不同的皮质区域对损伤的反应中是否有某些趋势呢？达马西欧指出，在人的颞区（头的两侧）靠近头后部的脑损伤与更靠近前部损伤的特点不同。靠近颞叶后部的损伤与概念性东西有关。如果损伤靠近前部，对概念的影响逐渐变小，直到海马附近，主要丧失的是与特定事件有关的记忆。这样，概念与事件记忆间的区别非常显著。可能在处理一般物体和事件的区域与仅仅处理其中一种的区域间有一种逐渐的转变。

73 达马西欧的建议与我对单个皮质区的功能的描述是一致的。对于每个皮质区而言，其他区域（通常是等级更低的）有输入到达它的中间各层；该皮质区把这些区域提取的特征组合构造成新的特征。

74 例如，当你沿视觉等级向上走时，你会从皮质 V1 区出发。V1 区处理相当简单的视觉特征（如有朝向的直线）。这些特征无时不出现。然后你到达处理诸如脸这类不那么频繁出现的复杂目标的区域，直到与海马相联系的皮质，这里检测的组合信号（包括视觉及其他信号）大多对应于唯一的事件。

75 至此，我们之前的讨论足以撑起两个普适要点：这些受损坏的视觉系统以一种奇怪而神秘的方式工作，它的行为与科学家所发现的关于猕猴和我们自己的视觉系统的连接方式和行为并不矛盾。

incompatible with what scientists have discovered about the wiring and behavior of the macaque's visual system and, by extension, our own.

76 But our task is to understand visual awareness. Awareness is the result of the many intricate processes needed to construct the visual image. Are there forms of brain damage that bear more directly on awareness itself? It turns out there are several.

77 The first is often referred to as "split brains." In its cleanest form it results when the corpus callosum—the large tract of nerve fibers that connect the cerebral cortex on one side of the head to that on the other side—is severed completely, together with a smaller fiber tract called the "anterior commissure". This surgical operation is performed to relieve certain cases of epilepsy that have failed to respond to other treatments. The corpus callosum can be lost due to other forms of brain damage, but there is usually some additional destruction elsewhere in the brain, so the interpretation of the results may not be so straightforward. There are also people who are born without the corpus callosum, but the brain usually develops in such a way as to compensate to some extent for early deficits, so again the results are not so dramatic as in the surgical cases.

78 The history of the subject is so odd that it is worth a brief mention. A distinguished American neurosurgeon reported in 1936 that no symptoms followed if the corpus callosum was cut. Another expert in the mid-fifties, reviewing the experimental results, wrote that "the corpus callosum is hardly connected with psychological functions at all." Karl Lashley (a clever and influential American neuroscientist who, curiously enough, was almost always wrong) went so far as to suggest, partly in jest, that the only thing the corpus callosum did was to keep the two hemispheres from collapsing into

76 然而我们的任务是理解视觉意识。它是构建视觉影像所必需的许多复杂处理的结果。是否有某些形式的脑损伤对意识本身有更直接的影响呢?现已发现确实有一些。

77 其中一种现象通常被称为“裂脑”。其最彻底的形式是胼胝体(连接大脑两侧皮质区的一大束神经纤维)以及称作“前连合”的一小束纤维被完全切除。在对癫痫病人的一般治疗失败后，为减轻其发作，会进行这种外科手术。其他形式的脑损伤也会导致病人失去胼胝体，但此时通常在脑其他部位也有额外损伤，因而无法像这样直截了当地解释结果。也有些人生来就没有胼胝体，但脑在发育过程中常能在某种程度上补偿早期的缺陷，因而结果并不如手术情况那样明显。

78 这个主题的历史十分奇特，因而值得作一简要叙述。一位著名的美国神经外科医生在 1936 年报告说，胼胝体被切除后并无症状。20 世纪 50 年代中期，另一位专家在回顾实验结果时写道：“胼胝体几乎不能与心理学功能联系到一起。”卡尔·拉什利(Karl Lashley，一位聪明而有影响的美国神经科学家。奇怪的是，他几乎总是错的)则走得更远。他曾开玩笑说，胼胝体的唯一功能是防止两个半球坍塌到一起(胼胝体显得有些硬，因此得名。“胼胝”有硬皮的意思)。我们现在知道这些观点是完全错误的。造成这种错误部分是由于胼胝体并不总被完全切除，但主要是因为检测手段不敏感或不恰当。

each other. (The corpus callosum appears to be somewhat hard, hence the name "callosum.") These opinions we now know to be gross errors, caused partly because the callosum was not always severed completely but mainly because the tests used were either insensitive or inappropriate.

79 The situation was dramatically changed by the work of Roger Sperry and his colleagues in the fifties and sixties. For this work Sperry was awarded a Nobel Prize in 1981. They showed clearly, by careful designed experiments, that a cat or a monkey whose brain had been split could be taught in such a way that one hemisphere learned one response while the other hemisphere learned another, or even a conflicting response to the same situation. As Sperry put it, "It is as if the animal had two separate brains."

80 Why is this? For most right-handed people only the left hemisphere can speak or communicate through writing. It also rules most of the capacity to deal with language, although the right hemisphere may understand spoken words to a limited extent and probably deals with the music of speech. When the callosum is cut, the left hemisphere sees only the right half of the visual field; the right hemisphere, the left half. Each hand is mainly controlled by the opposite hemisphere, although the other hemisphere can produce some of the coarser movements of the hand and arm. Except under special conditions, both hemispheres can hear what is being said.

81 Immediately after the operation the patient may experience various transient effects. For example, his hands may act at cross purposes, one buttoning up his shirt while the other follows unbuttoning. Such behavior usually subsides, and the patient appears comparatively normal. But careful testing reveals more.

79 罗杰·斯佩里(Roger Sperry)和同事们在20世纪五六十年代的工作使得情况明显改善。由于此项工作，斯佩里获得了1981年的诺贝尔奖。通过仔细设计的实验，他们清楚地表明，当一只猫或猴子的脑被分成两半时，可以教它的一侧半球学会一种反应，另一侧半球则学会另一种、甚至是对相同情况的完全矛盾的反应。正如斯佩里所说："这就好像动物有两个独立的脑。"①

80 为什么会这样呢？对大多数习惯于用右手的人而言，只有左半球能说话或通过写字进行交流。对于与语言相关的大多数能力也是如此，尽管右半球能在很有限的程度上理解口语，或许还能处理说话的音韵。当胼胝体被切除后，左半球只能看到视野右边的一半，而右半球则只能看到左边的一半。每只手主要是由对侧半球控制的，但同侧半球能控制手或手臂做某些比较简单的运动。除了特殊情况，每个半球都能听到说话。

81 刚进行完手术的病人可能经历各种瞬时效应。例如，他的两只手所做的目的正好相反，一只手扣上衬衣的扣子，另一只手则随后将其解开。这种行为通常会减弱，病人显得比较正常。但更细致的检查揭示了更多的东西。

① 这些在动物身上取得的结果导致人们对脑分裂的病人进行更加仔细的检查。这些工作特别是由斯佩里、约瑟夫·伯根(Joseph Bogen)、迈克尔·伽扎尼加(Michael Gazzaniga)、欧兰(Eran)、戴利亚·蔡德尔(Dahlia Zaidel)和他们的同事们开展的。

82 In the test, the patient is made to fix his gaze upon a screen onto which an image is flashed to one or the other side of his fixation point. This ensures that the visual information will reach only one of the two hemispheres (more elaborate methods are now available to do this).

83 When a picture is flashed into the patient's left (speaking) hemisphere, he can describe it the way a normal person can. This ability is not limited to speech. When asked, the patient can also point to objects with his right hand (largely controlled by the left hemisphere) without speaking. His right hand can also identify objects by touch even though he is prevented from seeing them.

84 If, however, a picture is flashed into the right (nonspeaking) hemisphere, the results are quite different. The left hand (largely controlled by this nonspeaking hemisphere) can point to and identify unseen objects by touch, as the right hand could do previously. But when the patient is asked to explain why his left hand behaved in that particular way, he will invent explanations based on what his left (speaking) hemisphere saw, not on what his right hemisphere knew. The experimenter can see that these explanations are false, since he knows what was really flashed into the nonspeaking hemisphere to produce the behavior. This is a good example of what is called "confabulation."

85 In short, one half of the brain appears to be almost totally ignorant of what the other half saw. A little information can sometimes leak across to the other side. While flashing a series of pictures into the right hemisphere of a woman, Michael Gazzaniga slipped in one of a nude. This made the patient blush. The left hemisphere was quite unaware of what the picture was but knew that it had produced a blush, so it said, "You do show some funny

82 在实验中，病人被要求把凝视点固定在一个屏幕上。屏幕上会有一个图像在他的凝视点的左侧或右侧闪烁。这样可以保证视觉信息仅到达两个半球中的一个。现在有更加精心设计的方法可以做到这一点。

83 当一个闪烁的图片到达能使用语言的左半球，他就能像正常人一样描述它。这种功能并不仅限于语言表达。病人也能按要求不说话而用右手指向目标(右手主要由左半球控制)。他还能不看一个物体而用右手识别它。

84 然而，如果闪烁图片到达了不能使用语言的右半球，结果则大不一样。左手主要由这个不能使用语言的半球控制，它能指向物体，也能通过触摸识别没看见的物体，这和右手所能做的是一样的。但当病人被问及为什么他的左手有这种特殊方式的行为时，他会依照能用语言表达的左半球所看见的场景虚构一个解释，但这并不是右半球所看见的。实验者知道真正闪烁进入那个不能使用语言的半球以产生行为的物体是什么，因而可以看出这些解释是错误的。这是一个解释“虚构症”①的很好的例子。

85 简单地说，看来脑的一半几乎完全忽略了另一半所看见的。只有极少的信息有时会泄露到对侧。在给一位妇女的右半球闪现一系列照片时，迈克尔·伽扎尼加(Michael Gazzaniga)加入了一张裸体照片。这使得病人有些脸红。她的左半球并不能察觉那些照片的内容，但知道它使她脸

① 虚构症(confabulation)，指患者用随意的编造来填补记忆中的空白。——译者注

pictures, don't you, Doctor?" After a while such a person may learn to make one side crosscue the other; for example, by signalling in some way with the left hand so that the speaking hemisphere can pick up the signal. In a normal subject, the detailed visual awareness in the right hemisphere can easily be transferred to the left hemisphere so that the person can describe it in words. When the corpus callosum is fully cut, this information cannot cross to the speaking hemisphere. The information is unable to pass via the various connections lower down in the brain.

86 Notice that I am not concerned here with how the two halves of the brain differ, except that language is normally on the left. I do not need to worry about whether the right-hand side has somewhat special properties, such as being rather better at recognizing faces, for example. Nor will I consider the extreme view expressed by some that while the left side is a "person," the right-hand side is merely an automaton. Obviously the right side lacks a well-developed language system and is therefore in some sense less "human" since language is a unique ability of human beings. Eventually we shall need to answer the question whether the right-hand side is more than an automaton, but I feel that can wait till we understand better the neural basis of awareness, to say nothing of the question of Free Will. The balance of professional opinion is strongly of the view that, apart from language, the cognition and motor capacities of the two sides, while not exactly the same, have the same general character.

87 Most split brain operations do not sever the intertectal commissure that connects the superior colliculus on one side to that on the other. The brain cannot use this intact pathway to transmit the information in visual awareness from one side to the other. For this reason it is unlikely that the superior colliculus is the seat of visual awareness, even though it is involved in visual attention.

红，因此她说：“医生，你是不是给我显示了一些很有趣的照片?”过了一会儿，病人学会了向另一侧半球提供一些交叉线索：例如，用左手以某种方式发信号从而使能用语言的半球能够识别该信号。对于正常人而言，右半球的详细的视觉意识能够很容易地传到左半球，因而能用语言描述它。胼胝体被完全切除后，这些信息无法传到能使用语言的半球。该信息无法通过脑中的各种低层次的连接传到对侧。

86 请注意，除了提到语言通常在左脑外，我并未涉及脑的两半有什么差异。我不必关心右侧脑是否有某些特殊能力，例如它十分擅长识别面孔。我也不必考虑某些人的一种极端的观点。他们认为左侧具有“人”的特性，而右侧仅仅是自动机。显然右侧缺乏发展完善的语言系统，因而从某种意义上说不那么具有“人类”的特点——因为语言是唯一标志人类的能力。事实上我们需要回答右侧是否高于自动机这个问题，但我觉得应该稍作等待，直到我们更好地理解了意识的神经机制，否则我们不能很好地作出回答，更不必说解答自由意志的问题了。折中的职业观点强调，除了语言外，两侧的感知和运动能力虽不完全相同，但一般特征是一致的。

87 大多数切开脑的手术并不切断两侧上丘的顶盖间连合。脑无法利用这个未触及的通路从一侧向另一侧传递视觉意识信息。因此，尽管上丘参与了视觉注意过程，它似乎不像是意识的位置。

88 Another fascinating phenomenon is known as "blindsight," studied extensively by the Oxford psychologist Larry Weiskrantz. Patients with blindsight can point to and differentiate between certain simple objects, while at the same time denying that they see them.

89 Blindsight is usually caused by fairly extensive damage to the primary visual cortex V1 (the striate cortex), in many cases only on one side of the head. In the test, a horizontal row of small lights is arranged so that when the patient fixes his gaze on a spot to one side of the lights, they all fall into the blind part of his visual field. After the sound of a warning buzzer one of the lights is lit for a short time. The patient is asked to point to where the light was, without moving his eyes or his head while the light is on. The patient normally demurs, saying that since he is blind there, there is no point in doing the test. With a little persuasion he may be coaxed to give it a try and "guess" where it was. The test is then repeated many times, sometimes with one light going on, sometimes another. And the surprised patient, in spite of denying that he can see anything, points to the active light fairly accurately, usually within 5 or 10 degrees.

90 Some patients can distinguish simple shapes, such as an X from an O, provided they are big enough, and some can also discriminate line orientation and flicker. There are claims that two patients adjusted their grasp so that it matched the shape and size of the object they were reaching for, at the same time denying they saw it. In some cases a patient's eyes can follow the movement of moving stripes, but this task may be handled by another part of his brain, such as the superior colliculus. A patient's pupils can respond to light changes, since changes of pupil size are involuntary and controlled by another small brain region.

88 另一个引人注目的现象被称为“盲视”。牛津的心理学家拉里·威斯克兰兹(Larry Weiskrantz)在这方面作了广泛的研究。盲视病人能指出并区分某些非常简单的物体，但同时又否认能看见它们①。

89 盲视通常是由于初级视觉 V1 区(纹状皮质)受到大面积损伤而引起的，在许多病例中损伤仅出现在头部的一侧。在实验中，一行小灯呈水平排列，使得病人在凝视这些灯光的一端时，它们全部落在视野的盲区。在一声警告的蜂鸣声之后，有一盏灯会短时间点亮，而此时病人不能转动眼睛或头。要求病人指出哪盏灯被点亮了，此时，病人通常对此表示异议，说既然他看不见那里的东西，没必要做这个实验。经过短暂的劝说之后，他会打算试一下并作“猜测”。实验会重复多次，有时这盏灯被点亮，有时则是另一盏被点亮。结果令病人大感惊讶，尽管他否认看见了任何东西，却能相当准确地指出亮的那盏灯，误差一般不超过5°~10°。②

90 有些病人还能区分简单的形状，比如 X 和 O，只要它们足够大。有些人还能鉴别直线的朝向和闪烁。有人声称有两个病人能调节手的形状，使之与即将触摸到的目标的形状和大小相匹配，同时却否认看到了这个物体。某些情况下病人的眼睛能跟踪运动条纹，但这个任务或许是由脑的其他部分(如上丘)完成的。病人的瞳孔也能对光强作出反应，因为瞳孔的大小不是随意的，而是由另一个小的脑区控制的。

① 他在猴子身上进行了大量的平行工作，但在这里我并不打算叙述它们。

② 实际上这个结果遭到了怀疑。例如，一种反对意见是，引起这种行为的原因是：眼睛把光散射到视网膜的其他位置，对应于病人可见的视野。但似乎并非如此，特别是现在表明照射到盲点的光不能产生这种效应(回想一下，在盲点没有光感受器，因此不会对光反应。另外，盲视病人的光感受器是完好的，并能检测信号。最初损伤的是视皮质)。进一步的实验已经回答了所有这些反对意见，目前对于盲视是个真实的现象已没什么可怀疑的了。

91 Thus, even with a badly damaged V1, the brain can detect some fairly simple visual stimuli and act on them although the affected patient will firmly deny his awareness of them.

92 What neural pathways are involved is still unclear. It was originally surmised that the information went through the superior colliculus, a part of the "old brain." It now seems unlikely that this is the whole story, since recent experiments have shown that the blindsight responses to wavelength involve the cones of the eye. The response to different wavelengths is similar to that of normal people, although a brighter light is needed. This makes it unlikely that the colliculus is the only pathway, since no color-sensitive neurons have been found there.

93 The problem is complicated in that the damage to cortical area V1 produces, in time, extensive cell death in the corresponding parts of the LGN (the thalamic relay), and that this in turn kills many of the retinal ganglion cells of the P type since, like hermits, they have no one to talk to. However, some P neurons remain, as do some neurons in the relevant regions of the LGN, presumably because they project to some undamaged places. There are direct though weak pathways from the LGN directly to areas in the cortex beyond V1, such as V4. These pathways may remain sufficiently intact to lead to a motor output (being able to point, for instance), but not enough for visual awareness. There is suggestive evidence that in some cases there are little islands of intact tissue within the damaged patch of area V1, so that V1 can still produce some effects in those regions, even though they may be small. Or it may simply turn out that an intact V1 is essential for awareness for another reason, and not merely because it normally produces a strong input to the higher visual areas. Whatever the reason, the patient can use some visual information while denying that he sees anything.

91 因此，尽管 V1 区受到了严重损坏，病人会坚决否认察觉到了这些刺激，但脑仍能探测到某些相当简单的视觉刺激，并能采取相应的行动。

92 目前还不清楚其中涉及的神经通路。最初猜测信息是通过"古脑"(old brain)的一部分即上丘传递的。现在看来远不止于此，因为最新的实验表明眼视锥细胞参与了盲视对光波长的反应。它们对不同波长的反应与正常人相似，只是所需的光更亮些。在上丘没发现对颜色敏感的神经元，因此它不会是唯一的通道。

93 这个问题很复杂，因为皮质 V1 区的损伤最终会导致侧膝体(丘脑的中继站)对应部位的细胞大量死亡，继而又将杀死大量的视网膜 P 型神经节细胞，因为就像隐士一样，它们没有可以交谈的对象①。然而，某些 P 型神经元保留了下来，就像侧膝体相关区域的一些神经元一样，可能是因为它们投射到了某些未受损害的部位。从侧膝体有直接但弱的通路到达 V1 区以上的皮质区，诸如 V4 区。这些通路可能保持足够完好，足以产生运动输出(例如，能够指出目标)，但尚不足以产生视觉意识。有些启发性的证据表明，在 V1 区损伤的部位中有一些未被触及的组织形成的小岛，因而 V1 区在这些区域仍能起到一定作用，虽然这种作用可能比较小②。或者最终发现由于别的原因，一个完整的 V1 区对意识是必需的，而不仅仅是因为通常它产生了到高级视觉区域的输入。不管这个理由是什么，病人在否认看见任何东西的同时确实能利用一些视觉信息。

① 如果一个神经元的所有输出只到达死亡的神经元，它本身往往也会死去。

② HC 是海马(hippocampus)的缩写。——译者注

94 Another interesting form of behavior has been found in victims of prosopagnosia. While hooked up to a lie detector and shown sets of both familiar and unfamiliar faces, the patients are unable to say which faces were familiar, yet the lie detector clearly showed that the brain was making such a distinction even though the patients were unaware of it. Here again we have a case where the brain can respond to a visual feature without awareness.

95 The hippocampus is a part of the brain that is not confined to vision but is concerned with a type of memory, it is marked HC. This has fewer layers than most neocortex. Because of its position near the top of the sensory processing hierarchy, one would be tempted to guess that here at last was the true seat of visual (and other) awareness. It receives input from many of the higher cortical areas and projects back to them. This elaborate one-way pathway is reentrant—that is, it returns very close to where it started—and this also might-suggest that this is where consciousness really resides, since the brain might use this pathway to reflect on itself.

96 The experimental evidence argues strongly against this otherwise attractive hypothesis. Hippocampal damage can be caused by an infection of the virus herpes encephalitis, which produces severe but sometimes rather limited damage. The virus appears to preter to attack the hippocampus and its associated cortex. The borders of the damage can be quite distinct. Because the damage can be located by an MRI scan and is not progressive, a patient can be followed over a period of years after the acute phase of the infection is over.

97 If you were to meet a person who had lost his hippocampus on both sides, plus the immediately adjacent cortical areas, you might not at first realize that he was in any way abnormal. It is striking to see a videotape of such a

94 另一种让人感兴趣的行为形式是在一些面容失认症患者身上发现的。当病人与测谎仪连起来并面对一组熟悉的和不熟悉的面孔时，他们无法说出哪些面孔是他们熟悉的，但是测谎器清晰地显示出脑正在作出这种鉴别，只是病人不知道罢了。这里我们再次遇到了这种情况，脑可以不觉察一个视觉特征却能作出反应。

95 海马是脑的一部分，实际上它并不仅限于视觉，而是与一种记忆类型有关，标志为 HC^2。它的层数比大多数新皮质少。因为它的位置靠近感觉处理等级的顶端，人们禁不住猜测这里终于是视觉(及其他)意识的真正位置。它从许多更高的皮质区接受输入并投射回去。这种复杂的单向通路是再进入的——它返回到离出发点很近的地方——这或许也暗示着它是意识的所在之处，因为脑可能使用这条通路去反映它自己。

96 这种假设看起来很吸引人，但是遭到了实验证据的强烈驳斥。海马损伤可能由一种病毒性疱疹脑炎感染造成，这种病会造成相当严重、但有时很有限的损坏。看来病毒易于攻击海马及与其相联系的皮质。损伤的边界会很清晰。由于损伤可用 MRI 扫描定位且不再发展，病人在感染严重期过后数年均可进行复查。

97 如果你碰巧遇到一个失去两侧海马以及邻近皮质区域的人，你并不会马上意识到他有何异常。看了这样一盘录像带你一定会感到吃惊：其中讲述了一个人，他能谈话、微笑、喝咖啡、下棋，等等，他几乎只有一个问题，那就是他不能记住大约一分钟以前发生的任何事件。在相互介绍

person: talking, smiling, drinking coffee, playing checkers, and so on. Almost his only problem is that he cannot remember any recent episode that happened more than a minute or so before. Introduced to you he will shake your hand, repeat your name, and make conversation. But if you go out of the room for a few minutes and then return, he will deny he had ever seen you before. His motor skills are preserved and he can learn new ones, which usually last for at least several years if not indefinitely, but he cannot remember the occasions when he learned them. His memory for categories is intact, but his memory for new events lasts only a very short time and is then almost totally lost. He may also be handicapped in his memory for episodes that took place before his brain damage. In short, he knows the meaning of the word *breakfast* and he knows how to eat his breakfast, but he has not the faintest idea of what he had for breakfast. If you ask him he will either tell you he can't remember or he will confabulate and describe what he thinks he might have eaten.

98 Although in a sense he has lost full human "consciousness," his short-term visual awareness appears to be unaltered, if it is impaired, it can only be in subtle ways that testing has not yet uncovered. The hippocampus and its closely associated cortical areas are thus not necessary for visual awareness. It is possible, however, that the information flowing in and then out of the hippocampus may normally reach consciousness, so it is sensible to keep an eye on the neural areas and pathways involved, as this may help to pin down the location of awareness in the brain.

99 The study of brain damage sometimes gives us results we can obtain in no other way. Unfortunately, this knowledge is often tantalizingly ambiguous because in most cases the damage is so messy. In spite of this limitation, in favorable cases the information can be decisive. At the least, the results of

时他会和你握手，复述你的名字，并进行交谈。但如果你暂时离开房间，过几分钟后再返回，他会否认见过你。他的运动技巧均被保留，还能学习新技术，并通常能保持数年甚至更长时间，只是他记不起来是什么时候学会这些技艺的。他对分类的记忆是完好的，但他对新事物的记忆仅能维持极短的时间，随后就几乎完全丧失了。他在回忆脑损伤前发生的事情时也有障碍。简而言之，他知道"早餐"一词的含义，也懂得如何吃早餐，但他对吃过什么东西几乎没任何印象。如果你问他，他或许会告诉你他不记得了，或者跟你瞎聊，并描述他认为他可能吃了些什么。

98 虽然从某种意义上说他失去了全部人类"意识"，但看来他的短时视觉意识并未改变。如果它受到了损伤，也只会是一种实验尚未揭示的细微方式。因此海马及其紧密相关的皮质区域并不是形成视觉意识所必需的。然而，流入和流出的信息通常有可能到达意识状态，因而有理由留意一下其中的神经区域和通路。这或许对找出脑中意识的位置有所帮助。①

99 对脑损伤的研究能得到一些其他方式无法得到的结果。遗憾的是，由于大多数情况下损伤是极复杂的，这些知识时常变得很模糊，令人着急。尽管有这些局限性，在顺利的情况下信息是明确的。脑损伤的结果至少能对脑的工作提供暗示，而这些可以用其他方法在人或动物身上探测到。在某些情况下，它证实了某些在猴子身上进行的实验所得到的结果在人身上也适用。

① 海马系统的确切功能及其神经元完成这些神经功能的精确方式，目前有很大争议。不过，虽然有人会在术语上对我所描述的大致情况吹毛求疵，它还是会被广泛接受的。

brain damage can suggest ideas about the workings of the brain that can be explored by other methods, either on man or on animals. In some cases it confirms for man what we have already learned from experiments on monkeys.

Chapter 14 Visual Awareness

"Philosophy is written in that great book that lies before our gaze—I mean the universe—but we cannot understand it if we do not first learn the language and grasp the symbols in which it is written."

—Galileo

100 Let us now survey the ground we have covered so far. The main theme of the book is the Astonishing Hypothesis—that each of us is the behavior of a vast, interacting set of neurons. Christof Koch and I suggested that the best way to approach the problem of consciousness was to study visual awareness, both in man and his near relations. However, the way people see things is not straightforward. Seeing is both a constructive and a complicated process. Psychological tests suggest that it is highly parallel but with a serial, "attentional" mechanism on top of the parallel one. Psychologists have produced several theories that try to explain the general nature of the visual processes, but none is much concerned with how the neurons in our brains behave.

101 The brain itself is made of neurons (plus various supporting cells). Each neuron, considered in molecular terms, is a complex object, often with a rather bizarre, irregular shape. Neurons are electrical signallers. They respond quickly to incoming electrical and chemical signals and dispatch fast electrochemical pulses down their axon, often over distances very many

第十四章　视觉觉知①

宇宙就像一部展现在我们眼前的伟大的著作。哲学就记载在这上面。但是如果我们不首先学习并掌握书写它们所用的语言和符号，我们就无法理解它们。

——伽利略

100 现在让我们总览一下到目前为止我们所涉及的领域。本书的主题是"惊人的假说"，即我们每个人的行为都不过是一个拥有大量相互作用的神经元群体活动的体现。克里斯托弗·科赫(Christof Koch)和我认为探索意识问题的最佳途径是研究视觉觉知，这包括研究人类及其近亲。然而，人们观看事物并不是一件直截了当的事情，它是一个具有建设性的、复杂的处理过程。心理学研究表明，它具有高度的并行性，又按照一定的顺序加工，而"注意"机制处于这些并行处理的顶端。心理学家提出过若干种理论试图来解释视觉过程的一般规律，但没有一种更多地涉及脑中神经元的行为。

101 脑本身是由神经元及大量支持细胞构成的。从分子角度考虑，每个神经元都是一个复杂的对象，常具有无规则的、异乎寻常的形状。神经元是电子信号装置。它们对输入的电学和化学信号快速地作出反应，

① 在《惊人的假说》中，consciousness 和 awareness 的意思都是意识，只是前者作为范围更广的、比较书面化的词，而后者则更多用于感觉系统(特别是视觉系统)，是比较口语化的词。在《惊人的假说》一书中的第一和第二部分，它们均译作"意识"，并不引起歧义。但在第三部分当中，作者以 visual awareness 作为 consciousness 研究的突破口，须区分这两个词。故在《惊人的假说》第三部分(第十四章至第十八章)中特将 awareness 按心理学译为觉知。——译者注

times greater than the diameter of their cell bodies. There are enormous numbers of them, of many distinct types, and they interact with each other in complicated ways.

102 The brain is not a general-purpose machine, like most modern computers. Each part, when fully developed, does a somewhat different and specific job, but, in almost any response, many parts interact together. This general picture is supported by studies of humans whose brains have been damaged and by modern methods of scanning the human brain from outside the head.

103 The visual system has far more distinct cortical areas than one might have expected. These areas are connected in a semihierarchical manner. A neuron in one of the lower cortical areas (i.e., those connected most closely to the eyes) is mainly interested in relatively simple aspects of only a small fragment of the visual scene, although even these neurons are influenced by the visual context of the fragment. The ones at the higher cortical levels respond best to more complex visual objects (such as faces or hands) and are not too fussy about exactly where these objects are in the visual scene. There appears to be no single cortical area whose activity corresponds to the global content of our visual awareness.

104 To understand how the brain works, we have to develop theoretical models that describe how sets of neurons interact with each other. At the moment, the neurons in these models are oversimplified. Modern computers, while many times faster than the ones available a generation ago, can simulate only a relatively smail number of these neurons and their interconnections. Nevertheless, these primitive models, of which there are several distinct types, often show surprising behavior, not unlike some of the behavior of the brain. They help provide us with new ways of thinking about how the brain might work.

并将它们的高速电化学脉冲沿轴突发送出去，其传送距离通常比细胞体直径还要大许多倍。脑中的这些神经元数量巨大，它们有许多不同的类型。这些神经元彼此具有复杂的连接。

102 与大多数现代计算机不同，脑不是一种通用机。在完全发育好以后，脑的每一部分完成某些不同的专门任务。另外，在几乎所有的反应中，都有许多部分相互作用。这种一般性观念得到了人脑研究结果的支持，这些研究包括对脑损伤者的研究以及使用现代扫描方法从头颅外进行的对人脑的研究。

103 视觉系统的不同的皮质区的数目比人们预料的要多得多。它们按一种近似等级的方式连接而成。在较低级的皮质区，神经元到眼睛的连接最短，它们主要对视野中一小块区域中的相对简单的特征敏感，尽管如此，这些神经元也受该区域所处的视觉环境影响。较高级皮质区的神经元则对复杂得多的视觉目标(如脸或手)有反应，对该物体在视野中的位置并不敏感。(目前看来)似乎并不存在单独的皮质区域与视觉觉知的全部内容相对应。

104 为了理解脑如何工作，我们必须发展出描述神经元集团间如何相互作用的理论模型。目前的这些模型对神经元进行了过分的简化。尽管现代计算机比其上一代在运算速度上快得多，但也只能对数目很少的一群这类简化神经元及其相互作用进行模拟。尽管如此，虽然这些不同类型的简化模型仍显得原始，却经常表现出一些令人吃惊的行为。这些行为与脑的某些行为有相似之处。它们为我们研究脑所可能采取的工作方式提供了新的途径。

105 This is the background, then, against which we have to approach the problem of visual awareness: how to explain what we see in terms of the activity of neurons. In other words, What is the "neural correlate" of visual awareness? Where are these "awareness neurons"—are they in a few places or all over the brain—and do they behave in any special way?

106 Let's start by looking once again at the ideas briefly outlined in a Chapter 2. Exactly what psychological processes are involved in visual awareness? If we can find where these different processes are situated in the brain, this knowledge may help us to locate the awareness neurons we are looking for.

107 Philip Johnson-Laird suggested that the brain has an operating system—as a modern computer has—and its actions correspond to consciousness. In his book *Mental Models* he puts this idea in a wider context. He suggests that the division between conscious and unconscious processes is a result of the very high degree of parallelism in the brain. Such parallel processing allows the organism to evolve special sensory, cognitive, and motor systems that operate rapidly, since many of their neurons can work at the same time (rather than one after another) as I have already described for the visual system. The overall control of all this activity by the more serial operating system enables decisions to be made rapidly and flexibly. A very rough analogy would be to an orchestral conductor (the operating system) controlling the parallel activities of all the members of an orchestra.

108 While this operating system can monitor the output of the neural systems it controls, he postulates that it does not have access to the details of their operations but only the results they present to it. By introspection we have access to only a limited amount of what is going on in our brains. We have no access to the many operations that lead up to the information given to the

105 以上阐述的是背景知识。在此基础上，我们着手解决视觉觉知问题，即如何从神经元活动的角度来解释我们所看见的事物。换句话说，视觉觉知的“神经关联”是什么？这些“觉知神经元”究竟位于何处？它们是集中在一小块地方还是分散在整个脑中？它们的行为是否有什么特别之处？

106 让我们首先回顾一下第二章①曾概述的各种观点。视觉觉知究竟包括哪种心理学处理过程呢？如果我们能够找出这些不同的处理过程在脑中的确切位置，或许会对定位我们所寻找的觉知神经元有所帮助。

107 菲力普·约翰逊-莱尔德认为，脑和现代计算机一样，具有一个操作系统。该操作系统的行为与意识相对应。他在著作《心理模型》(*Mental Models*)一书中，在更加广阔的背景下提出了这一思想。他认为，有意识和无意识过程的区别在于后者是脑中高度的并行处理的结果。正如我已在视觉系统中描述的那样，这种并行处理就是大量的神经元能够同时工作，而不是序列式一个接一个地处理信息。只有这才能使有机体有可能进化成具有特殊的、运转快速的感觉、认知及运动系统。而更为序列式的操作系统对所有这些活动进行全局控制，这样才能够快速、灵活地作出决定。粗略地打个比方，这就好像一个管弦乐队的指挥(相当于操作系统)控制着乐队所有成员同时演奏一样。

108 约翰逊-莱尔德假定，虽然这个操作系统可以监视它所控制的神经系统的输出，它能利用的只是它们传递给它的结果，而不是它们工作的细

① 此处的“第二章”是[英]弗朗西斯·克里克著，汪云九等译，湖南科学技术出版社 2018 年出版的《惊人的假说》的第二章，而非本导引的第二章。

brain's operating system. As he puts it, in introspection, "We tend to force intrinsically parallel notions into a serial straitjacket," since he envisages the operating system as operating largely in a serial manner. This is why introspection can be so misleading.

109 Johnson-Laird's ideas are clearly and forcibly expressed, but if we want to understand the brain in neural terms we have to identify the location and nature of this operating system. It may not necessarily have exactly the same properties as those tound in modern computers. The brain's operating system is probably not cleanly located in one special place. It is more likely to be distributed, in two senses: It may involve separate parts of the brain interacting together, and the active information in one of these parts may be distributed over many neurons, Johnson-Laird's description of the brain's operating system reminds one somewhat of the thalamus, but the neurons of the thalamus may be too tew to represent all the contents of visual awareness, although this could be tested. It seems more probable that some (but not all) neurons in the neocortex do this, probably under the influence of the thalamus.

110 At which stages in the various functional brain hierarchies should we look for the neural correlates of awareness? Johnson-Laird believes that the operating system is at the highest level of the processing hierarchies whereas, as we saw, Ray Jackendoff thinks that consciousness is more closely associated with the intermediate levels. Which idea is the more plausible?

111 Jackendoff's view of visual awareness is based on David Marr's idea of the $2^{1/2}$D sketch—roughly the viewer-centered representation of visible surfaces described in Chapter 6—rather than Marr's 3D (three dimensional) model. This is because humans directly experience only the presented side of objects

节。我们通过内省只能感觉到我们脑中所发生的情形的很少一部分。我们无法介入能产生信息并传给脑的操作系统的许多运作。因为他将操作系统视为主要是序列式的，所以他认为“在内省时，我们倾向于迫使本来是并行的概念进入序列式的狭窄束缚中”。这是使用内省法会出现错误的原因。

109 约翰逊-莱尔德的观点表达得很清楚，又很有说服力。但是，如果我们希望从神经的角度理解脑，还必须要识别该操作系统的位置和本质。它不一定与现代计算机的许多特性相一致。脑的操作系统可能并不是清晰地定位于某一特殊位置上。从两层意义上说，它更像是分布式的：它可能涉及脑中相互作用的若干分离的部分，而其中某一部分的活动信息又会分散到许多神经元。约翰逊-莱尔德对脑的操作系统的描述使人多少想起丘脑，但是丘脑的神经元太少了，以至于无法表达视觉觉知的全部内容(虽然这是可以验证的)。似乎更有可能的是，在丘脑的影响下新皮质的部分神经元(而不是全部神经元)可以表达视觉觉知。

110 我们寻找的觉知的神经关联会处于脑功能等级的哪个阶段呢？约翰逊-莱尔德认为，操作系统处在处理等级的最高层次，而雷·杰肯道夫认为觉知与中间层次有更多联系。究竟哪种观点更合理呢？

111 杰肯道夫关于视觉觉知的观点①是基于戴维·马尔(David Marr)的2.5维图而不是三维模型的思想的。这是由于人们直接感受到的只是视野中物体呈现的那一侧；物体后面看不见的部分则仅仅是推测。另外，

① 将杰肯道夫的观点归纳起来而不曲解他的意思，这并不容易。如果读者希望进一步理解，可以参阅他的书。我并不打算叙述他对音韵学、句法、语义等方面的论点以及他在音乐认知方面的见解。相反地，我将试图简化他的基本观点，特别是它们在视觉上的应用。

in the visual field; the presence of the invisible rear of an object is only an inference. On the other hand, he believes that visual understanding—what one is aware of—is determined by the 3D model together with "conceptual structures"—fancy words tor thoughts. This illustrates what he means by the intermediate-level theory of consciousness.

112 An example may make this clearer. If you look at a person whose back is turned to you, you can see the back of his head but not his face. Nevertheless, your brain infers that he has a face. We can deduce this because if he turned around and showed that the front of his head had no face, you would be very surprised. The viewer-centered representation corresponds to what you saw of the back of his head. It is *what you are vividly aware of*. What your brain infers about the front would come from some sort of 3D model representation. Jackendoff believes you are not directly conscious of this 3D model (nor of your thoughts, for that matter). Recall the old line: How do I know what I think till hear what I say.

113 I have put Jackendoff's penultimate version of his theory in the footnotes, as his wording is not easy to understand on a first reading. Applied to vision, what he means (if I understand him correctly) is that the "distinctions of form"—that is, the position, shape, color, motion, and so on of a visual object—are related to (caused by/supported by/projected from) a short-term memory representation that results from winner-take-all mechanisms (the "selection function"), and that this representation is "enriched" by attentional processing.

114 The value of Jackendoff's approach is that it warns us not to assume that the highest levels of the brain must necessarily be the only ones involved in visual awareness. The vivid representation in our brains of the scene directly

他相信对视觉输入的理解(即我们感觉到的是什么)是由三维模型和"概念结构"(conceptual structure，即思维的另一种堂皇的说法)决定的。以上就是他的意识的中间层次理论。

112 下面的例子有助于理解这个理论。如果你看见一个背对着你的人，你只能看见他的后脑勺，而看不见他的脸。然而，你的脑会推断出他有一张脸。我们会这样进行推理，因为如果他转过身来，表明他的头的正面并没有脸，你会感到十分惊讶。以观察者为中心的表象是与你所看见的他的头的后部相对应的。这是你所真实感觉到的。你的脑所作出的关于其正面的推断是从某种三维模型表象得到的。杰肯道夫认为你并不直接察觉这个三维模型(就此而言，同样你也没有直接察觉你自己的思想)。正如一句古语所说：未闻吾所言，安知吾所思?

113 由于初读杰肯道夫的著作①时不容易理解他的语言，我把他的理论的倒数第二种说法放在脚注中。② 如果我对他的理论的理解正确的话，他的观点应用于视觉即"形态上的差异"(包括一个视觉目标的位置、形状、颜色、运动等)是与一种短时记忆有关(或由它引起/支持/投射)的表象，这种表象是一种"胜者为王"机制(一种选择机制)的结果，而注意机制的作用使它更加丰富。

114 杰肯道夫的观点的价值在于，它提醒我们不要假设脑的最高层次必定是视觉觉知中涉及的唯一层次。我们面前的场景在脑中的栩栩如生的

① 想精确理解杰肯道夫的话的读者可以查阅他的著作(他的理论的最终版本，即理论八，还谈到了情感)。

② 他的原话是："每种觉知形式所表达的形态上的差异是由对应该形式的中间层次的结构引起/支持/投射的。该结构是短时记忆表象的匹配集的一部分，而这种表象是由选择机制指派的，并为注意处理所丰富。特别地，语言觉知是由音韵结构引起/支持/投射的，音乐觉知则对应于音乐表面，视觉感知来自2.5维图。"

before us may involve various intermediate levels. Other levels may be less vivid or, as he suggests, we may not be truly conscious of them at all.

115 This does not mean that information flows only from the surface representation to the 3D one; it almost certainly flows in both directions. When you imagine the front of the face in the example above, what you are aware of is a conscious surface representation generated by the unconscious 3D model. This distinction between the two types of representation will probably have to be refined as the subject develops, but it gives us a rough first idea of what it is we are trying to explain.

116 Exactly where these levels may be located in the cortex is not so clear. For vision, they might correspond more to the middle parts of the brain (such as the inferotemporal regions plus certain parietal regions) rather than to the frontal regions of the brain, but exactly which parts of the visual hierarchy (diagrammed in Fig.52) he is referring to remains to be discovered. (This is discussed more fully in Chapter 16.)

117 Having seen how some psychologists view the matter, let's now look at the problem from the point of view of a neuroscientist who knows about neurons, their connections, and the way they fire. What is the general character of the behavior of the neurons associated (or not) with consciousness? In other words, what is the "neural correlate" of consciousness? It is plausible that consciousness in some sense requires the activity of neurons. It may be correlated with a special type of activity of some of the neurons in the cortical system. Consciousness can undoubtedly take different forms, depending on which parts of the cortex are involved. Koch and I have hypothesized that there is only one basic mechanism (or a few) underlying them all. We expect that, at any moment in time, consciousness will

表象可能涉及了许多中间层次。其他层次可能不够生动，或者如他所推测，我们可能根本不能察觉它们(的活动)。

115 这并不意味着信息仅仅是从表面表象流向三维表象：几乎可以肯定双向流动是存在的。在上面的例子中，当你想象一张脸孔的正面时，你所感觉到的正是由无法感知的三维模型产生的可感知的表面表象。随着这一主题的发展，两种表象之间的区别或许还需进一步明确，但它对我们试图解释的问题给出了一种最初的、粗略的看法。

116 目前尚不清楚这些层次在皮质中的准确位置。就视觉而言，它们更可能对应于脑的中部(如下颞叶及某些顶区)，而不是脑的额区。但是杰肯道夫所指的究竟是视觉等级系统中哪个部分，仍有待于探索。

117 在看了一些心理学家对这个问题的观点之后，我们现在再从那些了解神经元、它们的连接以及发放方式的神经科学家的角度来看这个难题。与意识有关(或无关)的神经元的行为的一般特征是什么？换句话说，意识的“神经关联”是什么？从某种意义上说，神经元的活动对意识是必不可少的，这看起来是合理的。意识可能与皮层中某些神经元的一种特殊类型的活动有关。毫无疑问它会具有不同的形式，这取决于皮质的哪些部分参与了活动。科赫和我假设其中仅有一种(或少数几种)基本机制。我们认为，在任意时刻意识将会与瞬间的神经元集合的特定活动类型相对应。这些神经元正是具有相当潜力的候选者的集合中的一部分。因此，在神经水平上，这个问题为：

correspond to a particular type of activity in a transient set of neurons that are a fraction of a much larger set of potential candidates. The questions at the neural level then become:

- Where are these neurons in the brain?
- Are they of any particular neuronal type?
- What is special (if anything) about their connections?
- What is special (if anything) about the way they are firing?

118 How could we go about finding which neurons are involved in visual awareness? And are there any clues that might suggest the manner of their firing that corresponds to such awareness?

119 As we have seen, there are several hints from psychological theories. Awareness is likely to involve some form of attention, so we should study the mechanism the brain uses to attend to one visual object rather than another. Awareness is likely to involve some form of very short-term memory, so we should try to discover how neurons behave when storing and using such memories. Finally, we seem to be able to attend to more than one object at a time. This poses problems for some neuronal theories of awareness, so let us deal with this first.

120 What happens in the brain when we see an object? There are an almost infinite number of possible, different objects that we are capable of seeing. There cannot be a single neuron, often referred to as a "grandmother cell," for each one. The combinatorial possibilities for representing so many objects at all different values of depth, motion, color, orientation, and spatial location are simply too staggering. This does not preclude the existence of sets of somewhat specialized neurons responding to very specific and ecologically highly significant objects like the appearance of a face.

◆这些神经元在脑中位于何处？

◆它们是否属于某些特殊的神经元类型？

◆如果它们的连接具有特殊性，那其特殊性是什么？

◆如果它们的发放存在某些特殊方式，那其特殊方式是什么？

118 怎样去寻找那些与视觉觉知有关的神经元呢？是否存在某些线索暗示了与这种觉知相关的神经发放的模式呢？

119 正如我们已经看到的，心理学理论对我们有若干提示。某些形式的注意很有可能参与了觉知过程，因而我们应当研究脑选择性注意视觉目标的机制。觉知过程很有可能包括某些形式的极短时记忆，因而我们还应探索神经元储存和使用这种记忆时的行为。最后，我们似乎可以一次注意多个目标，这对觉知的某些神经理论提出了问题，因此我们从论述这个问题开始。

120 当我们看见一个物体时，脑子里究竟发生了些什么呢？我们会看到的可能存在的、不同的物体几乎是无限的。不可能对每个物体都存在一个相应的响应细胞(这种细胞常被称为“祖母细胞”)。表达如此多具有不同深度、运动、颜色、朝向及空间位置的物体，其可能的组合多得惊人。不过，这并不排除可能存在某些特异化的神经元集团，它们对相当特定的、生态上有重要意义的目标(如脸的外貌)有响应。

121 It seems probable that, at any moment, any particular object in the visual field is represented by the firing of a set of neurons. Because any object will have different characteristics (form, color, motion, etc.) that are processed in several different visual areas, it is reasonable to assume that seeing any one object often involves neurons in many different visual areas. The problem of how these neurons temporarily become active as a unit is often described as "the binding problem." As an object seen is often also heard, smelled, or felt, this binding must also occur across different sensory modalities.

122 Our experience of perceptual unity thus suggests that the brain in some way binds together, in a mutually coherent way, all those neurons actively responding to different aspects of a perceived object. In other words, if you are currently paying attention to a friend discussing some point with you, neurons that respond to the motion of his face, neurons that respond to its hue, neurons in your auditory cortex that respond to the words coming from his face, and possibly the memory traces associated with knowing whose face it is all have to be "bound" together, to carry a common label identifying them as neurons that jointly generate the perception of that specific face. (The brain can sometimes be tricked into making an incorrect binding, as when you hear the voice coming not from the ventriloquist but from his dummy.)

123 There are several types of binding. A neuron responding to a short line can be considered to be binding the set of points that make up that line. The inputs and the behavior of such a neuron are probably initially determined by our genes (and by developmental processes) that have evolved out of the experiences of our distant ancestors. Other forms of binding, such as that

121 似乎有可能的是，在任意时刻，视野中每个特定的物体均由一个神经元集团的发放来表达。① 由于每个物体具有不同的特征，如形状、颜色、运动等，这些特征由若干不同的视觉区域处理，因而有理由假设，看每一个物体时经常有许多不同视觉区域的神经元参与。这些神经元如何暂时地变成一个整体同时兴奋呢？这个问题常被称为“捆绑问题”(binding problem)。由于视觉过程常伴随听觉、嗅觉或触觉，这种捆绑必然也出现在不同感觉模块之间。②

122 我们都有这种体验，即对物体有整体知觉。这使我们认为，对于已看见的物体的不同特征，所有神经元都产生了积极的响应，而脑通过某种方式相互协调地把它们捆绑在一起。换句话说，如果你把注意力正集中在与你讨论某个观点的朋友上，那么，你脑中有些神经元对他的脸部运动反应，有些对脸的颜色反应，听觉皮层中的神经元则对他的讲的话有反应，还可能挖掘出储存这张脸属于哪个人的那些记忆痕迹，所有这些神经元都将捆绑在一起，以便携带相同的标记以表明它们共同生成了对那张特定的脸的认知(有时候脑也会受骗而作出错误的捆绑，比如把听到的口技表演者的声音当作被模仿物发出的)。

123 捆绑有若干种形式。一个对短线响应的神经元可以认为把组成该直线的各点捆绑在一起。这种神经元的输入和行为最初可能是由基因(及发育过程)确定的，这些基因是我们远古的祖先的经验进化的结果。另一种形式的捆绑，如对熟悉物体的识别，又如熟悉的字母表中的字母，

① 如果一个集团中的神经元空间上离得很近(意味着它们可能有某种相互连接)，接受有些相似的输入，并投射到多少有些相似的区域，那便不会引起任何特别的困难。在这种情况下，它们就像是单个神经网络中的神经元。令人遗憾的是，通常这种简单的神经网络每次只能处理一个目标。

② 现在还不能完全肯定捆绑问题如我所说的那样真实存在，或者脑通过某种未知的技巧绕了过去。

required for the recognition of familiar objects, such as the letters of a well-known alphabet, may be acquired by frequently repeated experience—that is, by being overlearned. This probably implies that many of the neurons involved have as a result become strongly connected together. These two forms of fairly permanent binding could produce neurons that, collectively, could respond to many objects (such as letters, numbers, and other familiar symbols), but there are not enough neurons in the brain to code for the almost infinite number of conceivable objects. The same is true of language. Each language has a large but limited number of words, but the number of possible well-formed sentences is almost infinite.

124 The binding we are especially concerned with is a third type, being neither determined during early development nor overlearned. It applies particularly to objects whose exact combination of features may be quite novel to us, such as seeing a new animal at the zoo. It is unlikely that all the actively involved neurons will be strongly connected together, at least in most cases. This binding must arise rapidly. By its very nature, this third type is largely transitory and must be able to bind visual features together into an almost infinite variety of possible combinations, although it may be able to do this for only a few combinations at any one time. If a particular stimulus is repeated frequently, this third type of transient binding may eventually build up the second, overlearned type of binding.

125 Unfortunately, we don't yet know how the brain expresses this third type of binding. What is especially unclear is whether, in focused awareness, we are conscious of only one object at a time, or whether our brains can deal with several objects simultaneously. We certainly appear to be aware of more than one object at once, but could this be an illusion? Does the brain really deal with several objects one after another in such rapid succession that they

可能从频繁的、重复性的体验中获得，也就是说，是通过反复学习得到的。这或许意味着参与某个过程的大量神经元最终彼此有紧密的连接。① 这两种形式的相当永久的捆绑可以产生一些神经元群体，它们作为整体可以对许多物体(如字母、数字及其他熟悉的符号)作出反应。但脑中不可能有足够多的神经元去编码几乎无穷数目的可感知的物体。对语言也是如此。每种语言都有大量的但数目有限的单词，而形式正确的句子的数目几乎是无限的。

124 我们最关心的是第三种形式的捆绑。它既不是由早期发育确定的，也不是由反复学习得到的。它特别适用于那些对我们而言比较新奇的物体，比如说我们在动物园里看见的一只新来的动物。在多数情况下，积极地参与该过程的神经元之间未必有较强的连接。这种捆绑必须能够快速实现。因此它主要是短暂的，并必须能够将视觉特征捆绑在一起构成几乎无限多种可能的组合，只不过也许在某一时刻它只能形成不多的几种组合。如果一种特定的刺激频繁地出现，这种第三种形式的瞬间的捆绑终将会建立起第二种形式的捆绑即反复学习获得的捆绑。

125 遗憾的是，我们并不了解脑如何表达第三种形式的捆绑。特别不清楚的是，在集中注意的觉知时，我们究竟每次仅仅感知一个物体，还是可以同时感知多个物体。表面上看，我们每次能感觉的绝不只一个物体，但这是否可能是错觉呢？脑真的能如此快速一个接一个地处理多个物体的信息，以致它们好像同时出现在我们脑海中吗？也许我们每次只能注意一个物体，但在注意之后，我们可以大致记住其中几个。

① 回忆一下，大多数皮质神经元具有成千上万的连接，其中很多在开始时很弱，这意味着只有当脑已经大致按正确方法构造好，才可能容易地、正确地进行学习。

appear to be simultaneous? Perhaps we can attend to only one object at a time but, having attended, can briefly "remember" several of them. Since we do not know for certain we have to consider all these possibilities. Let's assume first that the brain deals with one object at a time.

126 What sort of neural activity could correspond to binding? Of course the neural correlate of consciousness may involve only one particular type of neuron: for example, one sort of pyramidal cell in one particular cortical layer. The simplest idea would be that awareness occurs when some members of this special set of neurons fire at a very high rate (e.g., at about 400 or 500 Hertz), or that the firing is sustained for a reasonably long time. Thus "binding" would correspond to a relatively small fraction of cortical neurons, in several distinct cortical areas, firing very fast at the same time (or for a long period). This is likely to have two consequences: the rapidity or duration of firing would increase the effect produced on the neurons to which this active set of neurons projects, neurons corresponding to the implications (or the "meaning") of the object that one was aware of at that moment. In addition, the rapid (or sustained) firing might activate some form of very short term memory.

127 This simple idea will not work if the brain has to be aware of more than one object at precisely the same moment, and even with one object the brain may have to distinguish between figure and ground. To grasp this, suppose that in the visual field, near the center of vision, there is just a red circle and a blue square. Then some of the neurons corresponding to awareness would be firing fast (or for a sustained period), some signalling "red," some "blue," and others signalling "circle" and still others "square." How could the brain know which color to put with which shape? In other words, if awareness corresponded merely to rapid (or sustained) firing, the brain might easily confuse the attributes of different objects.

因为我们并不确切知道真相，所以我们必须考虑所有这些的可能性。让我们先假设脑每次只能处理一个物体。

126 究竟哪种类型的神经活动可能与捆绑有关呢？当然，意识的神经关联可能仅包含一种特殊类型的神经元，比如某个特殊皮质上的一种锥体细胞。一种最为简单的观点是，当这个特殊神经元集团的某些成员以一个相当高的频率发放（比如大约400或500赫兹），或维持一段适当长时期的发放，此时觉知便出现了。这样，捆绑仅对应于皮质神经元中相当小的一部分，它们在皮质中若干不同的区域同时高频发放（或都发放很长一段时间）。看起来这会有两个结果：这种快速或持续性的发放将增强这个兴奋的神经元集团对所投射到的神经元的影响，而这些被影响的神经元对应于此时脑所觉知的物体的“意义”。同时，这种快速的（或持续的）发放将激活某种形式的极短时记忆。

127 然而，如果脑能同时精确地觉知不只一个物体，那么这种观点就不能成立。即便脑每次只处理一个物体，它也必须区分目标和背景。为了理解这一点，不妨想象在一个视野中靠近视觉中央的地方，恰好有一个红色的圆和一个蓝色的方块。那么，对应于觉知的某些神经元将会快速发放（或持续发放一段时间），有些标识红色，有些标识蓝色，其他一些标识圆，当然还有一些标识方块。脑又怎样知道哪种颜色与哪种形状相互搭配呢？换句话说，如果觉知仅仅对应于快速（或持续）的发放，脑多半会将不同物体的属性混在一起。

128 There are several ways to get around this difficulty. An object may only enter vivid awareness if the brain "attends" to it. Perhaps the attentional mechanism can strengthen the firing of the neurons responding to one of the objects while weakening the activity of the neurons responding to other objects. If this were true, the brain would be able to deal with one object after another, as the attentional mechanism jumped from one object to the next. After all, this is what we do when we move our eyes. We attend first to one part of the visual field, then to another part, and so on. The attentional mechanism we need would have to be faster than this and operate between eye movements, when the eyes are stationary, since we can see several objects without moving our eyes.

129 A second alternative is that, in some manner, the attentional mechanism makes different neurons fire in somewhat different ways. The key idea here is that of correlated firing. It is based on the idea that what matters is not just the average rate of firing of a neuron but the exact moments at which each neuron fires. For simplicity, let us consider only two objects. The neurons associated with the properties of the first object will all fire at the same moment, in some sort of pattern. The neurons associated with the second object will also all fire together but at different moments from the first set.

130 An idealized example may make this clearer. Suppose the neurons of the first set fire very fast. Perhaps this set will also fire again, say 100 milliseconds later, and again 100 milliseconds after the second burst, and so on. Suppose the second set also produces a set of fast spikes every 100 milliseconds or so, but at times when the first set is silent. The other parts of the brain will not confuse the neurons in the first set with those in the second set, because the two sets never fire at the same moment (see Fig. 57).

128 有许多方法可以解决这个困难。或许只有当脑注意某个物体时才会形成对它的生动的觉知。或许注意机制使对被注意的物体反应的神经元的活动增强，同时削弱对其他物体反应的神经元的活动。倘若如此，脑只能随着注意机制从一个物体跳跃到另一个物体，一个接一个地进行处理，毕竟，当我们转动眼睛时，情形是这样的。我们先注意视野中的一部分区域，然后转而注意另一区域，如此下去。由于我们不动眼睛就能同时看见多个物体，故注意机制的速度必须比上述情况要快，并能在眼的两次转动之间工作。

129 另一种替代的解释是，注意机制以某种方式使不同的神经元以多少不同的方式发放。此时的关键在于相关发放。① 它基于这样一种观点，即重要的不仅仅在于神经元的平均发放率，更是每个神经元发放的精确时间。简单起见，让我们仅考虑两个物体。对第一个物体的特征反应的神经元都在同一时刻以某种模式发放，相应于第二个物体的神经元也都同时发放，但发放的时间与第一个神经元集团不同。

130 用一个理想化的例子可以把这个问题讲得更清楚。假设第一集团中的神经元发放很快。或许它们还会再次发放，比如说是在100毫秒以后。同样，在第二簇发放后过100毫秒又再次发放，如此下去。假设第二群神经元也同样每隔大约100毫秒发放一簇高速脉冲，但是只在第一群神经元处于静息状态的时候才发放。这样，脑中的其他部分不会把这两群神经元的发放混在一起，因为它们从不会同时发放。②（图57）

---

① 这一观点是克里斯托夫·冯·德·马尔斯博格(Christoph von der Malsburg)在1981年的一篇相当难懂的文章中提出的。此前，彼得·米尔纳(Peter Milner)及其他人也叙述过。

② 当然，一个群内轴突的脉冲并不必彼此精确同步。当电位变化沿接受脉冲的神经元的树突传向细胞体时，从时间上看，它们的效果会有所扩散。此外，当脉冲沿许多不同轴突传播的时间延迟也有不同。这样，一群神经元的发放时间只需在大约几毫秒范围内是同时的。

131 The basic idea here is that spikes arriving at a neuron at the same moment will produce a larger effect than the same number of spikes arriving at different times. The theoretical requirement is that the firing of the neurons in each set should be strongly correlated with each other, while at the same time firing of neurons in different sets should be weakly correlated, or not at all.

132 Let us now return to our main problem. This is to locate the "awareness" neurons and to discover what it is that makes their tiring symbolize what we see. This is like trying to solve a murder mystery. We know something about the victim (the nature of awareness) and we know various miscellaneous facts that may be related to the crime. Which approaches look the most promising and how should we follow them up?

图 57　时间轴上每根短的竖线表示一个神经元在某一时刻的发放。第一条水平线显示了标识“红色”的神经元的发放。下一条线则是标识“圆”的神经元的行为，等等。因为表示红色的神经元和表示圆的神经元大致在相同时间发放，而它们与表示蓝色的神经元的发放时间相差很大。脑因此可推断出圆是红色的而不是蓝色的。这种情况常被说成表示“红色”和“圆”的神经元的发放是相关的(表示“蓝色”和“方块”的神经元也是如此)，而互相关(比如“红色”和“方块”之间)为 0(为了说明问题，这个例子被大大地简化了)。

131　此处的基本观点是：同时到达一个神经元的许多脉冲将比不同时刻到达的同样数目的脉冲产生更大的效果。① 其理论要求是同一群神经元的发放有较强的关联，同时不同群的神经元之间关联较弱，甚至根本没有关联。②

132　让我们回到主要问题上，即定位“觉知”神经元并揭示使它们的发放象征着我们所看见的东西的机制是什么。这就像试图侦破一个神秘的谋杀案。我们了解受害者(觉知的本质)的一些线索，还知道可能与犯罪有关的许多杂乱的事实。哪方面进展看起来最有希望呢？由此下一步又该怎么做呢？

---

① 一种稍微详尽的理论引入了轴突传递过程中这种必然发生的时间延迟，使得离细胞体较远的突触比较近的略早接收到输入。这样，由于树突延迟时间上的小的差异，两个信号的最大效应将同时达到细胞体。更为详细的理论还考虑局部的抑制性神经元产生的抑制性效果的调节。所有这种定性的考虑应可通过小心的模拟定量化，如在计算机上模拟单个神经元在这种环境下的行为方式，并引入时间延迟等因素。

② 这种发放不太可能像图 57 表示的那样有规则。

133 The most direct set of clues would be any evidence that caught the suspect in the act. Can we find neurons whose behavior always correlates with the retevant visual percept? One way to do this would be to set up situations (such as viewing the Necker cube, described in Chapter 3) in which the visual information coming into the eyes remains the same but the percept changes. Which neurons change their firing, or style of firing, when the percept changes and which do not? If a particular neuron does not follow the percept, this provides it with an alibi. On the other hand if its firing does correlate with the percept we still have to decide whether it is the actual murderer or only an accomplice.

134 Suppose we try another tack. Can we pin down the crime to a particular town, or a particular district or apartment building? This would make our search more efficient. In our problem, can we say roughly where in the brain the awareness neurons for vision are likely to be located? Obviously we suspect the neocortex, although we cannot entirely neglect its close neighbors, the thalamus and the claustrum, nor even that older visual system, the superior colliculus, to say nothing of the corpus striatum and the cerebellum. Visual awareness is unlikely to reside in areas such as the auditory cortex, so we can confine most of our attention to the many visual cortical areas shown in Figure 48. Perhaps we might find evidence that some areas are more heavily involved than others.

135 This will not pin down the murderer but it may lead us in the right direction. Is the criminal likely to be any particular type of person—a powerful man, for example, a disturbed teenager, or a gang? In our case, what kind of neurons are likely to be involved? Excitatory neurons? Inhibitory neurons? Stellate cells, or pyramidal cells? If they are in the cortex, in which layer or layers are they to be found?

136 Another tack would be to see if there were any forms of communication that might give the game away. If it is the work of a gang, did they use a cellular

133 最直接的线索将是在现场捉住嫌疑犯。我们能否发现那些行为一直与视觉觉知有关的神经元呢？一种可能的办法是设置一种环境使进入眼睛的视觉信息保持不变，但知觉会发生变化。当知觉改变时，哪些神经元会改变其发放，或改变发放的方式，而哪些神经元不会改变？如果一个特定神经元的发放不随知觉改变，这就提供了一个“它不在现场”的证据。另外，如果它的发放确实与知觉有关，我们还需确定它是“真凶”还是“从犯”。

134 让我们换一种策略。我们能否将案发地点限定在某个特定的城镇、一个区或建筑物中的单元呢？这将使我们的搜索变得更有效。在我们的问题中，即我们能否大致说出视觉觉知神经元在脑中可能的定位呢？显然，我们推测它在新皮质。虽然我们不能完全忽略新皮质的紧密的近邻，如丘脑和屏状核，以及在进化上比较古老的视觉系统（older visual system）和上丘，更不能忽略纹状体和小脑。视觉觉知不太可能存在于听皮质等区域，因此我们可以将注意力主要集中在图 48[①] 所示的许多视觉皮质区域。或许我们能发现证据表明某些区域比其他区域被更紧密地牵涉视觉觉知。

135 这尚不足以找到凶手，但可能将我们引向正确的方向。罪犯可能是某种特殊类型的人，比如说，一名强壮的男子，一名心理失常的青少年，或者一群匪徒。在此处，可能涉及哪些类型的神经元呢？是兴奋性神经元？还是抑制性神经元？是星形细胞，还是锥体细胞？如果它们是在皮质上，那么在皮质中哪一层或哪些层才能找到它们呢？

136 另一种策略将是寻找它们之间是否有某些形式的通信联系，从而使之露出马脚。如果这是一帮匪徒所为，他们是否在汽车里使用了移动电

① 此处的图 48 是［英］弗朗西斯 · 克里克著，汪云九等译，湖南科学技术出版社 2018 年出版的《惊人的假说》中的图 48，而非本导引中的图。

telephone in a car? In neural terms, does awareness depend on some particular form of neural circuitry that only occurs in special places in the brain?

137 Perhaps one should look for a motive for the crime. What benefit was the murder to the murderer? Did he profit financially and, if so, where was the money sent? If we could look there, we might be able to track backwards to the murderer. In neural terms, to what parts of the brain is the visual information dispatched? And how are these parts connected to the visual areas of the cortex?

138 Alternatively, one could ask if there is any special behavior that might lead us to the suspect. This might be correlated firing between groups of neurons, or perhaps rhythmical or patterned firing of one sort or another. If a gang is suspected, who is likely to be the leader, who decides what the gang should do? We believe that awareness often involves the brain making decisions as to which interpretation is the most plausible. There may be a winner-take-all mechanism involving certain sets of neurons. If we could spot such a mechanism the neuronal nature of the winners might point us toward the awareness neurons. Was some particular weapon used? As mentioned earlier, we strongly suspect that very short term memory may be an essential feature of awareness. Also that some form of attentional mechanism may help to produce vivid awareness, so anything we can learn about how they work neuronally might lead us in the right direction.

139 In short, there are a number of experimental approaches that might conceivably lead us to the neurons and to the behavior that we are looking for. At this stage we can't afford to neglect any lead that looks even remotely promising, since we have a difficult problem to solve. Let us now examine in more detail the nature of these different approaches.

话？用神经学的术语说，觉知是否依赖于仅仅出现在脑中特定位置上的某些特别形式的神经回路呢？

137 或许有人会寻找犯罪的动机。凶手犯罪能得到某种利益吗？他是否能得到经济上的好处呢？倘若如此，赃款被运到哪里去了呢？如果我们能在那里找到的话，我们或许就能追查到凶手。用神经的术语讲，视觉信息被发送到脑中的哪些部位了？这些部位又是如何与皮质视觉区域连接的呢？

138 此外，有人会问是否有某些特殊的行为将我们引至嫌疑犯。这或许是神经元群之间的相关发放，或许是这种或那种形式的节律或模式发放。如果我们怀疑是一群匪徒所为，谁最有可能是头目呢？谁决定了匪徒们的行动？我们相信，觉知过程中经常涉及脑对哪种解释最为合理进行判断。这可能是一种包含某些神经元集团的一种"胜者为王"机制。如果我们能发现这种机制，那么胜者的神经本质也许能将我们指向觉知神经元。其作案时是否用过什么特殊的武器呢？正如前面所述，我们很有把握地猜测极短时记忆是觉知的本质特征。同时，某些形式的注意机制或许协助产生生动的觉知，因此，我们所知道的关于这些在神经水平的工作的任何知识都将把我们引向正确的方向。

139 简单地说，通过大量的实验手段能从观念上将我们引导到所寻找的神经元及它们的行为。现阶段，因为我们要解决的问题十分困难，我们不能忽略任何哪怕看起来只有很少希望的线索，现在，让我们更仔细地检查这些不同的途径的本质。

四、科学方法

《几何原本》导引

《自然科学经典导引》为什么要讲数学？为什么要谈《几何原本》？很多人可能都会问这两个问题。

关于这两个问题，直接和简短的回答是：数学是自然科学的基本工具；《几何原本》的公理化方法从根本上影响和推动了数学、自然科学的发展。

数学是自然科学的基本工具，这应该是大家的共识，这里无需赘言。大家可能更好奇：《几何原本》真的有这么大的影响力吗？

爱因斯坦的评价，应该更有说服力：

“我们推崇古代希腊是西方科学的摇篮。在那里，世界第一次目睹了一个逻辑体系的奇迹，这个逻辑体系如此精密地一步一步推进，以致它的每一个命题都是绝对不容置疑的——我这里说的是欧几里得几何。”①

《几何原本》的意义不只在数学本身，更重要的是它体现出的理性思维方式对后世的深远影响；同样地，阅读《几何原本》不是为了让大家学习数学知识，而是为了进一步明白什么是科学思维。

---

① ［美］阿尔伯特·爱因斯坦：《爱因斯坦文集(第一卷)》，许良英，范岱年编译，商务印书馆1976年版，第313页。

该书英文版书名是 *Elements*，故也称《原本》。作者是古希腊数学家欧几里得(约公元前 325 年—公元前 270 年)，生平不详。现存的欧几里得的各种画像也是后人的想象。他活跃于托勒密一世(公元前 323—前 283 年)时期的亚历山大里亚，而且很有可能曾在柏拉图学院学习。

《原本》的第一个中译本是 1607 年由利玛窦和徐光启合译的，定名为《几何原本》(以下简称《原本》)，当时翻译了前 6 卷。后续部分于 1857 年由伟烈亚力和李善兰译出。

该书有以下几个突出的特点：

(1)成书时间早。《原本》完成于 2300 多年前，一般认为，欧几里得是总结了当时已有的数学成果，用自己的方法给出证明。《原本》一书，使我们不禁赞叹，人类的思辨水平在 2300 多年前就达到了如此的高度。

(2)内容丰富。《原本》全书共 13 卷，中译本的正文也有 599 页，约 64 万字之多(见《几何原本》，兰纪正、朱恩宽译，译林出版社 2014 年版)。内容涵盖了平面几何(第 1~6 卷)、数论(第 7~9 卷)、无理数(第 10 卷)、立体几何(第 11~13 卷)。

(3)影响力大。《原本》以 5 个公设、5 个公理为最基本的依据，言必有据地推演出 467 个命题，构建了一座数学知识的大厦。《原本》被称为“最成功和最有影响的教科书”，是发行量仅次于《圣经》的书籍①。《原本》的公理化体系和演绎证明方法，极大地影响了数学、科学的发展，对整个人类的思维方式有着极大的影响。它的逻辑严谨性直到 19 世纪才被超越。2300 年后的今天，其前六卷仍然是学校教科书《平面几何》的基本内容，被认为是所有受教育的人都应该学习的知识。

在传统方法中，公理是一个不证自明的命题，一般比较简单或者符合直觉，其真实性被视为是理所当然的，且被当作演绎及推导其他结论的起点。公设、公理有何不同？词典 *The Concise Oxford Dictionary of Mathematics* 的词条“postulate”给出了一个极简短的回答：

① Carl. B. Boyer：《数学史》，秦传安译，中央编译出版社 2012 年版，第 136 页。

postulate = AXIOM.

即“公设 = 公理”。两者是同义词，现在一般统一称为公理。

什么是公理化方法？公理化方法是一种从少数几个基本概念、公理出发，通过逻辑的演绎推理得出一系列命题，从而建立整个理论体系的方法。古希腊哲学家曾经探索世界的本原——世界的物质形式如此多样，是否存在几个基本元素构造出复杂多样的大千世界？我们知道，柏拉图对此给出的答案是四元素“火”“气”“水”“土”。类似地，面对已有的各种数学成果，构造或演绎它们的本原在哪里，是否有几个最基本的前提，能推演出当时的整个数学。由公理出发，逻辑地演绎一门学科，成为后来数学、物理等学科著书立说的范本，牛顿的《自然哲学之数学原理》就是按照这个公理化体系完成的。

篇幅所限，本课程选录了《原本》第 I 卷的前 20 个命题，再加上命题 47（勾股定理）。我们以其中的两个命题为例，简要说明《原本》的逻辑演绎特点。例如：

命题 2：由一个所给定的点（作为端点）作一线段等于已知线段。

这个问题就是把一个长度进行移动。《原本》完成命题 2 使用了多少基本依据呢？1 次公设 1，1 次公设 2，2 次公设 3，1 次公理 1，1 次公理 3，2 次定义 15，1 次命题 1。

为什么要这么麻烦，直接用圆规比划过去不就完了吗？但是，这样不完美，因为这会多一个公设，即需要假定圆规是可以固定的。而这是多余的，因为它可以由已有的前提推导出来。也就是说，欧几里得选择的公设、公理是希望尽可能少，这是一个很重要的初衷。

命题 20 的内容是“在任何三角形中，任意两边之和大于第三边”。

如下图，需要证明 $BA+AC>BC$。

这个命题据说曾经遭到一些人的嘲笑：这是驴子也知道的事情，难道还需要证明吗？假如驴子在点 B，它要去吃点 C 的食物，肯定是直接沿直线由 B 走到 C，不可能傻到从 B 先到 A，再走到 C 啊。

欧几里得证明这个结论使用到了公理 5、命题 5、命题 19。但是完成命题

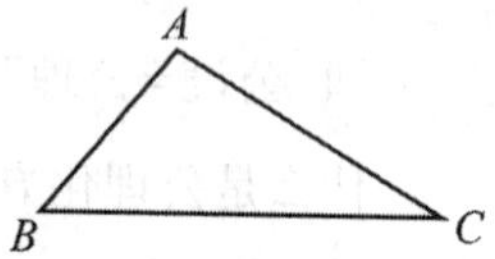

19 又使用了命题 18。逐步追溯下去，不考虑公理和公设，完成命题 20 的证明需要命题 1，2，3，4，5，7，8，9，10，11，13，15，16，18，19 的支撑。

欧几里得为什么要大费周章证明驴通过本能就能知道的东西？事实上，这个看似无聊、不必要的证明，正是《原本》所体现的“言必有据”的理性思维方式。直觉不是总是可靠的，我们在这个直觉上不犯错误，不能保证我们在每一个直觉认识上不犯错误。数学家 E. T. Bell 有一句名言：“Obvious” is the most dangerous word in mathematics。要阐明一个观点或结论，必须要有严谨的逻辑推导，确保其正确性。这种一丝不苟的严谨逻辑思维，影响了两千多年来人们研究数学、探索自然科学的方法；促进了近代科学的发展；甚至成为你我思维的基因。换句话说，今天我们每一个人思考、看待世界的方式，多多少少在受到欧几里得的影响。

为什么我们要读《原本》？我们是希望大家看到《原本》的公理化方法所承载和传达的理念。同学们经常问一个问题：学习数学有什么用啊？我的看法是，数学不完全是为了用；它有时候就像一个智力游戏，是凭借兴趣驱动的逻辑思维过程。数学如果只考虑怎么用，恐怕就不会有跨越式的发展。科学研究也是如此。我们应该鼓励在好奇心驱使下的研究，这样才可能使得科学朝着未知的世界发展。传说欧几里得曾经把钱退给学生，让他离开，因为学生想在学习中获取实利。这个故事大家都知道，可能很多人只是一笑而过，但其中的深意正是要认识数学乃至科学的“无功利性”。认识数学的“无用性”、科学的“无功利性”，这也是“自然科学经典导引”课程的一个重要目标。

关于《原本》在科学发展中的作用，请让我再次引用爱因斯坦的评价：

“西方科学的发展是以两个伟大成就为基础的，那就是：希腊哲学家发明的形式逻辑体系（在欧几里得几何学中），以及通过系统的实验发现有可能找出因果关系（在文艺复兴时期）”。①

① ［美］阿尔伯特·爱因斯坦：《爱因斯坦文集（第一卷）》，许良英，范岱年编译，商务印书馆 1976 年版，第 574 页。

这个评价一方面肯定了《原本》所承载的公理化方法的重要性，另一方面也不至于过分夸大其在科学研究中的地位。今天看来，《原本》也有很多不完美的地方，这些不完美，正好促使后来的数学家不断探索新的更高形式的统一。下一章我们要讲到的非欧几何，就是在这个背景下产生的。

最后，让我们做一个有趣的假设：如果欧几里得穿越到今天，他和今天的大学生相比，谁更厉害？我的看法是，欧几里得即使不知道现在的微积分、大学物理，没听说过DNA，不会用手机、电脑，但是他的思维水平、逻辑能力，做研究的功力、毅力，仍然应该远超常人。他依然值得我们敬佩和学习。谁更厉害，不是看知道东西的多与少，是取决于思维能力与水平、创新创造的能力和水平。这也正好是我们这个通识课要传达的一个理念。那就让我们坐下来，阅读经典，和欧几里得对话，看看他的《原本》能否给我们打开新的“脑洞”。

（黄正华）

Elements

Euclid

Book I

DEFINITIONS

(1) A **point** is that which has no part.

(2) A **line** is breadthless length.

(3) The extremities of a line are points.

(4) A **straight line** is a line which lies evenly with the points on itself.

(5) A **surface** is that which has length and breadth only.

(6) The extremities of a surface are lines.

(7) A **plane surface** is a surface which lies evenly with the straight lines on itself.

(8) A **plane angle** is the inclination to one another of two lines in a plane which meet one another and do not lie in a straight line.

(9) And when the lines containing the angle are straight, the angle is called **rectilineal**.

(10) When a straight line set up on a straight line makes the adjacent angles equal to one another, each of the equal angles is **right**, and the straight line standing on the other is called a **perpendicular** to that on which it stands.

(11) An **obtuse angle** is an angle greater than a right angle.

(12) An **acute angle** is an angle less than a right angle.

(13) A **boundary** is that which is an extremity of anything.

(14) A **figure** is that which is contained by any boundary or boundaries.

(15) A **circle** is a plane figure contained by one line such that all the straight

---

本文选自 Sir Thomas L. Health 著 *The Thirteen Books of Euclid's Elements*。

几何原本

欧几里得

第I卷

定义

(1) **点**是没有部分的.

(2) **线**只有长度而没有宽度.

(3) **一线**①的两端是点.

(4) **直线**是它上面的点一样地平放着的线.

(5) **面**只有长度和宽度.

(6) 面的边缘是线.

(7) **平面**是它上面的线一样地平放着的面.

(8) **平面角**是在一平面内但不在一条直线上的两条相交线相互的倾斜度.

(9) 当包含角的两条线都是直线时，这个角叫做**直线角**.

(10) 当一条直线和另一条直线交成的邻角彼此相等时，这些角的每一个叫做**直角**，而且称这一条直线**垂直**于另一条直线.

(11) 大于直角的角叫做**钝角**.

(12) 小于直角的角叫做**锐角**.

(13) **边界**是物体的边缘.

(14) **图形**是被一个边界或几个边界所围成的.

(15) **圆**是由一条线围成的平面图形，其内有一点与这条线上的点连接成的所有线段都相等.

本文选自[古希腊]欧几里得：《几何原本》，兰纪正，朱恩宽译，梁宗巨，张毓新，徐伯谦校订，译林出版社2014年版，第1-17页，第39-40页。

① 不一定是直线。

lines falling upon it from one point among those lying within the figure are equal to one another.

(16) And the point is called the **centre** of the circle.

(17) A **diameter** of the circle is any straight line drawn through the centre and terminated in both directions by the circumference of the circle, and such a straight line also bisects the circle.

(18) A **semicircle** is the figure contained by the diameter and the circumference cut off by it. And the centre of the semicircle is the same as that of the circle.

(19) **Rectilineal figures** are those which are contained by straight lines, **trilateral** figures being those contained by three, **quadrilateral** those contained by four, and **multilateral** those contained by more than four straight lines.

(20) Of trilateral figures, an **equilateral triangle** is that which has its three sides equal, an **isosceles triangle** that which has two of its sides alone equal, and a **scalene triangle** that which has its three sides unequal.

(21) Further, of trilateral figures, a **right-angled triangle** is that which has a right angle, an **obtuse-angled triangle** that which has an obtuse angle, and an **acute-angled triangle** that which has its three angles acute.

(22) Of quadrilateral figures, a **square** is that which is both equilateral and right-angled; an **oblong** that which is right-angled but not equilateral; a **rhombus** that which is equilateral but not right-angled; and a **rhomboid** that which has its opposite sides and angles equal to one another but is neither equilateral nor right-angled. And let quadrilaterals other than these be called **trapezia**.

(23) **Parallel** straight lines are straight lines which, being in the same plane and being produced indefinitely in both directions, do not meet one another in either direction.

POSTULATES

Let the following be postulated:

(1) To draw a straight line from any point to any point.

(16) 而且把这个点叫做**圆心**.

(17) 圆的**直径**是任意一条经过圆心的直线在两个方向被圆周截得的线段，且把圆二等分.

(18) **半圆**是直径和由它截得的圆周所围成的图形．而且半圆的心和圆心相同.

(19) **直线形**是由线段围成的，**三边形**是由三条线段围成的，**四边形**是由四条线段围成的，**多边形**是由四条以上线段围成的.

(20) 在三边形中，三条边相等的，叫做**等边三角形**；只有两条边相等的，叫做**等腰三角形**；各边不等的，叫做**不等边三角形**.

(21) 此外，在三边形中，有一个角是直角的，叫做**直角三角形**；有一个角是钝角的，叫做**钝角三角形**；有三个角是锐角的，叫做**锐角三角形**.

(22) 在四边形中，四边相等且四个角是直角的，叫做**正方形**；角是直角，但四边不全相等的，叫做**长方形**；四边相等，但角不是直角的，叫做**菱形**；对角相等且对边也相等，但边不全相等且角不是直角的，叫做**斜方形**；其余的四边形叫做**不规则四边形**.

(23) **平行直线**是在同一平面内的一些直线，向两个方向无限延长，在不论哪个方向它们都不相交。

公设

(1) 由任意一点到另外任意一点可以画直线.

(2) 一条有限直线可以继续延长.

(2) To produce a finite straight line continuously in a straight line.

(3) To describe a circle with any centre and distance.

(4) That all right angles are equal to one another.

(5) That, if a straight line falling on two straight lines make the interior angles on the same side less than two right angles, the two straight lines, if produced indefinitely, meet on that side on which are the angles less than the two right angles.

COMMON NOTIONS

(1) Things which are equal to the same thing are also equal to one another.

(2) If equals be added to equals, the wholes are equal.

(3) If equals be subtracted from equals, the remainders are equal.

(4) Things which coincide with one another are equal to one another.

(5) The whole is greater than the part.

PROPOSITION

Proposition 1.

On a given finite straight line to construct an equilateral triangle.

Let *AB* be the given finite straight line. Thus it is required to construct an equilateral triangle on the straight line *AB*.

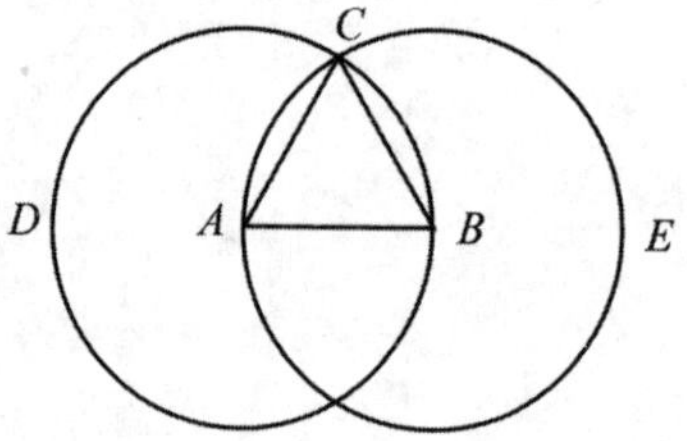

With centre *A* and distance *AB* let the circle *BCD* be described; [Post.3]

again, with centre *B* and distance *BA* let the circle *ACE* be described; [Post.3]

and from the point *C*, in which the circles cut one another, to the points *A*, *B* let the straight lines *CA*, *CB* be joined. [Post.1]

Now, since the point *A* is the centre of the circle *CDB*, *AC* is equal to *AB*. [Def.15]

Again, since the point *B* is the centre of the circle *CAE*, *BC* is equal to *BA*. [Def.15]

But *CA* was also proved equal to *AB*; therefore each of the straight lines *CA*,

(3)以任意点为心及任意的距离①可以画圆.

(4)凡直角都彼此相等.

(5)同平面内一条直线和另外两条直线相交，若在某一侧的两个内角的和小于二直角，则这二直线经无限延长后在这一侧相交②。

公理

(1)等于同量③的量彼此相等.

(2)等量加等量，其和仍相等.

(3)等量减等量，其差仍相等.

(4)彼此能重合的物体是全等④的.

(5)整体大于部分.

命题

命题 1

在一个给定的有限直线上作一个等边三角形。

设 AB 是所给定的有限直线.

那么，要求在线段 AB 上作一个等边三角形。

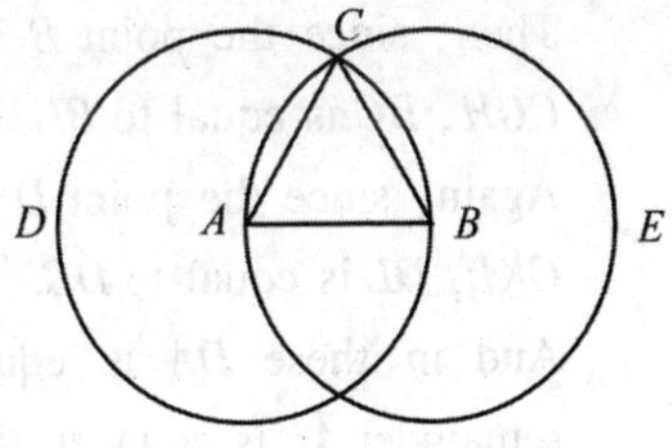

以 A 为心，且以 AB 为距离画圆 BCD；[公设 3]

再以 B 为心，且以 BA 为距离画圆 ACE；[公设 3]

由两圆的交点 C 到 A、B 连线 CA，CB. [公设 1]

因为，点 A 是圆 CDB 的圆心，AC 等于 AB. [定义 15]

① 到此原文中无“半径”二字出现，此处“距离”即圆的半径。

② 这就是大家提到的欧几里得第 5 公设，即现行平面几何中的平行公理的原始等价命题。

③ 这里的“量”与公理 4 中的“物体”在原文中是同一个字 thing。

④ 为了区别面积相等与图形相等，译者将图形“相等”译为“全等”。

CB is equal to *AB*.

And things which are equal to the same thing are also equal to one another; [C.N. 1] therefore *CA* is also equal to *CB*.

Therefore the three straightlines *CA*, *AB*, *BC* are equal to one another.

Therefore the triangle *ABC* is equilateral; and it has been constructed on the given finite straight line *AB*.

(Being) what it was required to do.

Proposition 2.

To place at a given point (asan extremity) a straight line equal to a given straight line.

Let *A* be the given point, and *BC* the given straight line.

Thus it is required to place at the point *A* (as an extremity) a straight line equal to the given straight line *BC*.

From the point *A* to the point *B* let the straight line *AB* be joined; [Post.1] and on it let the equilateral triangle *DAB* be constructed. [I. 1]

Let the straight lines *AE*, *BF* be produced in a straight line with *DA*, *DB*; [Post.2] with centre *B* and distance *BC* let the circle *CGH* be described; [Post.3] and again, with centre *D* and distance *DG* let the circle *GKL* be described. [Post.3]

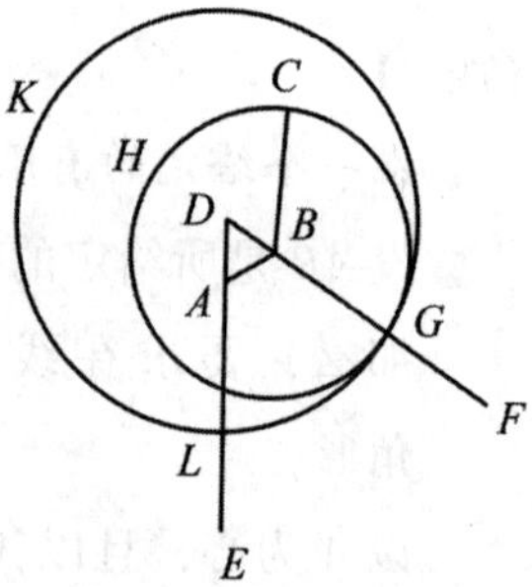

Then, since the point *B* is the centre of the circle *CGH*, *BC* is equal to *BG*.

Again, since the point *D* is the centre of the circle *GKL*, *DL* is equal to *DG*.

And in these *DA* is equal to *DB*; therefore the remainder *AL* is equal to the remainder *BG*. [C.N.3]

But *BC* was also proved equal to *BG*; therefore each of the straight lines *AL*, *BC* is equal to *BG*.

And things which are equal to the same thing are also equal to one another; [C.N.1] therefore *AL* is also equal to *BC*.

Therefore at the given point *A* the straight line *AL* is placed equal to the given straight line *BC*.

(Being) what it was required to do.

又点 B 是圆 CAE 的圆心，BC 等于 BA. [定义 15]

但是，已经证明了 CA 等于 AB；所以线段 CA，CB 都等于 AB.

而且等于同量的量彼此相等； [公理 1]

因此，CA 也等于 CB.

三条线段 CA，AB，BC 彼此相等.

所以三角形 ABC 是等边的，即在已知有限直线 AB 上作出了这个三角形.

这就是所要求作的.

命题 2

由一个所给定的点（作为端点）作一线段等于已知线段.

设 A 是所给定的点，BC 是已知线段，

那么，要求由点 A（作为端点）作一线段等于已知线段 BC.

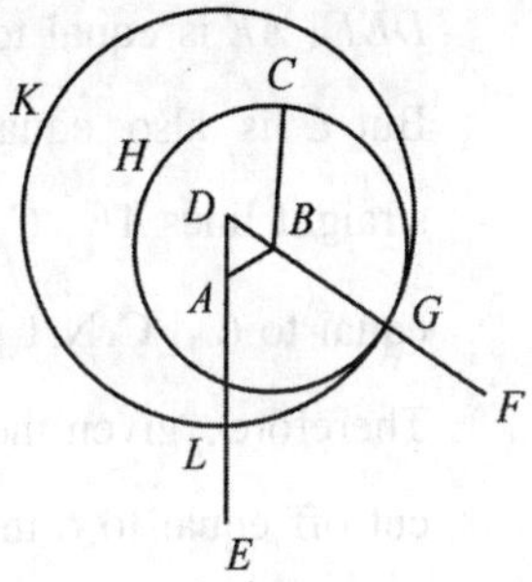

由点 A 到点 B 连线段 AB， [公设 1]

而且在 AB 上作等边三角形 DAB，[I. 1]①

延长 DA，DB 成直线 AE，BF， [公设 2]

以 B 为心，以 BC 为距离画圆 CGH. [公设 3]

再以 D 为心，以 DG 为距离画圆 GKL. [公设 3]

因为点 B 是圆 CGH 的心，故 BC 等于 BG. [定义 15]

且点 DA 是圆 GKL 的心，故 DL 等于 DG. [定义 15]

又 DA 等于 DB，所以余量 AL 等于余量 BG. [公理 3]

但已证明了 BC 等于 BG，所以线段 AL、BC 的每一个都等于 BG. 又因等于同量的量彼此相等. [公理 1]

所以，AL 也等于 BC.

从而，由给定的点 A 作出了线段 AL 等于已知线段 BC.

这就是所要求作的.

① [I. 1] 表示第 1 卷，第 1 个命题，此后均如此。

Proposition 3.

Given two unequal straight lines, to cut off from the greater a straight line equal to the less.

Let *AB*, *C* be the-two given unequal straight lines, and let *AB* be the greater of them.

Thus it is required to cut off from *AB* the greater a straight line equal to *C* the less.

At the point *A* let *AD* be placed equal to the straight line *C*; [I. 2] and with centre *A* and distance *AD* let the circle *DEF* be described. [Post.3]

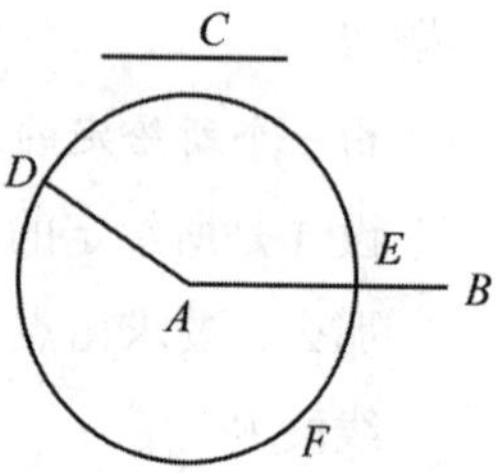

Now, since the point *A* is the centre of the circle *DEF*, *AE* is equal to *AD*. [Def.15]

But *C* is also equal to *AD*. Therefore each of the straight lines *AE*, *C* is equal to *AD*; so that *AE* is also equal to *C*. [C.N.1]

Therefore, given the two straight lines *AB*, *C*, from *AB* the greater *AE* has been cut off equal to *C* the less.

(Being) what it was required to do.

Proposition 4.

If two triangles have the two sides equal to two sides respectively, and have the angles contained by the equal straight lines equal, they will also have the base equal to the base, the triangle will be equal to the triangle, and the remaining angles will be equal to the remaining angles respectively, namely those which the equal sides subtend.

Let *ABC*, *DEF* be two triangles having the two sides *AB*, *AC* equal to the two sides *DE*, *DF* respectively, namely *AB* to *DE* and *AC* to *DF*, and the angle *BAC* equal to the angle *EDF*.

I say that the base *BC* is also equal to the base *EF*, the triangle *ABC* will be

命题 3

已知两条不相等的线段，试由大的上边截取一条线段使它等于另外一条.

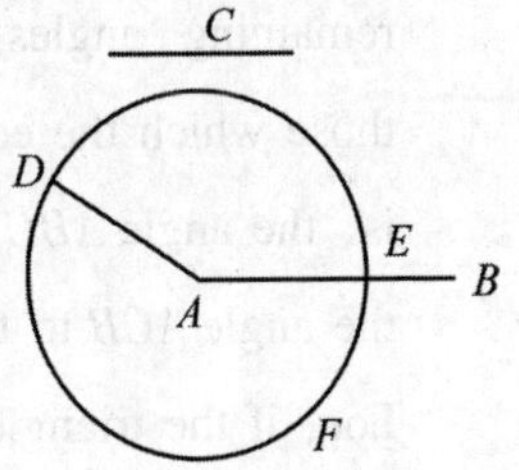

设 AB，C 是两条不相等的线段，且 AB 大于 C.

这样要求由较大的 AB 上截取一段等于较小的 C.

由点 A 取 AD 等于线段 C，［I.2］

且以 A 为心，以 AD 为距离画圆 DEF.［公设 3］

因为点 A 是圆 DEF 的圆心，故 AE 等于 AD.

［定义 15］

但 C 也等于 AD. 所以线段 AE，C 的每一条都等于 AD；这样，AE 也等于 C.

［公理 1］

所以，已知两条线段 AB，C，由较大的 AB 上截取了 AE 等于较小的 C.

这就是所要求作的.

命题 4

如果两个三角形中，一个的两边分别等于另一个的两边，而且这些相等的线段所夹的角相等. 那么，它们的底边等于底边，三角形全等于三角形，这样其余的角也分别等于相应的角，即那些等边所对的角.

设 ABC，DEF 是两个三角形，两边 AB，AC 分别等于边 DE，DF，即 AB 等于 DE，且 AC 等于 DF，以及角 BAC 等于角 EDF.

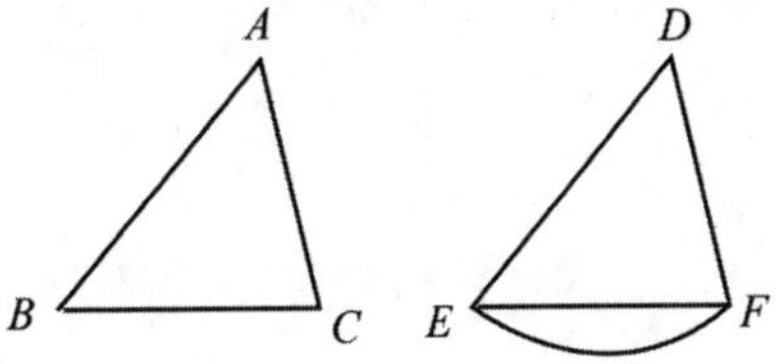

我断言底 BC 也等于底 EF，三角形 ABC 全等于三角形 DEF，其余的角分别等于其余的角，即这些等边所对的角，也就是角 ABC 等于角 DEF，且角 ACB 等于角 DFE.

equal to the triangle *DEF*, and the remaining angles will be equal to the remaining angles respectively, namely those which the equal sides subtend, that is, the angle *ABC* to the angle *DEF*, and the angle *ACB* to the angle *DFE*.

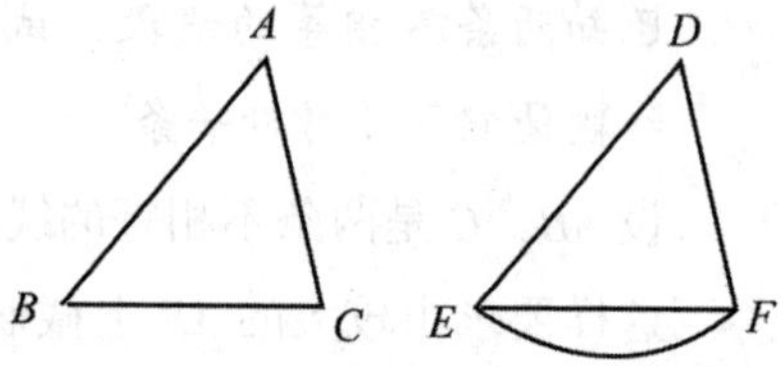

For, if the triangle *ABC* be applied to the triangle *DEF*, and if the point *A* be placed on the point *D* and the straight line *AB* on *DE*, then the point *B* will also coincide with *E*, because *AB* is equal to *DE*.

Again, *AB* coinciding with *DE*, the straight line *AC* will also coincide with *DF*, because the angle *BAC* is equal to the angle *EDF*; hence the point *C* will also coincide with the point *F* because *AC* is again equal to *DF*.

But *B* also coincided with *E*; hence the base *BC* will coincide with the base *EF*. [For if, when *B* coincides with *E* and *C* with *F*, the base *BC* does not coincide with the base *EF*, two straight lines will enclose a space: which is impossible. Therefore the base *BC* will coincide with *EF*] and will be equal to it. [C.N.4]

Thus the whole triangle *ABC* will coincide with the whole triangle *DEF*, and will be equal to it.

And the remaining angles will also coincide with the remaining angles and will be equal to them, the angle *ABC* to the angle *DEF*, and the angle *ACB* to the angle *DFE*.

Therefore etc.

(Being) what it was required to prove.

如果移动三角形 ABC 到三角形 DEF 上，若点 A 落在点 D 上且线段 AB 落在 DE 上，因为 AB 等于 DE. 那么，点 B 也就与点 E 重合.

又，AB 与 DE 重合，因为角 BAC 等于角 EDF，线段 AC 也与 DF 重合.

因为 AC 等于 DF，故点 C 也与点 F 重合.

但是，点 B 也与点 E 重合，故底 BC 也与底 EF 重合.

[事实上，当 B 与 E 重合且 C 与 F 重合时，底 BC 不与底 EF 重合．则二条直线就围成一块空间：这是不可能的，所以底 BC 就与 EF 重合]二者就相等. [公理 4]

这样，整个三角形 ABC 与整个三角形 DEF 重合，于是它们全等.

且其余的角也与其余的角重合，于是它们都相等，即角 ABC 等于角 DEF，且角 ACB 等于角 DFE.

这就是所要证明的.

Proposition 5.

In isosceles triangles the angles at the base are equal to one another, and, if the equal straight lines be produced further, the angles under the base will be equal to one another.

Let *ABC* be an isosceles triangle having the side *AB* equal to the side *AC*; and let the straight lines *BD*, *CE* be produced further in a straight line with *AB*, *AC*. [Post.2]

I say that the angle *ABC* is equal to the angle *ACB*, and the angle *CBD* to the angle *BCE*.

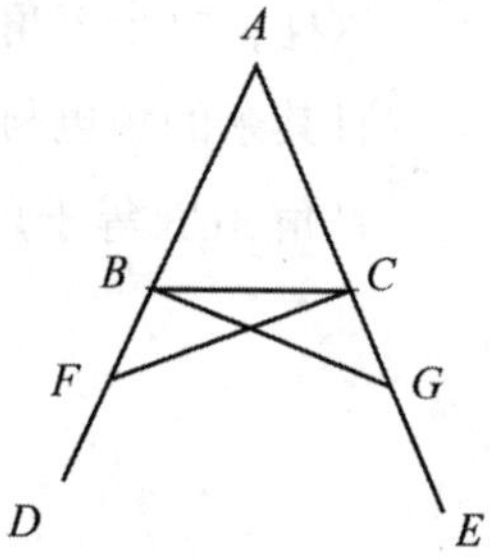

Let a point *F* be taken at random on *BD*; from *AE* the greater let *AG* be cut off equal to *AF* the less; [I. 3] and let the straight lines *FC*, *GB* be joined. [Post.1]

Then, since *AF* is equal to *AG* and *AB* to *AC*, the two sides *FA*, *AC* are equal to the two sides *GA*, *AB*, respectively; and they contain a common angle, the angle *FAG*.

Therefore the base *FC* is equal to the base *GB*, and the triangle *AFC* is equal to the triangle *AGB*, and the remaining angles will be equal to the remaining angles respectively, namely those which the equal sides subtend, that is, the angle *ACF* to the angle *ABG*, and the angle *AFC* to the angle *AGB*. [I. 4] and, since the whole *AF* is equal to the whole *AG*, and in these *AB* is equal to *AC*, the remainder *BF* is equal to the remainder *CG*.

But *FC* was also proved equal to *GB*; therefore the two sides *BF*, *FC* are equal to the two sides *CG*, *GB* respectively; and the angle *BFC* is equal to the angle *CGB*, while the base *BC* is common to them; therefore the triangle *BFC* is also equal to the triangle *CGB*, and the remaining angles will be equal to the remaining angles respectively, namely those which the equal sides subtend; therefore the angle *FBC* is equal to the angle *GCB*, and the angle *BCF* to the angle *CBG*.

命题 5

在等腰三角形中，两底角彼此相等，并且若向下延长两腰，则在底以下的两个角也彼此相等.

设 *ABC* 是一个等腰三角形，边 *AB* 等于边 *AC*，且延长 *AB*，*AC* 成直线 *BD*，*CE*. ［公设 2］

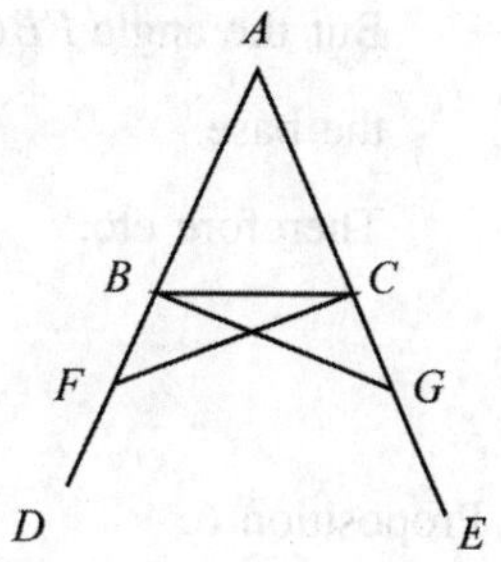

我断言角 *ABC* 等于角 *ACB*，且角 *CBD* 等于角 *BCE*.

在 *BD* 上任取一点 *F*，且在较大的 *AE* 上截取一段 *AG* 等于较小的 *AF*， ［I. 3］

连接 *FC* 和 *GB*. ［公设 1］

因为 *AF* 等于 *AG*，*AB* 等于 *AC*，两边 *FA*、*AC* 分别等于边 *GA*、*AB*，且它们包含着公共角 *FAG*.

所以底 *FC* 等于底 *GB*，且三角形 *AFC* 全等于三角形 *AGB*，其余的角也分别相等，即相等的边所对的角，也就是角 *ACF* 等于角 *ABG*，角 *AFC* 等于角 *AGB*. ［I. 4］

又因为，整体 *AF* 等于整体 *AG*，且在它们中的 *AB* 等于 *AC*，余量 *BF* 等于余量 *CG*. ［公理 3］

但是已经证明了 *FC* 等于 *GB*；

所以，两边 *BF*、*FC* 分别等于两边 *CG*、*GB*，且角 *BFC* 等于角 *CGB*.

这里底 *BC* 是公用的；所以，三角形 *BFC* 也全等于三角形 *CGB*；又，其余的角也分别相等，即等边所对的角.

所以角 *FBC* 等于角 *GCB*，且角 *BCF* 等于角 *CBG*.

因此，由于已经证明了整个角 *ABG* 等于角 *ACF*，且角 *CBG* 等于角 *BCF*，其余的角 *ABC* 等于其余的角 *ACB*. ［公理 3］

又它们都在三角形 *ABC* 的底边以上.

从而，也就证明了角 *FBC* 等于角 *GCB*，且它们都在三角形的底边以下.

证完

Accordingly, since the whole angle ABG was proved equal to the angle ACF, and in these the angle CBG is equal to the angle BCF, the remaining angle ABC is equal to the remaining angle ACB; and they are at the base of the triangle ABC.

But the angle FBC was also proved equal to the angle GCB; and they are under the base.

Therefore etc.

Q.E.D.

Proposition 6.

If in a triangle two angles be equal to one another, the sides which subtend the equal angles will also be equal to one another.

Let ABC be a triangle having the angle ABC equal to the angle ACB;

I say that the side AB is also equal to the side AC.

For, if AB is unequal to AC, one of them is greater.

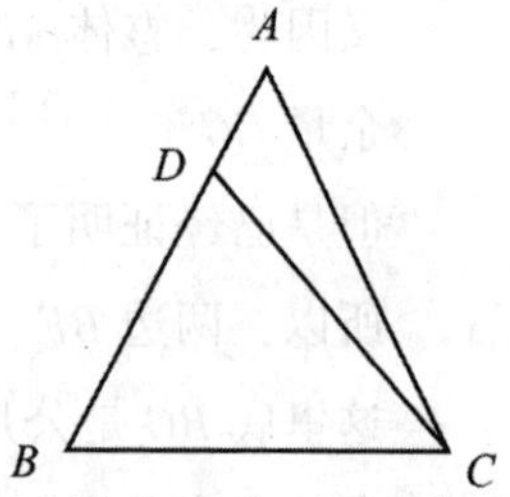

Let AB be greater; and from AB the greater let DB be cut off equal to AC the less;

let DC be joined.

Then, since DB is equal to AC, and BC is common, the two sides DB, BC are equal to the two sides AC, CB respectively; and the angle DBC is equal to the angle ACB; therefore the base DC is equal to the base AB, and the triangle DBC will be equal to the triangle ACB, the less to the greater; which is absurd.

Therefore AB is not unequal to AC; it is therefore equal to it.

Therefore etc.

Q.E.D.

命题 6

如果在一个三角形中，有两角彼此相等，则等角所对的边也彼此相等.

设在三角形 ABC 中，角 ABC 等于角 ACB.

我断言边 AB 也等于边 AC.

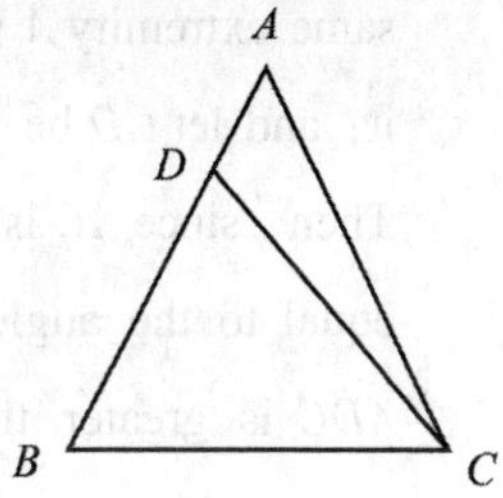

因为，若 AB 不等于 AC，其中必有一个较大，设 AB 是较大的；由 AB 上截取 DB 等于较小的 AC；

[I. 3]

连接 DC.

那么，因为 DB 等于 AC 且 BC 公用，两边 DB、BC 分别等于边 AC、CB，且角 DBC 等于角 ACB.

所以，底 DC 等于底 AB，且三角形 DBC 全等于三角形 ACB，即小的等于大的；这是不合理的.

所以，AB 不能不等于 AC，从而它等于它.

证完

Proposition 7.

Given two straight lines constructed on a straight line (from its extremities) and meeting in a point, there cannot be constructed on the same straight line (from its extremities), and on the same side of it, two other straight lines meeting in another point and equal to the former two respectively, namely each to that which has the same extremity with it.

For, if possible, given two straight lines *AC*, *CB* constructed on the straight line *AB* and meeting at the point *C*, let two other straight lines *AD*, *DB* be constructed on the same straight line *AB*, on the same side of it, meeting in another point *D* and equal to the former two respectively, namely each to that which has the same extremity with it, so that *CA* is equal to *DA* which has the same extremity *A* with it, and *CB* to *DB* which has the same extremity *B* with it; and let *CD* be joined.

Then, since *AC* is equal to *AD*, the angle *ACD* is also equal to the angle *ADC*; [I. 5] therefore the angle *ADC* is greater than the angle *DCB*; therefore the angle *CDB* is much greater than the angle *DCB*.

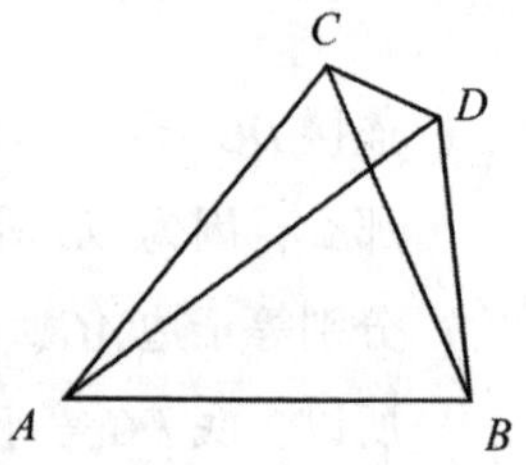

Again, since *CB* is equal to *DB*, the angle *CDB* is also equal to the angle *DCB*.

But it was also proved much greater than it: which is impossible.

Therefore etc.

Q.E.D.

Proposition 8.

If two triangles have the two sides equal to two sides respectively, and have also the base equal to the base, they will also have the angles equal which are contained by the equal straight lines.

Let *ABC*, *DEF* be two triangles having the two sides *AB*, *AC* equal to the two sides *DE*, *DF* respectively, namely *AB* to *DE*, and *AC* to *DF*; and let them have

命题 7

设在给定的线段上(从它的两个端点)作出相交于一点的二线段，则不可能在该线段(从它的两个端点)的同侧作出相交于另一点的另外二条线段，使得作出的二线段分别等于前面二线段，即每个交点到相同端点的线段相等.

因为，如果可能的话，在给定的线段 *AB* 上作出交于点 *C* 的两条线段 *AC*、*CB*. 设在 *AB* 同侧能作另外两条线段 *AD*、*DB* 相交于另外一点 *D*. 而且这二线段分别等于前面二线段，即每个交点到相同的端点. 这样 *CA* 等于 *DA*，它们有相同的端点 *A*，且 *CB* 等于 *DB*，它们也有相同的端点 *B*，连接 *CD*.

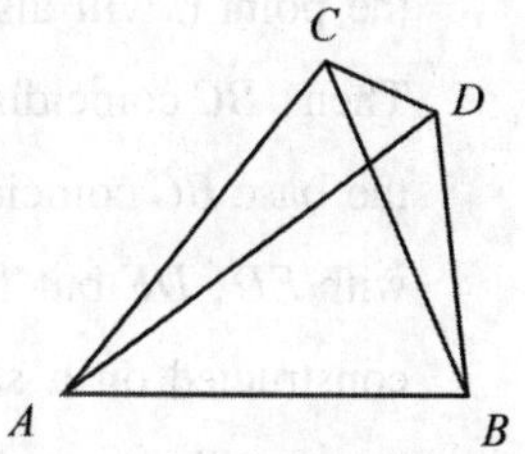

因为，*AC* 等于 *AD*，角 *ACD* 也等于角 *ADC*， [I.5]

所以，角 *ADC* 大于角 *DCB*，所以角 *CDB* 比角 *DCB* 更大.

又，因为 *CB* 等于 *DB*，且角 *CDB* 也等于角 *DCB*. 但是已被证明了它更大于它：这是不可能的.① 证完

命题 8

如果两个三角形的一个有两边分别等于另一个的两边，并且一个的底等于另一个的底，则夹在等边中间的角也相等.

设 *ABC*、*DEF* 是两个三角形，两边 *AB*、*AC* 分别等于两边 *DE*、*DF*，即 *AB* 等于 *DE*，且 *AC* 等于 *DF*. 又设底 *BC* 等于底 *EF*.

我断言角 *BAC* 等于角 *EDF*.

---

① 还可假设点 *D* 在△*ABC* 内，或点 *C* 在△*ABD* 内，或点 *D* 在 *AC* 上，或点 *C* 在 *BD* 上时，均可引出矛盾. 说明以上情况不可能存在. 因而，命题成立.

the base BC equal to the base EF; I say that the angle BAC is also equal to the angle EDF.

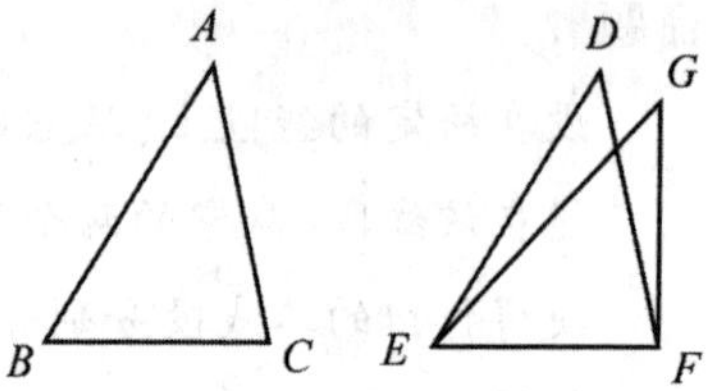

For, if the triangle ABC be applied to the triangle DEF, and if the point B be placed on the point E and the straight line BC on EF, the point C will also coincide with F, because BC is equal to EF.

Then, BC coinciding with EF, BA, AC will also coincide with ED, DF; for, if the base BC coincides with the base EF, and the sides BA, AC do not coincide with ED, DF but fall beside them as EG, GF, then, given two straight lines constructed on a straight line (from its extremities) and meeting in a point, there will have been constructed on the same straight line (from its extremities), and on the same side of it, two other straight lines meeting in another point and equal to the former two respectively, namely each to that which has the same extremity with it.

But they cannot be so constructed. [I. 7]

Therefore it is not possible that, if the base BC be applied to the base EF, the sides BA, AC should not coincide with ED, DF; they will therefore coincide, so that the angle BAC will also coincide with the angle EDF, and will be equal to it.

If therefore etc.

Q.E.D.

Proposition 9.

To bisect a given rectilineal angle.

Let the angle BAC be the given rectilineal angle.

Thus it is required to bisect it.

Let a point D be taken at random on AB; let AE be cut off from AC equal to AD; [I. 3] let DE be joined, and on DE let the equilateral triangle DEF be constructed; let AF be joined.

I say that the angle BAC has been bisected by the straight line AF.

若移动三角形 ABC 到三角形 DEF，且点 B 落在点 E 上，线段 BC 在 EF 上，点 C 也就和 F 重合.

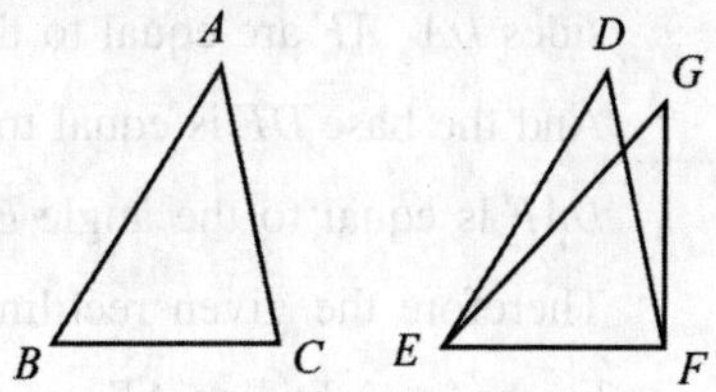

因为，BC 等于 EF.

故 BC 和 EF 重合，BA、AC 也和 ED、DF 重合.

因为，若底 BC 与底 EF 重合，而边 BA、AC 不与 ED、DF 重合且落在它们旁边的 EG、GF 处.

那么，在给定的线段(从它的端点)以上有相交于一点的已知两条线段，这时，在同一线段(从它的端点)的同一侧作出了交于另一点的另外两条线段，它们分别等于前面二线段，即每一交点到同一端点的连线.

但是，不能作出后二线段. [I.7]

如果把底 BC 移动到底 EF，边 BA、AC 和 ED、DF 不重合：这是不可能的，因此，它们要重合. 这样一来，角 BAC 也重合于角 EDF，即它们相等.

证完

命题 9

二等分一个给定的直线角.

设角 BAC 是一个给定的直线角，要求二等分这个角.

设在 AB 上任意取一点 D，在 AC 上截取 AE 等于 AD； [I.3]

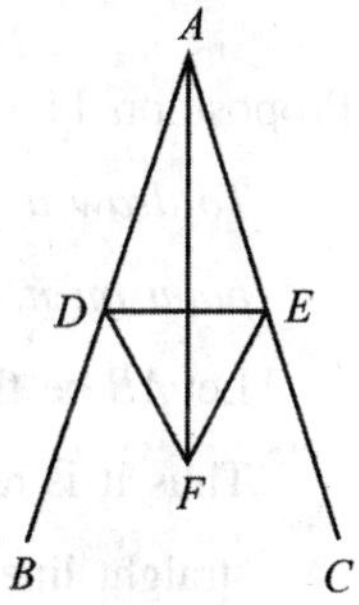

连接 DE，且在 DE 上作一个等边三角形 DEF，连接 AF.

我断言角 BAC 被 AF 所平分.

因为，AD 等于 AE，且 AF 公用，两边 DA、AF 分别等于两边 EA、AF.

又底 DF 等于底 EF；

For, since AD is equal to AE, and AF is common, the two sides DA, AF are equal to the two sides EA, AF respectively. And the base DF is equal to the base EF; therefore the angle DAF is equal to the angle EAF. [I. 8]

Therefore the given rectilineal angle BAC has been bisected by the straight line AF.

Q.E.F.

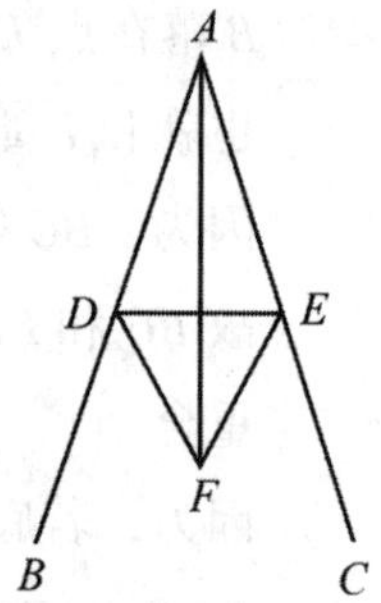

Proposition 10.

To bisect a given finite straight line.

Let AB be the given finite straight line.

Thus it is required to bisect the finite straightline AB.

Let the equilateral triangle ABC be constructed on it, [I. 1] and let the angle ACB be bisected by the straight line CD; [I. 9]

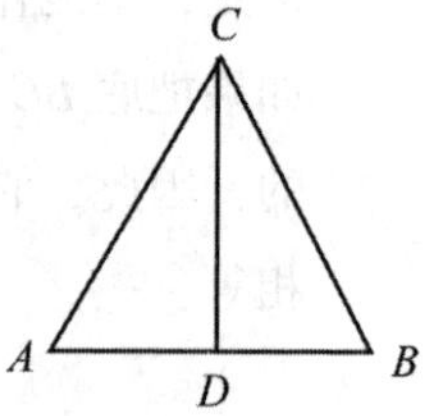

I say that the straight line AB has been bisected at the point D.

For, since AC is equal to CB, and CD is common, the two sides AC, CD are equal to the two sides BC, CD respectively; and the angle ACD is equal to the angle BCD; therefore the base AD is equal to the base BD. [I. 4]

Therefore the given finite straight line AB has been bisected at D.

Q.E.F.

Proposition 11.

To draw a straight line at right angles to a given straight line from a given point on it.

Let AB be the given straight line, and C the given point on it.

Thus it is required to draw from the point C a straight line at right angles to the straight line AB.

Let a point D be taken at random on AC; let CE be made equal to CD; [I. 3]

所以，角 *DAF* 等于角 *EAF*. [I. 8]

从而，直线 *AF* 二等分已知直线角 *BAC*.

作完

命题 10

二等分已知有限直线.

设 *AB* 是已知有限直线，那么，要求二等分有限直线 *AB*.

设在 *AB* 上作一个等边三角形 *ABC*. [I. 1]

且设直线 *CD* 二等分角 *ACB*. [I. 9]

我断言线段 *AB* 在点 *D* 被二等分.

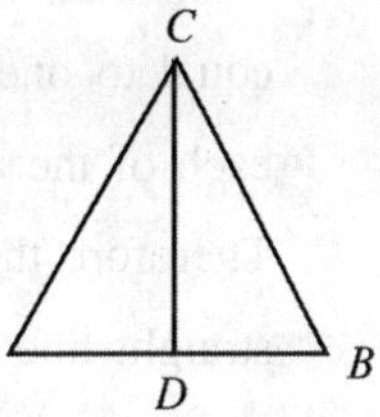

事实上，由于 *AC* 等于 *CB*，且 *CD* 公用；两边 *AC*、*CD* 分别等于两边 *BC*、*CD*；且角 *ACD* 等于角 *BCD*.

所以，底 *AD* 等于底 *BD*. [I. 4]

从而，将给定的有限直线 *AB* 二等分于点 *D*.

作完

命题 11

由给定的直线上一已知点作一直线和给定的直线成直角.

设 *AB* 是给定的直线，*C* 是它上边的已知点. 那么，要求由点 *C* 作一直线和直线 *AB* 成直角.

设在 *AC* 上任意取一点 *D*，且使 *CE* 等于 *CD*， [I. 3]

在 *DE* 上作一个等边三角形 *FDE*，连接 *FC*. [I. 1]

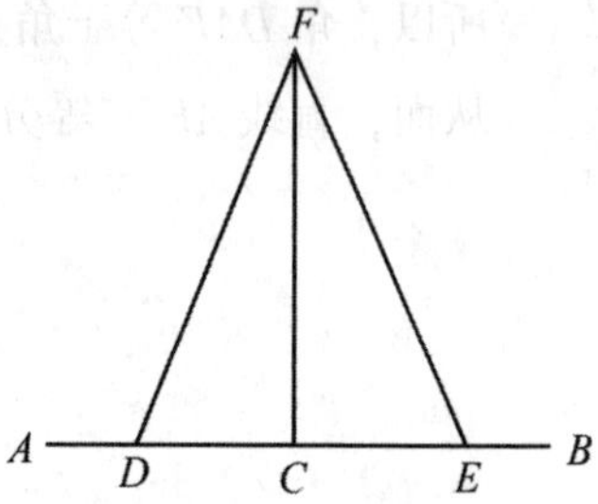

on *DE* let the equilateral triangle *FDE* be constructed, [I. 1] and let *FC* be joined; I say that the straight line *FC* has been drawn at right angles to the given straight line *AB* from *C* the given point on it.

For, since *DC* is equal to *CE*, and *CF* is common, the two sides *DC*, *CF* are equal to the two sides *EC*, *CF* respectively; and the base *DF* is equal to the base *FE*; therefore the angle *DCF* is equal to the angle *ECF*; [I. 8] and they are adjacent angles.

But, when a straight line set up on a straight line makes the adjacent angles equal to one another, each of the equal angles is right; [Def. 10] therefore each of the angles *DCF*, *FCE* is right.

Therefore the straight line *CF* has been drawn at right angles to the given straight line AB from the given point *C* on it.

Q.E.F.

Proposition 12.

To a given infinite straight line, from a given point which is not on it, to draw a perpendicular straight line.

Let *AB* be the given infinite straight line, and *C* the given point which is not on it;

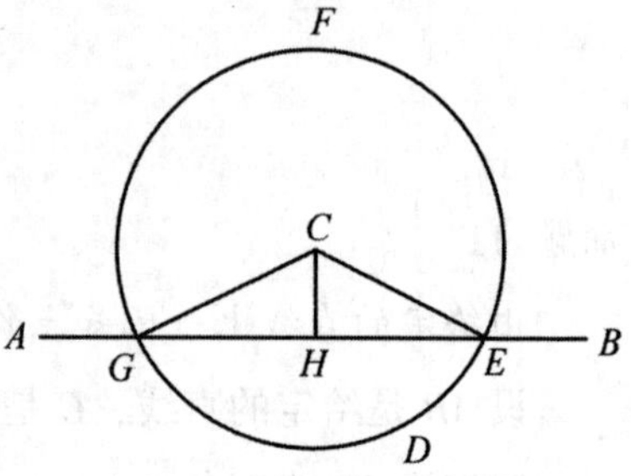

thus it is required to draw to the given infinite straight line *AB*, from the given point *C* which is not on it, a perpendicular straight line.

For let a point *D* be taken at random on the other side of the straight line *AB*, and with centre *C* and distance *CD* let the circle *EFG* be described; [Post.3]

let the straight line *EG* be bisected at *H*, [I.10] and let the straight lines *CG*,

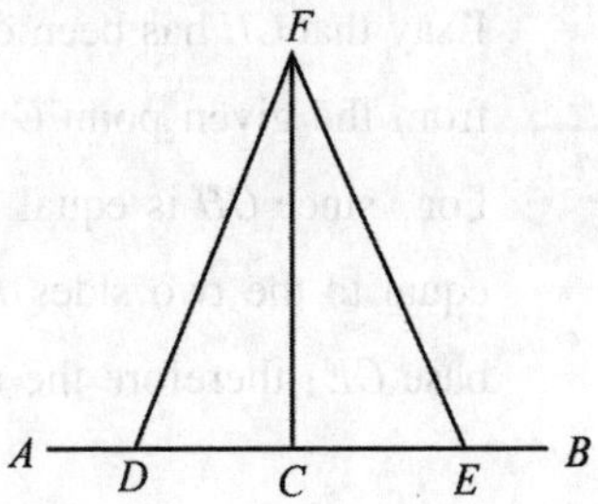

我断言直线 *FC* 就是由已知直线 *AB* 上的已知点 *C* 作出的和 *AB* 成直角的直线.

因为，由于 *DC* 等于 *CE*，且 *CF* 公用；两边 *DC*、*CF* 分别等于两边 *EC*、*CF*；且底 *DF* 等于底 *FE*，

所以，角 *DCF* 等于角 *ECF*， [I.8]

它们又是邻角．但是，当一条直线和另一条直线相交成相等的邻角时，这些等角的每一个都是直角． [定义 10]

所以，角 *DCF*，角 *FCE* 每一个都是直角.

从而，由给定的直线 *AB* 上的已知点 *C* 作出的直线 *CF* 和 *AB* 成直角.

作完

命题 12

由给定的无限直线外一已知点作该直线的垂线.

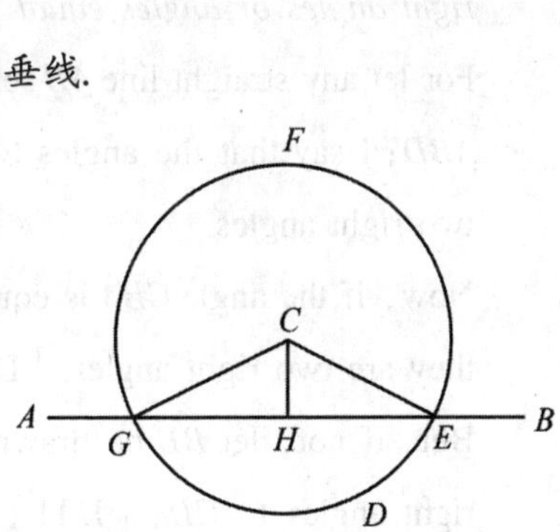

设 *AB* 为给定的无限直线，且设已知点 *C* 不在它上．要求由点 *C* 作无限直线 *AB* 的垂线.

设在直线 *AB* 的另一侧任取一点 *D*，且以点 *C* 为心，以 *CD* 为距离作圆 *EFG*. [公设 3]

设线段 *EG* 被点 *H* 二等分， [I.10]

连接 *CG*，*CH*，*CE*. [公设 1]

我断言 *CH* 就是由不在已知无限直线 *AB* 上的已知点 *C* 所作该直线的垂线.

因为 *GH* 等于 *HE*，且 *HC* 公用；两边 *GH*、*HC* 分别等于两边 *EH*，*HC*；且

CH, *CE* be joined. [Post.1]

I say that *CH* has been drawn perpendicular to the given infinite straight line *AB* from the given point *C* which is not on it.

For, since *GH* is equal to *HE*, and *HC* is common, the two sides *GH*, *HC* are equal to the two sides *EH*, *HC* respectively; and the base *CG* is equal to the base *CE*; therefore the angle *CHG* is equal to the angle *EHC*.

[I.8]

And they are adjacent angles.

But, when a straight line set up on a straight line makes the adjacent angles equal to one another, each of the equal angles is right, and the straight line standing on the other is called a perpendicular to that on which it stands.

[Def. 10]

Therefore *CH* has been drawn perpendicular to the given infinite straight line *AB* from the given point *C* which is not on it.

Q.E.F.

Proposition 13.

If a straight line set up on a straight line make angles, it will make either two right angles or angles equal to two right angles.

For let any straight line *AB* set up on the straight line *CD* make the angles *CBA*, *ABD*; I say that the angles *CBA*, *ABD* are either two right angles or equal to two right angles.

Now, if the angle *CBA* is equal to the angle *ABD*, they are two right angles. [Def.10]

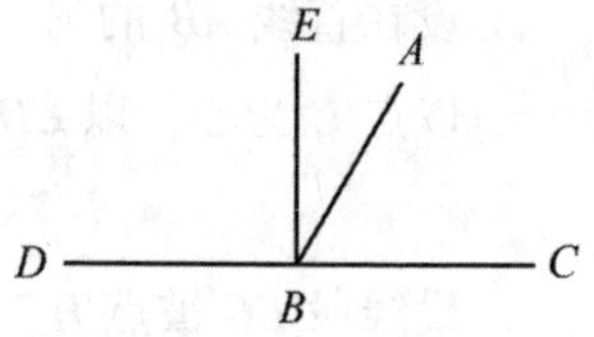

But, if not, let *BE* be drawn from the point *B* at right angles to *CD*; [I. 11] therefore the angles *CBE*, *EBD* are two right angles.

Then, since the angle *CBE* is equal to the two angles *CBA*, *ABE*, let the angle *EBD* be added to each; therefore the angles *CBE*, *EBD* are equal to the three angles *CBA*, *ABE*, *EBD*. [C.N.2]

底 CG 等于 CE.

所以，角 CHG 等于角 EHC. [I. 8]

且它们是邻角.

但是，当两条直线相交成相等的邻角时，每一个角都是直角，而且称一条直线垂直于另一条直线. [定义 10]

所以，由不在所给定的无限直线 AB 上的已知点 C 作出了 CH 垂直于 AB.

作完

命题 13

一条直线和另一条直线所交成的角，或者是两个直角或者它们的和等于两个直角.

设任意直线 AB 在直线 CD 的上侧和它交成角 CBA，ABD.

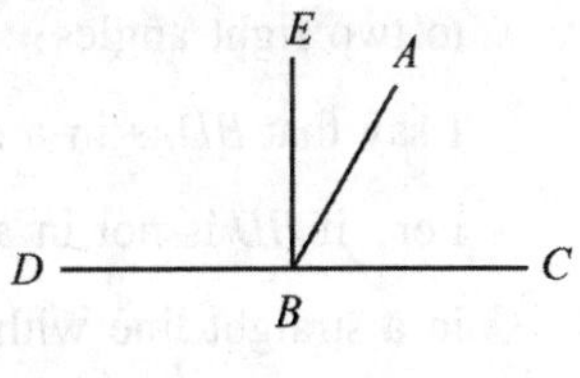

我断言角 CBA，ABD 或者都是直角或者其和等于两个直角.

现在，若角 CBA 等于角 ABD，那么它们是两个直角. [定义 10]

但是，假若不是，设 BE 是由点 B 所作的和 CD 成直角的直线. [I. 11]

于是角 CBE，EBD 是两个直角.

这时因为角 CBE 等于两个角 CBA，ABE 的和，给它们各加上角 EBD；则角 CBE，EBD 的和就等于三个角 CBA，ABE，EBD 的和. [公理 2]

再者，因为角 DBA 等于两个角 DBE，EBA 的和，给它们各加上角 ABC；则角 DBA，ABC 的和就等于三个角 DBE，EBA，ABC 的和. [公理 2]

Again, since the angle *DBA* is equal to the two angles *DBE*, *EBA*, let the angle *ABC be added to each*; *therefore the angles DBA*, *ABC* are equal to the three angles *DBE*, *EBA*, *ABC*. [C.N.2]

But the angles *CBE*, *EBD* were also proved equal to the same three angles; and things which are equal to the same thing are also equal to one another; [C.N.1] therefore the angles *CBE*, *EBD* are also equal to the angles *DBA*, *ABC*.

But the angles *CBE*, *EBD* are two right angles; therefore the angles *DBA*, *ABC* are also equal to two right angles.

Therefore etc.

Q.E.D.

Proposition 14.

If with any straight line, and at a point on it, two straight lines not lying on the same side make the adjacent angles equal to two right angles, the two straight lines will be in a sraight line with one another.

For with any straight line *AB*, and at the point *B* on it, let the two straight lines *BC*, *BD* not lying on the same side make the adjacent angles *ABC*, *ABD* equal to two right angles;

I say that *BD* is in a straight line with *CB*.

For, if *BD* is not in a straight line with *BC*, let *BE* be in a straight line with *CB*.

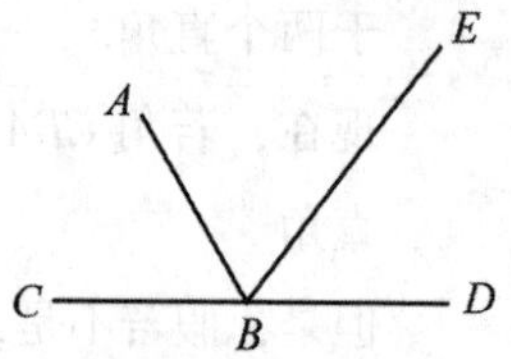

Then, since the straight line *AB* stands on the straight line *CBE*, the angles *ABC*, *ABE* are equal to two right angles. [I. 13]

But the angles *ABC*, *ABD* are also equal to two right angles; therefore the angles *CBA*, *ABE* are equal to the angles *CBA*, *ABD*. [Post. 4 and C.N.1]

Let the angle *CBA* be subtracted from each; therefore the remaining angle *ABE* is equal to the remaining angle *ABD*, [C.N.3]

但是，角 *CBE*，*EBD* 的和也被证明了等于相同的三个角的和．而等于同量的量彼此相等， [公理 1]

故角 *CBE*，*EBD* 的和也等于角 *DBA*，*ABC* 的和．但是角 *CBE*，*EBD* 的和是两个直角.

所以，角 *DBA*，*ABC* 的和也等于两个直角.

证完

命题 14

如果过任意直线上一点有两条直线不在这一直线的同侧，且和直线所成邻角和等于二直角. 则这两条直线在同一直线上.

因为，过任意直线 *AB* 上面一点 *B*，有两条不在 *AB* 同侧的直线 *BC*，*BD* 成邻角 *ABC*，*ABD*，其和等于两个直角.

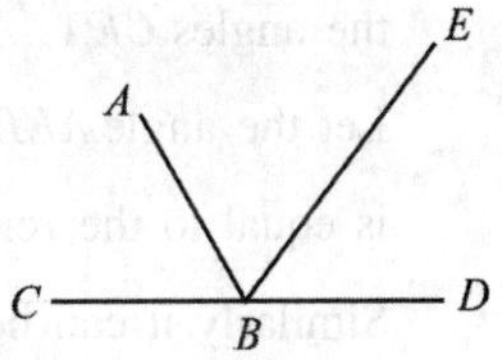

我断言 *BD* 和 *CB* 在同一直线上.

事实上，如果 *BD* 和 *BC* 不在同一直线上，设 *BE* 和 *CB* 在同一直线上．因为，直线 *AB* 位于直线 *CBE* 之上，角 *ABC*，*ABE* 的和等于两个直角. 但角 *ABC*，*ABD* 的和也等于两个直角. [I. 13]

所以，角 *CBA*，*ABE* 的和等于角 *CBA*，*ABD* 的和.

[公设 4 和公理 1]

由它们中各减去角 *CBA*，于是余下的角 *ABE* 等于余下的角 *ABD*.

[公理 3]

这时，小角等于大角：这是不可能的.

the less to the greater: which is impossible.

Therefore *BE* is not in a straight line with *CB*.

Similarly we can prove that neither is any other straight line except *BD*.

Q.E.D.

Proposition 15.

If two straight lines cut one another, they make the vertical angles equal to one another.

For let the straight lines *AB*, *CD* cut one another at the point *E*; I say that the angle *AEC* is equal to the angle *DEB*, and the angle *CEB* to the angle *AED*.

For, since the straight line *AE* stands on the straight line *CD*, making the angles *CEA*, *AED*, the angles *CEA*, *AED* are equal to two right angles. [I. 13]

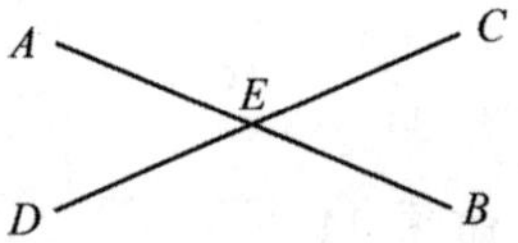

Again, since the straight line *DE* stands on the straight line *AB*, making the angles *AED*, *DEB*, the angles *AED*, *DEB* are equal to two right angles. [I. 13]

But the angles *CEA*, *AED* were also proved equal to two right angles; therefore the angles *CEA*, *AED* are equal to the angles *AED*, *DEB*. [Post.4 and C.N.1]

Let the angle *AED* be subtracted from each; therefore the remaining angle *CEA* is equal to the remaining angle *BED*. [C. N. 3]

Similarly it can be proved that theangles *CEB*, *DEA* are also equal.

Therefore etc.

[Porism. From this it is manifest that, if two straight lines cut one another, they will make the angles at the point of section equal to four right angles.]

Q.E.D.

所以，*BE* 和 *CB* 不在一直线上.

类似地，我们可证明除 *BD* 外再没有其他的直线和 *CB* 在同一直线上.

所以，*CB* 和 *BD* 在同一直线上.

证完

命题 15

如果两直线相交，则它们交成的对顶角相等.

设直线 *AB*，*CD* 相交于点 *E*.

我断言角 *AEC* 等于角 *DEB*，且角 *CEB* 等于角 *AED*.

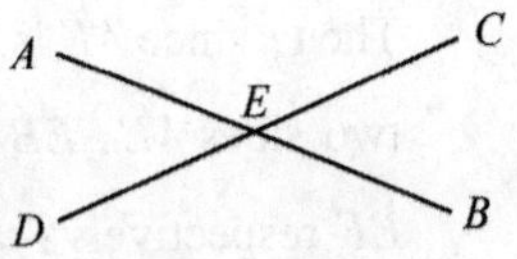

事实上，因为直线 *AE* 位于直线 *CD* 上侧，而构成角 *CEA*，*AED*；角 *CEA*，*AED* 的和等于两个直角.

又，因为直线 *DE* 位于直线 *AB* 的上侧，构成角 *AED*，*DEB*；角 *AED*，*DEB* 的和等于两个直角. [I. 13]

但是，已经证明了角 *CEA*，*AED* 的和等于两个直角.

故角 *CEA*，*AED* 的和等于角 *AED*，*DEB* 的和. [公设 4 和公理 1]

由它们中各减去角 *AED*，则其余的角 *CEA* 等于其余的角 *BED*. [公理 3]

类似地，可以证明角 *CEB* 也等于角 *DEA*.

证完

[推论 很明显，若两条直线相交，则在交点处所构成的角的和等于四个直角.]

Proposition 16.

In any triangle, if one of the sides be produced, the exterior angle is greater than either of the interior and opposite angles.

Let *ABC* be a triangle, and let one side of it *BC* be produced to *D*; I say that the exterior angle *ACD* is greater than either of the interior and opposite angles *CBA*, *BAC*.

Let *AC* be bisected at *E*, [I. 10] and let *BE* be joined and produced in a straight line to *F*; let *EF* be made equal to *BE*, [I. 3] let *FC* be joined, [Post. 1] and let *AC* be drawn through to *G*. [Post.2]

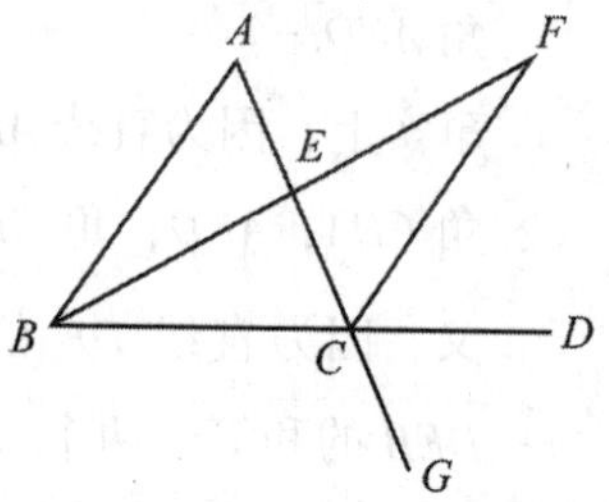

Then, since *AE* is equal to *EC*, and *BE* to *EF*, the two sides *AE*, *EB* are equal to the two sides *CE*, *EF* respectively; and the angle *AEB* is equal to the angle *FEC*, for they are vertical angles. [I. 15]

Therefore the base *AB* is equal to the base *FC*, and the triangle *ABE* is equal to the triangle *CFE*, and the remaining angles are equal to the remaining angles respectively, namely those which the equal sides subtend; [I. 4] therefore the angle *BAE* is equal to the angle *ECF*.

But the angle *ECD* is greater than the angle *ECF*; [C.N.5] therefore the angle *ACD* is greater than the angle *BAE*.

Similarly also, if BC be bisected, the angle *BCG*, that is, the angle *ACD*, [I. 15] can be proved greater than the angle ABC as well.

Therefore etc.

Q.E.D.

命题 16

在任意的三角形中，若延长一边，则外角大于任何一个内对角.

设 ABC 是一个三角形，延长边 BC 到点 D.

我断言外角 ACD 大于内角 CBA，BAC 的任何一个.

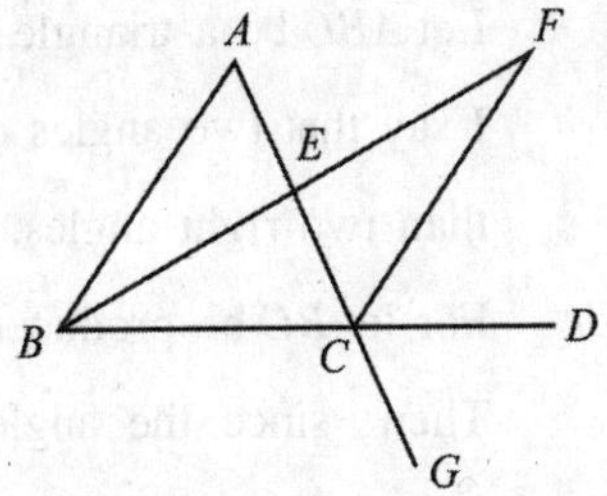

设 AC 被二等分于点 E，[I. 10]

连接 BE 并延长至点 F，使 EF 等于 BE，[I. 3]

连接 FC，[公设 1]

延长 AC 至 G. [公设 2]

那么，因为 AE 等于 EC，BE 等于 EF，两边 AE，EB 分别等于两边 CE，EF. 又角 AEB 等于角 FEC，因为它们是对顶角. [I. 15]

所以，底 AB 等于底 FC，且三角形 ABE 全等于三角形 CFE，余下的角也分别等于余下的角，即等边所对的角. [I. 4]

所以，角 BAE 等于角 ECF.

但是，角 ECD 大于角 ECF. [公理 5]

所以，角 ACD 大于角 BAE.

类似地也有，若 BC 被平分，角 BCG，也就是角 ACD. [I. 15]

可以证明它大于角 ABC.

证完

Proposition 17.

In any triangle two angles taken together in any manner are less than two right angles.

Let *ABC* be a triangle;

I say that two angles of the triangle *ABC* taken together in any manner are less than two right angles.

For let *BC* be produced to *D*. [Post.2]

Then, since the angle *ACD* is an exterior angle of the triangle *ABC*, it is grater than the interior and opposite angle *ABC*. [I.16]

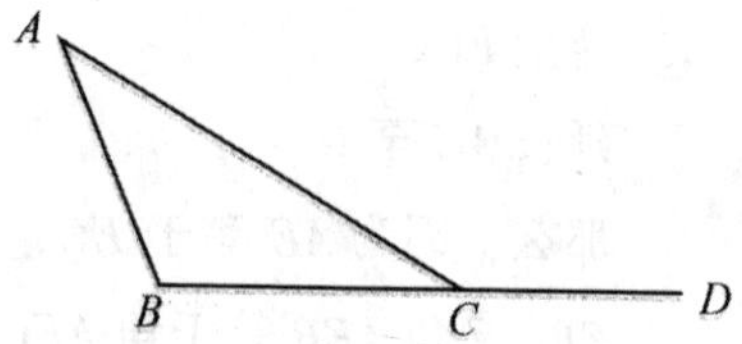

Let the angle *ACB* be added to each; therefore the angles *ACD*, *ACB* are greater than the angles *ABC*, *BCA*.

But the angles *ACD*, *ACB* are equal to two right angles.

Therefore the angles *ABC*, *BCA* are less than two right angles.

Similarly we can prove that the angles *BAC*, *ACB* are also less than two right angles, and so are the angles *CAB*, *ABC* as well.

Therefore etc.

Q.E.D.

Proposition 18.

In any triangle the greater side subtends the greater angle.

For let *ABC* be a triangle having the side *AC* greater than *AB*; I say that the angle *ABC* is also greater than the angle *BCA*.

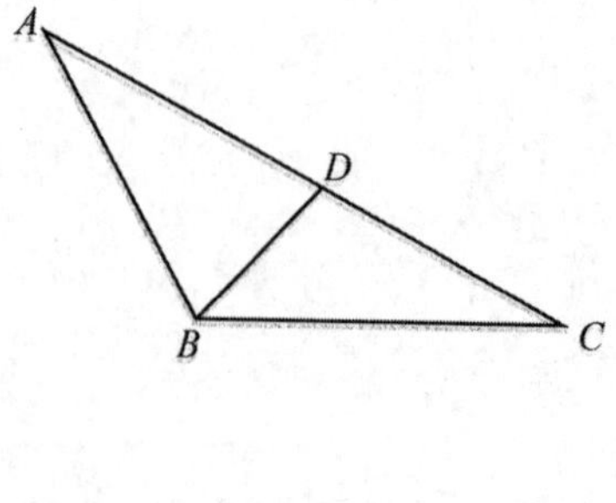

For, since *AC* is greater than *AB*, let *AD* be made equal to *AB* [I. 3], and let *BD* be joined.

Then, since the angle *ADB* is an exterior

命题 17

在任何三角形中，任意两角之和小于两直角.

设 ABC 是一个三角形，我断言三角形 ABC 的任意两个角的和小于两个直角.

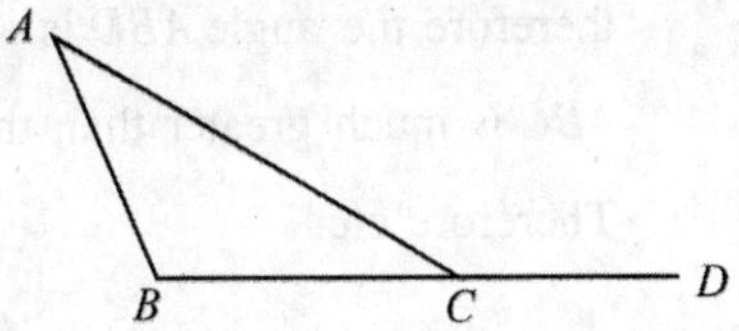

将 BC 延长至 D. [公设 2]

于是角 ACD 是三角形 ABC 的外角，它大于内对角 ABC.

把角 ACB 加在它们各边，则角 ACD，ACB 的和大于角 ABC，BCA 的和.

但是角 ACD，ACB 的和等于两个直角. [I. 13]

所以，角 ABC，BCA 的和小于两个直角.

类似地，我们可以证明角 BAC，ACB 的和也小于两个直角；角 CAB，ABC 的和也是这样.

证完

命题 18

在任何三角形中大边对大角.

设在三角形 ABC 中边 AC 大于 AB.

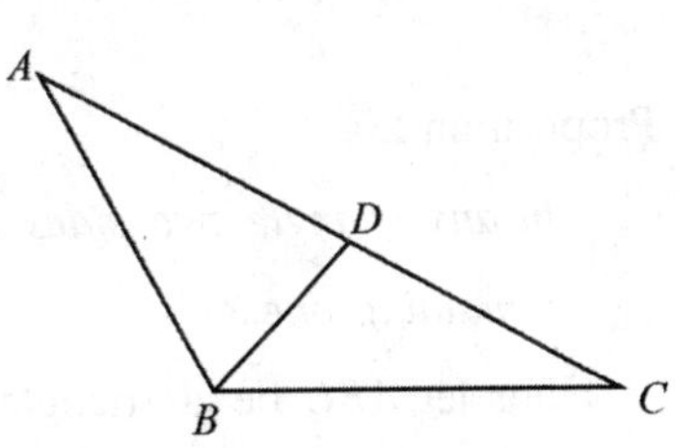

我断言角 ABC 也大于角 BCA.

事实上，因为 AC 大于 AB，取 AD 等于 AB. [I. 3]

连接 BD. 那么，因为角 ADB 是三角形 BCD 的外角，它大于内对角 DCB. [I. 16]

angle of the triangle BCD, it is greater than the interior and opposite angle DCB. [I. 16]

But the angle ADB is equal to the angle ABD, since the side AB is equal to AD; therefore the angle ABD is also greater than the angle ACB; therefore the angle ABC is much greater than the angle ACB.

Therefore etc.

Q.E.D.

Proposition 19.

In any triangle the greater angle is subtended by the greater side.

Let ABC be a triangle having the angle ABC greater than the angle BCA; I say that the side AC is also greater than the side AB.

For, if not, AC is either equal to AB or less.

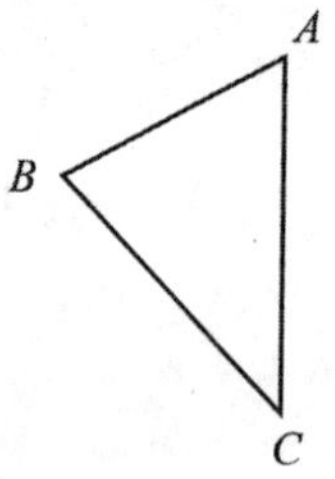

Now AC is not equal to AB; for then the angle ABC would also have been equal to the angle ACB; [I. 5] but it is not; therefore AC is not equal to AB.

Neither is AC less than AB, for then the angle ABC would also have been less than the angle ACB; [I. 18] but it is not; therefore AC is not less than AB.

And it was proved that it is not equal either.

Therefore AC is greater than AB.

Therefore etc.

Q.E.D.

Proposition 20.

In any triangle two sides taken together in any manner are greater than the remaining one.

For let ABC be a triangle; I say that in the triangle ABC two sides taken together in any manner are greater than the remaining one, namely

BA, AC greater than BC,

但是角 ADB 等于角 ABD，

这是因为，边 AB 等于 AD.

所以，角 ABD 也大于角 ACB，从而，角 ABC 比角 ACB 更大.

证完

命题 19

在任何三角形中，大角对大边.

设在三角形 ABC 中，角 ABC 大于角 BCA.

我断言边 AC 也大于边 AB.

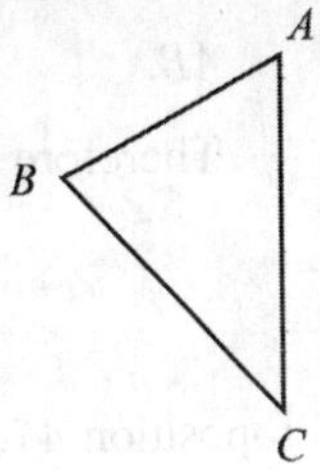

因为，假若不是这样，则 AC 等于或小于 AB. 现在设 AC 等于 AB；那么，角 ABC 也等于角 ACB，[I. 5]

但它是不等的. 所以，AC 不等于 AB.

AC 也不能小于 AB，因为这样角 ABC 也小于角 ACB，[I. 18]

但是，它不是这样的. 所以，AC 不小于 AB，

已经证明了一个不等于另外一个. 从而，AC 大于 AB.

证完

命题 20

在任何三角形中，任意两边之和大于第三边.

设 ABC 为一个三角形.

我断言在三角形 ABC 中，任意两边之和大于其余一边，即

AB, BC greater than AC,

BC, CA greater than AB.

For let BA be drawn through to the point D, let DA be made equal to CA, and let DC be joined.

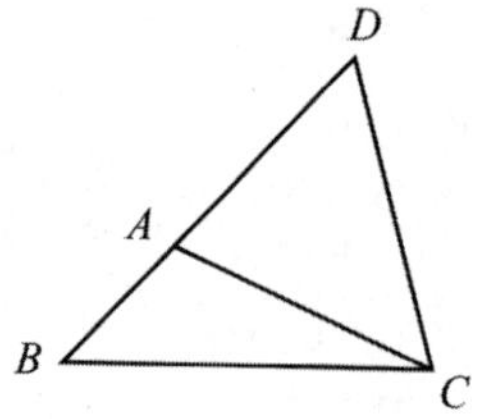

Then, since DA is equal to AC, the angle ADC is also equal to the angle ACD; [I. 5] therefore the angle BCD is greater than the angle ADC. [C.N.5]

And, since DCB is a triangle having the angle BCD greater than the angle BDC, and the greater angle is subtended by the greater side, [I. 19] therefore DB is greater than BC.

But DA is equal to AC; therefore BA, AC are greater than BC.

Similarly we can prove that AB, BC are also greater than CA, and BC, CA than AB.

Therefore etc.

Q.E.D.

Proposition 47.

In right-angled triangles the square on the side subtending the right angle is equal to the squares on the sides containing the right angle.

Let ABC be a right-angled triangle having the angle BAC right; I say that the square on BC is equal to the squares on BA, AC.

For let there be described on BC the square BDEC, and on BA, AC the squares GB, HC; [I. 46] through A let AL be drawn parallel to either BD or CE, and let AD, FC be joined.

Then, since each of the angles BAC, BAG is right, it follows that with a straight line BA, and at the point A on it, the two straight lines AC, AG not lying on the same side make the adjacent angles equal to two right angles; therefore CA is in a straight line with AG. [I. 14]

For the same reason BA is also in a straight line with AH.

And, since the angle DBC is equal to the angle FBA : for each is right: let the angle ABC be added to each; therefore the whole angle DBA is equal to the

BA，AC 之和大于 BC，

AB，BC 之和大于 AC，

BC，CA 之和大于 AB.

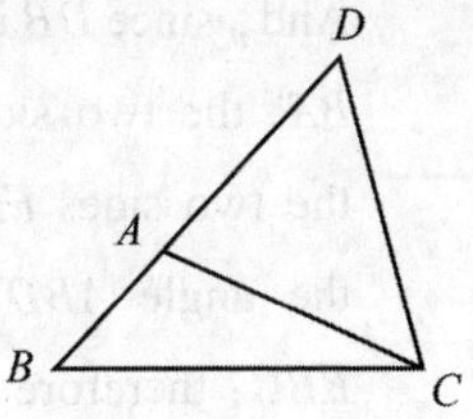

事实上，延长 BA 至点 D，使 DA 等于 CA，连接 DC.

则因 DA 等于 AC，

角 ADC 也等于角 ACD； [I. 5]

所以，角 BCD 大于角 ADC. [公理 5]

由于 DCB 是三角形，它的角 BCD 大于角 BDC，而且较大角所对的边较大， [I. 19]

所以 DB 大于 BC.

但是 DA 等于 AC，

故 BA，AC 的和大于 BC.

类似地，可以证明 AB，BC 的和也大于 CA；BC，CA 的和也大于 AB.

证完

命题 47

在直角三角形中，直角所对的边上的正方形等于夹直角两边上正方形的和.

设 ABC 是直角三角形，已知角 BAC 是直角.

我断言 BC 上的正方形等于 BA，AC 上的正方形的和.

事实上，在 BC 上作正方形 BDEC，且在 BA，AC 上作正方形 GB，HC. [I. 46]

过 A 作 AL 平行于 BD 或 CE，连接 AD，FC.

因为角 BAC，BAG 的每一个都是直角，在一条直线 BA 上的一个点 A 有两条直线 AC，AG 不在它的同一侧所成的两邻角的和等于二直角，于是 CA 与 AG 在同一条直线上. [I. 14]

whole angle *FBC*. [C.N. 2]

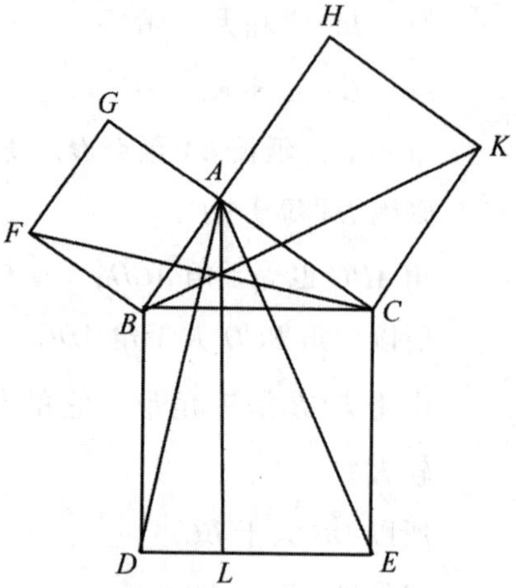

And, since *DB* is equal to *BC*, and *FB* to *BA*, the two sides *AB*, *BD* are equal to the two sides *FB*, *BC* respectively, and the angle *ABD* is equal to the angle *FBC*; therefore the base *AD* is equal to the base *FC*, and the triangle *ABD* is equal to the triangle *FBC*. [I. 4]

Now the parallelogram *BL* is double of the triangle *ABD*, for they have the same base *BD* and are in the same parallels *BD*, *AL*. [I. 41]

And the square *GB* is double of the triangle *FBC*, for they again have the same base *FB* and are in the same parallels *FB*, *GC*. [I. 41]

[But the doubles of equals are equal to one another.]

Therefore the parallelogram *BL* is also equal to the square *GB*.

Similarly, if *AE*, *BK* be joined, the parallelogram *CL* can also be proved equal to the square *HC*; therefore the whole square *BDEC* is equal to the two squares *GB*, *HC*. [C.N. 2]

And the square *BDEC* is described on *BC*, and the squares *GB*, *HC* on *BA*, *AC*.

Therefore the square on the side *BC* is equal to the squares on the sides *BA*, *AC*.

Therefore etc.

Q. E. D.

同理，*BA* 也与 *AH* 在同一条直线上.

又因角 *DBC* 等于角 *FBA*：因为每一个角都是直角：给以上两角各加上角 *ABC*；

所以，整体角 *DBA* 等于整体角 *FBC*.

［公理 2］

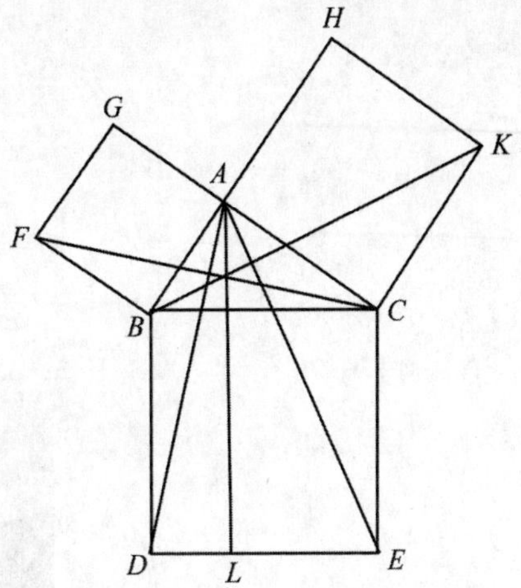

又因为 *DB* 等于 *BC*，*FB* 等于 *BA*；两边 *AB*，*BD* 分别等于两边 *FB*，*BC*.

又角 *ABD* 等于角 *FBC*；所以底 *AD* 等于底 *FC*，且三角形 *ABD* 全等于三角形 *FBC*. ［I. 4］

现在，平行四边形 *BL* 等于三角形 *ABD* 的二倍，因为它们有同底 *BD* 且在平行线 *BD*，*AL* 之间. ［I. 41］

又正方形 *GB* 是三角形 *FBC* 的二倍，因为它们又有同底 *FB* 且在相同的平行线 *FB*，*GC* 之间. ［I. 41］

［但是，等量的二倍仍然是彼此相等的.］

故平行四边形 *BL* 也等于正方形 *GB*.

类似地，若连接 *AE*、*BK*，也能证明平行四边形 *CL* 等于正方形 *HC*.

故整体正方形 *BDEC* 等于两个正方形 *GB*，*HC* 的和. ［公理 2］

而正方形 *BDEC* 是在 *BC* 上作出的，正方形 *GB*，*HC* 是在 *BA*，*AC* 上作出的.

所以，在边 *BC* 上的正方形等于边 *BA*，*AC* 上正方形的和.

证完

《科学与假设》导引

18 世纪不仅是启蒙心智的时代，也是一个信仰科学的时代。哈雷彗星在 1758 年的重现使人类对科学产生“惊人的”、“非凡的”和“异乎寻常”的敬畏感。一方面，人们开始思考：如何更好地认识世界、认识世界有无确定的规则等问题。这些思考深深地嵌入了 19 世纪科学发展的进程之中。以弗朗西斯·培根、惠威尔及穆勒等人所倡导的归纳主义，孔德和斯宾塞为代表的实证主义成为科学认识方法论的重要观点，并对促进近代科学发展起了巨大的作用。另一方面，早在 18 世纪末，康德不满意经验论的归纳主义阶梯，提出了不同的观点：不是从经验上升到理论，而是以先天的“感性直观的纯形式”和“知性的纯粹概念或纯粹范畴”去组织后天经验，以构成绝对可靠的“先验综合知识”。虽然时至今日，经验论和先验论都有着众多的追随信徒，但二者对于科学理论体系的阐释均存在一定的局限性。庞加莱吸取了二者的精华，强调在从事实过渡到定律以及由定律提升为原理时，科学家应充分享有发挥能动性的自由，进而提出了经验约定论，在科学哲学发展进程中写下了惊艳的一笔。

亨利·庞加莱(Jules Henri Poincaré，1854—1912)也被一些研究者译为“彭加勒”，该名字对于读者而言并不陌生。耳熟能详的“庞加莱猜想”就是亨利·庞加莱于1904年所提出。伟大的数学家是人们对庞加莱的第一印象。但事实上，亨利·庞加莱一生的工作和生活所涉猎领域远非数学家的称谓所能概括。他于1854年4月29日生于法国南锡，父亲莱昂·庞加莱是一位医生兼南希医科大学教授；二叔父安托万·庞加莱曾为国家道路桥梁部的检查官；堂弟雷蒙·庞加莱曾于1913年至1920年担任法兰西第三共和国第九届总统。虽然出身名门，但亨利·庞加莱(下文简称：庞加莱)的童年是不幸的。幼时的庞加莱形象思维和语言能力发育迟缓，语言与意识不协调，识别图形的能力欠佳。5岁时又患白喉病，从此留下了喉头麻痹症。但庞加莱表现出惊人的学习天赋，读书速度超快且过目不忘。虽视力不佳，上课时难以看清老师的板书，但听觉记忆能力超群。庞加莱中学时便展现出惊人的数学天赋，两次荣获法国公立中学生数学竞赛头等奖，1873年被高等工科学校作为第一名录取，并先后在巴黎综合理工大学、南锡矿业大学和巴黎大学学习。庞加莱1881年任巴黎大学教授，先后讲授数学分析、光学、电学、流体平衡、电学中的数学、天文学、热力学等课程，1887年当选为法国科学院院士，后任院长，1906年当选法兰西学院院士，获得法国学者的最高荣誉。

19世纪数学的发展一开始就在“数学家之王”高斯(1777—1855)身影的笼罩之下，他的研究遍及所有数学领域。然而，庞加莱走出了高斯光环所照耀之处，他在数学的四个主要领域(算术、代数、几何、分析)都做出了开创性的成就，也是“相对论”的开创先驱，被誉为“对于数学和它的应用具有全面知识的最后一个人”。为追求真理，庞加莱一直奋斗到生命的最后一息。临终前三周，即1912年6月26日，庞加莱抱病在法国道德教育联盟成立大会上发表了最后一次公开演讲。他说，“人生就是持续的斗争”“如果我们偶尔享受到相对的宁静，那正是我们先辈顽强地进行了斗争。假使我们的精力、我们的警惕松懈片刻，我们就将失去先辈为我们赢得的斗争成果。”

除了在专业领域做出惊人的探索，庞加莱也通过自己对科学发展过程的观察、概括、总结和反思，提出了“经验约定论”、“关系实在论”等科学哲学的经典思想。这些论述集中呈现于四本著作：1902 年出版的《科学与假设》(*Science and Hypothesis*)、1905 年出版的《科学的价值》(*The Value of Science*)、1913 年出版的《科学与方法》(*Science and Method*)以及去世前撰写并由友人于1913 年整理出版的短文汇编《最后的沉思》(*Last Essays*)。庞加莱的科学哲学思想在狭隘经验论和极端先验论中保持了必要的张力，强调科学家既要注意约定的实验根源和实验的指导作用，又要大胆假设和自由创造。正如庞加莱在《最后的沉思》所言："两个十分遥远的系统，当它们的距离无限增加时，它们之间的相互引力趋近于零。经验告诉我们，这近似地为真；经验不能够告诉我们，这完全为真，因为两个系统之间的距离总是有限的。但是，没有任何东西妨碍我们假定这完全为真；即使经验与该原理似乎不符，也没有任何东西妨碍我们。让我们设想当距离增加而相互之间的引力减小，此后引力又开始增加的情况。没有任何东西妨碍我们承认，对更大的距离而言，引力在减小，并最终趋于零。只有把目前所考虑的原理本身作为约定，这才能使它免受经验的冲击。约定是经验向我们提示的，但我们却可以自由地采用它。"

1902 年出版的 *Science and Hypothesis*(《科学与假设》)是由庞加莱已发表的短论、讲演、书评、科学著作的序言串接而成，共分为四编：第一编主要讲数学领域的归纳法及推理；第二编主要讲几何学，从欧式几何讲到非欧几何以及几何学空间；第三编主要讲力学，包括经典力学、相对运动和绝对运动以及热力学问题；第四编主要讲物理学，包括物理学的假设、光学、电学以及概率演算等问题。虽然该书涉及众多领域，但却真正做到了"雅俗共赏"。一方面，正如我国科技哲学家李醒民教授所言"在法国的公园和咖啡馆，经常可以看到普通工人和店员手捧庞加莱的小册子聚精会神地阅读"；另一方面，据《爱因斯坦文集(第一卷)》记载，在"奥林比亚科学院"时期(1902 年 3 月至 1905 年 11 月)，爱因斯坦和他的挚友索洛文、哈比希特曾利用晚上的业余时间一起研读过《科学与假设》，有时念一页或半页，有时只念一句话，立即就会引起强烈的争论，当问题比较重要时，争论可以延续数日之久。《自然科学经典导引》选择了 *Science and Hypothesis* 的第二部分的第三章“非欧几何学”和第四部分的第九章“物理学的假设”。一方面，这两章是庞加莱经验约定论思想的集中体现和系统阐释；另一方面，透过挑选的章节，读者可以一窥科学边界拓展的全过程。选编的章节致力于回答两个关键问题：(1)什么是假设？(2)科学的进步与假设有何关系？它们是 *Science and Hypothesis* 的核心，理解了这两个问题就基本把握了庞加莱经验约定论的精髓。

对于“什么是假设”，庞加莱并没有给假设下一个明确、完备而清晰的定义，只是提出“一切概括都是假设”。但一切假设是否均为概括呢？庞加莱用假设的外延阐释对该问题进行了系统回答。庞加莱指出，第一类假设是“隐蔽的定义”，“是纯属自然的，人们难以摆脱。我们很难假设遥远物体的作用就是可以忽略，还有小规模的运动是遵循线性法则以及结果是关于起因的函数。”第二类假设是中性假设，是一种能够帮助科学家计算、理解图像或是坚定信念的辅助性技巧或手段，例如“我们在大多数问题的分析里也就假设在推算开始，物质要么是连续的，或者相反是由原子构成的。假设也有可能是相反的，但结果最终还是一样的，也只是相反的路走得更复杂一些，仅此而已。”第三类假设是概括，是对事物之间关系的描述，“是真正的归纳总结，这些就是必须要实验来验证或者证实。不管结果是对是错，都不能说一点用都没有。”

假设与科学进步的关系是什么呢？正如庞加莱在引言中所说：“人们略加思索，便可以察觉假设占据着多重大的地位，数学家没有它便不能工作，更不必说实验家了”。正文章节中随处可见假设的证明促进科学进步的例子。例如，对于“约定性”的假设(第一种假设)，选编的第三章“非欧几何学”认为：“我们对于实验结果的选择都是在可能的结果中根据事实情况决定的。但是我们不受拘束，也不能有任何矛盾的地方。所以，我们的教条就可以一直不倒；尽管证明它们的实验定理说的都不是绝对的。”因此，约定无所谓对错，只是出于便利性的选择而已，询问欧式几何为真还是非欧几何为真没有任何意义，因为“一种几何学不能说比另一个更有正确性，只能说是更易于解释定理”。

故当罗巴切夫斯基几何和黎曼几何抛弃欧式几何“通过一给定点只能引一条直线与已知直线平行”的公设(约定)而通过进一步的演绎分析后，新的几何大厦就建立起来了。对于“概括性”的假设(第三类假设)，它的证明就是科学进步的充要条件，它们应当“总是尽可能早地、尽可能经常地受到证实”。即使它们被证伪，被科学家“毫无保留地放弃”，科学家也应当“满足开心，因为这也是他获得新发现的偶然机会”，因为“考虑一切可能发生在那现象的情况，如果在验证时情况不对，那就是因为有没预料到的情况或者一些我们无法理解的情况出现，这就是我们发现未知的新事物的绝好机会。”同时，被证伪的假设“比起我们运用的猜想还有更多的价值。不仅仅是因为我们之前通过严密的实验得来。而且没有了之前的猜想，我们只能凭机遇来进行试验，那什么都无法得出。我们什么伟大的想法都无法得出，除了我们知道的事件越来越多，但都是没经过任何推理思考的原始数据。”

庞加莱的《科学与假设》为我们认识科学进步的过程、认识科学家在科学进步中扮演的角色提供了很好的呈现。一方面，庞加莱既看到了科学的客观性，又看到了科学家的主观能动性，事实不以人的意志为转移，但对于事实的选择、事实的概括以及概括(假设)的证明则依赖科学家的努力；另一方面，庞加莱既看到了科学的永恒性，也看到了科学理论的相对性，科学进步是一个逼近真理的过程，任何的理论、假设都要接受证明，是否能够被证明(无论证实还是证伪)是科学理论重要的特征，科学就是在不断的证明中进步，证明正是科学家从柏拉图“洞喻”中的地洞走向地面看到“真理之光”的过程。值得关注的是，庞加莱既看到了对假设证实的功用，更看到了证伪的价值，没有证伪，科学家发现不了未知，证实是“人们只不过多编入了一个事实”，只有证实和证伪携手并进，科学才能够进步。

(王传毅)

Science and Hypothesis

Henri Poincaré

Chapter 3 The Non-Euclidean Geometries

1 Every conclusion supposes premises; these premises themselves either are self-evident and need no demonstration, or can be established only by relying upon other propositions, and since we can not go back thus to infinity, every deductive science, and in particular geometry, must rest on a certain number of undemonstrable axioms. All treatises on geometry begin, therefore, by the enunciation of these axioms. But among these there is a distinction to be made: Some, for example, "Things which are equal to the same thing are equal to one another," are not propositions of geometry, but propositions of analysis. I regard them as analytic judgments *a priori*, and shall notconcern myself with them.

2 But I must lay stress upon other axioms which are peculiar to geometry. Most treatises enunciate three of these explicitly:

1° Through two points can pass only one straight;

2° The straight line is the shortest path from one point to another;

3° Through a given point there is not more than one parallel to a given straight.

3 Although generally a proof of the second of these axioms is omitted, it would be possible to deduce it from the other two and from those, much more numerous, which are implicitly admitted without enunciating them, as I shall explain further on.

本文选自亨利·庞加莱著 *Science and Hypothesis*。

科学与假设

亨利 · 庞加莱

第三章　非欧几里得几何学

1　任何结论都假设先有前提，这些前提，要么本身是自明的而无须证明，要么只能在其他命题的基础上才能成立。既然人们不能这样追溯到无穷，所有的演绎科学，特别是几何学，都必须建立在一定数量的不可证明的公理之上才行。因此，所有关于几何学的论述都是从阐明这些公理开始的。然而这些公理之间也存在区别，其中一些，例如，"等于同量的量彼此相等"，就不是几何学命题，而是分析学命题。我认为它们是一种先验分析判断，对此我不再去关注它们。

2　但是，我必须强调那些专属于几何学的公理。在大多数专著中，对以下三个公理都有明确的阐述：

1° 经过两点只能作一条直线；

2° 直线是两点间的最短距离；

3° 通过一给定点，只能引一条直线与一给定的直线平行。

3　虽然人们一般会省去对第二公理的证明，但它可以从另外两个公理，以及那些不言而喻的、数量更多的公理中演绎出来，我将在后续篇幅里说明。

本文由本导引主编桑建平根据亨利 · 庞加莱著 *Science and Hypothesis* 的英译版的第三章、第九章翻译而成。

4 It was long sought in vain to demonstrate likewise the third axiom, known as *Euclid's Postulate.* What vast effort has been wasted in this chimeric hope is truly unimaginable. Finally, in the first quarter of the nineteenth century, and almost at the same time, a Hungarian and a Russian, Bolyai and Lobachevski, established irrefutably that this demonstration is impossible; they have almost rid us of inventors of geometries "sans postulatum"; since then the Académie des Sciences receives only about one or two new demonstrations a year.

5 The question was not exhausted; it soon made a great stride by the publication of Riemann's celebrated memoir entitled: *Ueber die Hypothesen welche der Geometrie zu Grunde liegen.* This paper has inspired most of the recent works of which I shall speak further on, and among which it is proper to cite those of Beltrami and of Helmholtz.

6 THE BOLYAI-LOBACHEVSKI GEOMETRY.—If it were possible to deduce Euclid's postulate from the other axioms, it is evident that in denying the postulate and admitting the other axioms, we should be led to contradictory consequences; it would therefore be impossible to base on such premises a coherent geometry.

7 Now this is precisely what Lobachevski did.

8 He assumes at the start that: *Through a given point can be drawn two parallels to a given straight.* And he retains besides all Euclid's other axioms.

9 From these hypotheses he deduces a series of theorems among which it is impossible to find any contradiction, and he constructs a geometry whose faultless logic is inferior in nothing to that of the Euclidean geometry.

4 长期以来，人们一直试图证明第三公理，即所谓的欧几里得公设，但结果都是徒劳的。人们无法想象为追求这一幻想所付出的努力。结果在19世纪初期的25年，几乎是同时，两位科学家，俄国的罗巴切夫斯基(Lobatchewsky)和匈牙利的鲍耶(Bolyai)，无可辩驳地证明了这是不可能的。他们几乎让我们摆脱了“无公设”的几何发明家，从此法国科学院(Académie des Sciences)每年只收到一两个新的证明。

5 但问题并没有就此结束，不久黎曼(Riemann)就发表了著名论文，题为“几何学之基本假设”(*Ueber die Hypothesen welche der Gometrie zum Grunde liegen*)，这一问题才有了巨大进展。这篇论文激发了我将进一步讨论的大多数近代专著，在这些著作中引用贝尔特拉米(Beltrami)和亥姆霍兹(Helmholtz)的著作是恰当的。

6 **鲍耶-罗巴切夫斯基几何学** 如果可以从其他几个公理出发导出欧几里得公设，那么很明显，如果拒绝公设而保留其他公理，就会导致矛盾的结果。因此，不可能在这些前提下建立融洽的几何学。

7 然而，罗巴切夫斯基恰恰就是这么做的。

8 他一开始就假定：通过一给定点可以引两条直线与给定的直线平行。并且他保留了欧几里得的所有其他公理。

9 从这些假设出发，他演绎出一系列的定理，这些定理之间找不到任何矛盾，他构造了一个逻辑上和欧几里得几何学一样无懈可击的几何学。

10 The theorems are, of course, very different from those to which we are accustomed, and they can not fail to be at first a little disconcerting.

11 Thus the sum of the angles of a triangle is always less than two right angles, and the difference between this sum and two right angles is proportional to the surface of the triangle.

12 It is impossible to construct a figure similar to a given figure but of different dimensions.

13 If we divide a circumference into n equal parts, and draw tangents at the points of division, these n tangents will form a polygon if the radius of the circle is small enough; but if this radius is sufficiently great they will not meet.

14 It is useless to multiply these examples; Lobachevski's propositions have no relation to those of Euclid, but they are not less logically bound one to another.

15 RIEMANN'S GEOMETRY.—Imagine a world uniquely peopled by beings of no thickness (height); and suppose these "infinitely flat" animals are all in the same plane and can not get out. Admit besides that this world is sufficiently far from others to be free from their influence. While we are making hypotheses, it costs us no more to endow these beings with reason and believe them capable of creating a geometry. In that case, they will certainly attribute to space only two dimensions.

10 然而，这些定理与人们所习惯的那些定理是截然不同的，起初会使人感到有些困惑。

11 例如，一个三角形的内角之和总是小于两直角，这个和与两直角的差则与三角形的面积成正比。

12 要构造出与给定图形相似但大小不等的图形是不可能的。

13 如果一圆周被分为 n 等份，并在各分点引切线，那么当圆的半径足够小时，这 n 条切线就会形成一个多边形，但如果半径足够大，它们就不会相交。

14 不需要再多举这样的例子。罗巴切夫斯基的命题与欧几里得的命题毫无关系，但它们在逻辑上依然是相互联系的。

15 **黎曼几何学** 想象一个没有厚度(高度)的动物所生存的世界，假设这些"无限扁平"的动物都在同一个平面上，而不可能走出这个平面。进一步假定，这个世界与其他世界的距离已经足够遥远，可以不受那些世界的影响。当做这些假设时，人们不妨再赋予这些动物以理性，并相信它们能够创造出几何学。在这种情况下，它们肯定认为空间是二维的了。

16 But suppose now that these imaginary animals, while remaining without thickness, have the form of a spherical, and not of a plane, figure, and are all on the same sphere without power to get off. What geometry will they construct? First it is clear they will attribute to space only two dimensions; what will play for them the rôle of the straight line will be the shortest path from one point to another on the sphere, that is to say, an arc of a great circle; in a word, their geometry will be the spherical geometry.

17 What they will call space will be this sphere on which they must stay, and on which happen all the phenomena they can know. Their space will therefore be *unbounded* since on a sphere one can always go forward without ever being stopped, and yet it will be *finite*; one can never find the end of it, but one can make a tour of it.

18 Well, Riemann's geometry is spherical geometry extended to three dimensions. To construct it, the German mathematician had to throw overboard, not only Euclid's postulate, but also the first axiom: *Only one straight can pass through two points.*

19 On a sphere, through two given points we can draw *in general* only one great circle (which, as we have just seen, would play the rôle of the straight for our imaginary beings); but there is an exception: if the two given points are diametrically opposite, an infinity of great circles can be drawn through them.

20 In the same way, in Riemann's geometry (at least in one of its forms), through two points will pass in general only a single straight; but there are exceptional cases where through two points an infinity of straights can pass.

16 其次，再假设这些想象中的动物，虽然没有厚度，但它们的形状是球形的而不是平面的，它们都在同一个球面上，它们无法逃离。它们会建立什么样的几何学呢？首先，很明显它们只属于二维空间。对它们来说，直线是球面上一点到另一点的最短路径，也就是说一个大圆的圆弧。总之，它们的几何学将是球面几何学。

17 它们所说的空间就是它们所处的球面，在这个球面上发生着它们所熟悉的一切现象。因此，它们的空间将是无界的，因为在一个球面上，它们可以永远向前走而不会停下来；但这空间将是有限的，虽然永远也找不到它的尽头，但却可以完成整个旅程。

18 黎曼几何学是推广到三维的球面几何。德国数学家为了构建这种几何学，不仅要抛弃欧几里得公设，甚至连第一公理，即通过两点只能作一条直线也要抛弃掉。

19 通过球面上的两给定点，通常只能画一个大圆(正如我们前面所提到的，这个大圆对于人们想象中的动物来说是一条直线)。但有一个例外：如果给定的两个点在直径的两端，则可以作无穷多个大圆。

20 同样在黎曼几何学(至少其中一种)中，通过两点一般只能引一条直线，但也有例外，通过两点可以引无数条直线。

21 There is a sort of opposition between Riemann's geometry and that of Lobachevski.

Thus the sum of the angles of a triangle is:

Equal to two right angles in Euclid's geometry;

Less than two right angles in that of Lobachevski;

Greater than two right angles in that of Riemann.

22 The number of straights through a given point that can be drawn coplanar to a given straight, but nowhere meeting it, is equal:

To one in Euclid's geometry;

To zero in that of Riemann;

To infinity in that of Lobachevski.

23 Add that Riemann's space is finite, although unbounded, in the sense given above to these two words.

24 THE SURFACES OF CONSTANT CURVATURE.—One objection still remained possible. The theorems of Lobachevski and of Riemann present no contradiction; but however numerous the consequences these two geometers have drawn from their hypotheses, they must have stopped before exhausting them, since their number would be infinite; who can say then that if they had pushed their deductions farther they would not have eventually reached some contradiction?

25 This difficulty does not exist for Riemann's geometry, provided it is limited to two dimensions; in fact, as we have seen, two-dimensional Riemannian geometry does not differ from spherical geometry, which is only a branch of ordinary geometry, and consequently is beyond all discussion.

21 在黎曼几何学与罗巴切夫斯基几何学之间存在着某种对立的地方。例如，三角形的内角之和：
在欧几里得几何学中等于两直角；
在罗巴切夫斯基几何学中小于两直角；
在黎曼几何学中大于两直角。

22 过给定点引与给定直线共面但在任何地方都不相交的直线的数目是：
在欧几里得几何学中有一条；
在黎曼几何学中一条也没有；
在罗巴切夫斯基几何学中有无穷多条。

23 此外，黎曼空间虽然无界，但却是有限的，这与前面提到的这两个词的意义相同。

24 **常曲率面** 然而，某种反对意见依然可能存在。罗巴切夫斯基的定理与黎曼的定理之间并不矛盾；但是无论这两位几何学家从他们的假设中推导出多少结果，他们势必在穷尽所有结果之前就停下来，不然其数目会无限；谁又能说，如果他们将演绎继续下去，最终就不会得出某种矛盾的结论呢？

25 这种困难对于黎曼几何学来说并不会存在，只要把它限定在二维即可。事实上，正如我们已经看到的，二维黎曼几何学与球面几何学并无差别，球面几何学只是普通几何学的一个分支，因此不需要讨论。

26 Beltrami, in correlating likewise Lobachevski's two-dimensional geometry with a branch of ordinary geometry, has equally refuted the objection so far as it is concerned.

27 Here is how he accomplished it. Consider any figure on a surface. Imagine this figure traced on a flexible and inextensible canvas applied over this surface in such a way that when the canvas is displaced and deformed, the various lines of this figure can change their form without changing their length. In general, this flexible and inextensible figure can not be displaced without leaving the surface; but there are certain particular surfaces for which such a movement would be possible; these are the surfaces of constant curvature.

28 If we resume the comparison made above and imagine beings without thickness living on one of these surfaces, they will regard as possible the motion of a figure all of whose lines remain constant in length. On the contrary, such a movement would appear absurd to animals without thickness living on a surface of variable curvature.

29 These surfaces of constant curvature are of two sorts: Some are of *positive curvature*, and can be deformed so as to be applied over a sphere. The geometry of these surfaces reduces itself therefore to the spherical geometry, which is that of Riemann.

30 The others are of *negative curvature*. Beltrami has shown that the geometry of these surfaces is none other than that of Lobachevski. The two-dimensional geometries of Riemann and Lobachevski are thus correlated to the Euclidean geometry.

26 贝尔特拉米通过证明罗巴切夫斯基的二维几何学只是普通几何学的一个分支，也同样驳斥了上述反对意见。

27 他的论证是这样的：考虑一面上有任一图形。想象这个图形被画在一个可弯曲而不可伸缩的画布上，且画布紧贴在这一面上，当画布被移动和变形时，图形的不同线条在不改变长度的情况下改变它们的形状。一般这个可弯曲而不可伸缩的图形不可能在不离开表面的情况下被移动。但在某些特殊表面，这样的移动是可能的，它们就是常曲率面。

28 如果我们继续上面所做的比较，想象没有厚度的动物生活在某个这样的表面上，它们会认为，在所有线条的长度都保持恒定的情形下，图形的移动是可能的。相反，对于生活在可变曲率表面上而没有厚度的动物来说，这样的移动是荒谬的。

29 这种常曲率面可分为两种：一种是正曲率的，能够变形而紧贴在球面上。所以，这种面上的几何归属于球面几何，这就是黎曼几何学。

30 另一种是负曲率的，贝尔特拉米已经证明这些面的几何就是罗巴切夫斯基几何学。因此，黎曼和罗巴切夫斯基的二维几何学与欧几里得几何学相关。

31 INTERPRETATION OF NON-EUCLIDEAN GEOMETRIES.—So vanishes the objection so far as two-dimensional geometries are concerned.

32 It would be easy to extend Beltrami's reasoning to three-dimensional geometries. The minds that space of four dimensions does not repel will see no difficulty in it, but they are few. I prefer therefore to proceed otherwise.

33 Consider a certain plane, which I shall call the fundamental plane, and construct a sort of dictionary, by making correspond each to each a double series of terms written in two columns, just as correspond in the ordinary dictionaries the words of two languages whose significance is the same:

Space: Portion of space situated above the fundamental plane.

Plane: Sphere cutting the fundamental plane orthogonally.

Straight: Circle cutting the fundamental plane orthogonally.

Sphere: Sphere.

Circle: Circle.

Angle: Angle.

Distance between two points: Logarithm of the cross ratio of these two points and the intersections of the fundamental plane with a circle passing through these two points and cutting it orthogonally. Etc., Etc.

34 Now take Lobachevski's theorems and translate them with the aid of this dictionary as we translate a German text with the aid of a German-English dictionary. *We shall thus obtain theorems of the ordinary geometry*. For example, that theorem of Lobachevski: "the sum of the angles of a triangle is less than two right angles" is translated thus: "If a curvilinear triangle has for sides circlearcs which prolonged would cut orthogonally the fundamental plane, the sum of the angles of this curvilinear triangle will be less than two right angles." Thus, however far the consequences of Lobachevski's

31 **非欧几里得几何学的诠释** 如果以二维几何学为限，就不会有反对意见了。

32 很容易把贝尔特拉米的论证推广到三维几何学，若人的心智对四维空间不畏缩也就不会觉得有什么困难，但这样的人极少。因此，我还是采取别的方法来叙述吧。

33 考虑某一平面，我将其称之为基本平面，然后再编撰一部词典，把两组术语一一对应地写成两列，就像在普通字典中，两种语言中有相同意义的词汇相对应一样：

空间：在基本平面上方的空间部分。

平面：与基本平面相交成直角的球面。

直线：与基本平面相交成直角的圆。

球：球。

圆：圆。

角：角。

两点之间的距离：基本平面与经过此两点的正交圆的交点，以及此两点之交比的对数，等等。

34 现在，借助这本词典来翻译罗巴切夫斯基的定理，就像使用德英词典来翻译德语文本一样。这样人们就得到普通几何学的定理。例如，罗巴切夫斯定理：“三角形的内角之和小于两直角”，可译为：“如果一曲线三角形的边延长后是与基本平面正交的圆弧线，则该曲线三角形的内角之和小于两直角。”因此，无论如何推广罗巴切夫斯基假设的结果，它们永远不会导致矛盾。事实上，如果罗巴切夫斯基的两个定理是矛盾的，那么这两个定理经词典翻译后也是矛盾的。由于这些译文是普通几何学的

hypotheses are pushed, they will never lead to a contradiction. In fact, if two of Lobachevski's theorems were contradictory, it would be the same with the translations of these two theorems, made by the aid of our dictionary, but these translations are theorems of ordinary geometry and no one doubts that the ordinary geometry is free from contradiction. Whence comes this certainty and is it justified? That is a question I can not treat here because it would require to be enlarged upon, but which is very interesting and I think not insoluble.

35 Nothing remains then of the objection above formulated. This is not all. Lobachevski's geometry, susceptible of a concrete interpretation, ceases to be a vain logical exercise and is capable of applications; I have not the time to speak here of these applications, nor of the aid that Klein and I have gotten from them for the integration of linear differential equations.

36 This interpretation moreover is not unique, and several dictionaries analogous to the preceding could be constructed, which would enable us by a simple "translation" to transform Lobachevski's theorems into theorems of ordinary geometry.

37 THE IMPLICIT AXIOMS.—Are the axioms explicitly enunciated in our treatises the sole foundations of geometry? We may be assured of the contrary by noticing that after they are successively abandoned there are still left over some propositions common to the theories of Euclid, Lobachevski and Riemann. These propositions must rest on premises the geometers admit without enunciation. It is interesting to try to disentangle them from the classic demonstrations.

定理，而没有人对普通几何学无矛盾表示怀疑。这种确定性从何而来？是否合理？这是一个我不能在此处理的问题，因为说起来话就长了，但它却是一个非常有趣的问题，我认为它不是不能解决的。

35 因此，前面所提出的反对意见也就不存在了，但这还不是全部。罗巴切夫斯基几何学可以被具体地加以诠释，不再是一个无用的逻辑练习，并能够应用。在这里我无闲暇讨论这些应用，也没有时间讲克莱因(Klein)和我由于线性微分方程的积分而从中得到的帮助。

36 此外，这种诠释并非唯一，可以编撰一些类似于上述的词典，这将使我们能够通过简单的翻译将罗巴切夫斯基的定理转换成普通几何学的定理。

37 **隐含的公理** 在我们的专著中所明确阐述的公理是几何学的唯一基础吗？当我们试着依次把这些公理抛弃后，仍然存在一些欧几里得、罗巴切夫斯基和黎曼几何学中共有的命题，就可以确信并非如此。这些命题必须建立在几何学家不加阐述而公认的前提上。尝试将它们从经典的证明中清理出来是一件有趣的事。

38 Stuart Mill has claimed that every definition contains an axiom, because in defining one affirms implicitly the existence of the object defined. This is going much too far; it is rare that in mathematics a definition is given without its being followed by the demonstration of the existence of the object defined, and when this is dispensed with it is generally because the reader can easily supply it. It must not be forgotten that the word existence has not the same sense when it refers to a mathematical entity and when it is a question of a material object. A mathematical entity exists, provided its definition implies no contradiction, either in itself, or with the propositions already admitted.

39 But if Stuart Mill's observation can not be applied to all definitions, it is none the less just for some of them. The plane is sometimes defined as follows: The plane is a surface such that the straight which joins any two of its points is wholly on this surface.

40 This definition manifestly hides a new axiom; it is true we might change it, and that would be preferable, but then we should have to enunciate the axiom explicitly.

41 Other definitions would suggest reflections not less important.

42 Such, for example, is that of the equality of two figures; two figures are equal when they can be superposed; to superpose them one must be displaced until it coincides with the other; but how shall it be displaced? If we should ask this, no doubt we should be told that it must be done without altering the shape and as a rigid solid. The vicious circle would then be evident.

43 In fact this definition defines nothing; it would have no meaning for a being living in a world where there were only fluids. If it seems clear to us, that is

38 斯图亚特·密尔(Stuart Mill)断言，所有定义都含有公理，因为通过定义，人们隐含地肯定了所定义对象的存在。像这样说未免太过分了。在数学中，若给出一个定义，很少不去证明所定义对象的存在，如果没有证明所定义对象的存在，通常是因为读者可以很容易地补充它。不要忘记，“存在”这个词，当它指的是一个数学实体时，它的意思和它指的是一个物质对象时的意思是不一样的。一个数学实体的存在，前提是它的定义中没有隐含矛盾，无论是它本身，还是与先前公认的命题之间都没有矛盾。

39 如果斯图亚特·密尔的观察不能适用于所有的定义，那么对其中某些定义却是适用的。平面有时是这样定义的：

平面是这样一种面，将面上任意两点连接起来的直线完全位于该面上。

40 该定义显然隐藏着一个新的公理。的确，人们可以改变它，这也许会更好一些，但必须明确地阐明该公理。

41 其他定义也可能引起并非不重要的思考。

42 例如，两个图形全等的问题：两个图形可以重叠时是全等的。要重叠它们，必须移动其中一个，直到与另一个重叠。但如何移动它呢？如果问这个问题，那么人们无疑会被告知，它应该在不改变其形状的情况下完成，就像一个刚体被移动了一样。这样一来，就会出现循环论证了。

43 事实上，这个定义就没有给出什么内容。对于居住在只有流体的世界里的动物来说，是毫无意义的。如果人们认为这似乎是清楚的，那是因为

because we are used to the properties of natural solids which do not differ much from those of the ideal solids, all of whose dimensions are invariable.

44 Yet, imperfect as it may be, this definition implies an axiom.

45 The possibility of the motion of a rigid figure is not a self-evident truth, or at least it is so only in the fashion of Euclid's postulate and not as an analytic judgment *a priori* would be.

46 Moreover, in studying the definitions and the demonstrations of geometry, we see that one is obliged to admit without proof not only the possibility of this motion, but some of its properties besides.

47 This is at once seen from the definition of the straight line. Many defective definitions have been given, but the true one is that which is implied in all the demonstrations where the straight line enters:

48 "It may happen that the motion of a rigid figure is such that all the points of a line belonging to this figure remain motionless while all the points situated outside of this line move. Such a line will be called a straight line." We have designedly, in this enunciation, separated the definition from the axiom it implies.

49 Many demonstrations, such as those of the cases of the equality of triangles, of the possibility of dropping a perpendicular from a point to a straight, presume propositions which are not enunciated, for they require the admission that it is possible to transport a figure in a certain way in space.

我们习惯了天然固体的性质，这些性质与理想固体的性质差别不大，而理想固体自身的大小是不变的。

44 尽管这个定义可能并不完美，但它仍隐含着一个公理。

45 一个刚性图形运动的可能性不是一个不言自明的真理，或至少它只能像欧几里得公设那样，而不是像先验分析判断那样。

46 而且，当研究几何学的定义和证明时，我们发现，人们在不加证明的情况下，不仅要承认这种运动的可能性，而且还要承认它的某些性质。

47 这首先表现在直线的定义中。对于直线，已有很多有缺陷的定义，而真正的定义却隐含在用到直线的所有证明中的那一种：

48 “一个刚性图形的运动可能是这样的：属于该图形上某线的所有点都静止不动，而该线外的所有点都在运动。这样的一条线被称为直线。”在这一阐述中，我们有意地把定义同它所蕴含的公理分开。

49 很多证明，例如有关三角形全等的证明以及从一点向一直线引垂线的证明，都假设有些可省略叙述的命题，因为它们需要承认在空间中以某种方式移动图形是可能的。

50 THE FOURTH GEOMETRY.—Among these implicit axioms, there is one which seems to me to merit some attention, because when it is abandoned a fourth geometry can be constructed as coherent as those of Euclid, Lobachevski and Riemann.

51 To prove that a perpendicular may always be erected at a point *A* to a straight *AB*, we consider a straight *AC* movable around the point *A* and initially coincident with the fixed straight *AB*; and we make it turn about the point *A* until it comes into the prolongation of *AB*.

52 Thus two propositions are presupposed: First, that such a rotation is possible, and next that it may be continued until the two straights come into the prolongation one of the other.

53 If the first point is admitted and the second rejected, we are led to a series of theorems even stranger than those of Lobachevski and Riemann, but equally exempt from contradiction.

54 I shall cite only one of these theorems and that not the most singular: *A real straight may be perpendicular to itself.*

55 LIE'S THEOREM.—The number of axioms implicitly introduced in the classic demonstrations is greater than necessary, and it would be interesting to reduce it to a minimum. It may first be asked whether this reduction is possible, whether the number of necessary axioms and that of imaginable geometries are not infinite.

56 A theorem of Sophus Lie dominates this whole discussion. It may be thus enunciated: Suppose the following premises are admitted:

50 **第四种几何学** 在这些隐含的公理中，有一公理是值得注意的，当人们抛弃它时，可以构建一个像欧几里得、罗巴切夫斯基和黎曼几何学一样融洽的第四种几何学。

51 为了证明在 A 点总可以向直线 AB 引垂线，考虑一条围绕 A 点可转动的直线 AC，并且最初与固定的直线 AB 重合。然后，使 AC 绕着 A 点转动，直到它位于 AB 的延长线上为止。

52 因此，人们假设有两个命题：第一，这样的转动是可能的；第二，这样的转动可以继续下去，直到这两条直线互为延长线时为止。

53 如果承认第一个命题而否认第二个命题，就会得到一系列比罗巴切夫斯基和黎曼的定理更为奇异的定理，但同样没有矛盾。

54 从这些定理中仅选一个叙述，而它并非是最奇异的：一条实直线可以垂直于它自身。

55 **李定理** 隐含地引入到经典证明中的公理的数量比所需要的多，把它们减少到最小也许是有趣的。首先可能会问这种简化是否可能，以及必要的公理数目和可能的几何学的数目是否不是无限的。

56 索弗斯·李(Sophus Lie)的定理对这个问题的讨论是极其重要的。它可以用以下方式来阐述，假设承认以下前提：

1° Space has n dimensions;

2° The motion of a rigid figure is possible;

3° It requires p conditions to determine the position of this figure in space.

57 *The number of geometries compatible with these premises will be limited.*

58 I may even add that if n is given, a superior limit can be assigned to p.

59 If therefore the possibility of motion is admitted, there can be invented only a finite (and even a rather small) number of three-dimensional geometries.

60 RIEMANN'S GEOMETRIES.—Yet this result seems contradicted by Riemann, for this savant constructs an infinity of different geometries, and that to which his name is ordinarily given is only a particular case.

61 All depends, he says, on how the length of a curve is defined. Now, there is an infinity of ways of defining this length, and each of them may be the starting point of a new geometry.

62 That is perfectly true, but most of these definitions are incompatible with the motion of a rigid figure, which in the theorem of Lie is supposed possible. These geometries of Riemann, in many ways so interesting, could never therefore be other than purely analytic and would not lend themselves to demonstrations analogous to those of Euclid.

63 ON THE NATURE OF AXIOMS.—Most mathematicians regard Lobachevski's geometry only as a mere logical curiosity; some of them, however, have gone farther. Since several geometries are possible, is it certain ours is the true one? Experience no doubt teaches us that the sum of the angles of a

1° 空间是 n 维的；

2° 刚性图形的运动是可能的；

3° 确定该图形在空间中的位置需要 p 个条件。

57 符合这些前提的几何学的数目是有限的。

58 甚至可以补充说，如果给定 n，则可以指定 p 的上限。

59 因此，如果承认前面所提到的运动的可能性，则只能发明有限数目的(甚至是相当有限的)三维几何学。

60 **黎曼几何学** 然而，这个结果似乎与黎曼的理论相矛盾，因为这位学者发明了无数的几何学，而通常所称的黎曼几何学只是这些几何学中的一个特例。

61 他说，这一切都取决于如何定义曲线的长度。现在，有无数种定义这个长度的方法，每一种都可能成为新几何学的起点。

62 这是完全正确的，但是这些定义中的大多数都与刚性图形的运动不相容，而在李氏定理中，刚性图形的运动是可能的。这些黎曼几何学虽然如此有趣，但永远不过是纯粹分析性的，也不可能像欧几里得几何学那样加以证明。

63 **论公理的本性** 大多数数学家认为罗巴切夫斯基的几何学仅仅是一种逻辑上的奇巧，然而，部分数学家则更进一步。如果存在多种几何学，那我们的几何学一定是正确的吗？毫无疑问，经验告诉我们三角形的内角

triangle is equal to two right angles; but this is because the triangles we deal with are too little; the difference, according to Lobachevski, is proportional to the surface of the triangle; will it not perhaps become sensible when we shall operate on larger triangles or when our measurements shall become more precise? The Euclidean geometry would thus be only a provisional geometry.

64 To discuss this opinion, we should first ask ourselves what is the nature of the geometric axioms.

65 Are they synthetic *a priori* judgments, as Kant said?

66 They would then impose themselves upon us with such force that we could not conceive the contrary proposition, nor build upon it a theoretic edifice. There would be no non-Euclidean geometry.

67 To be convinced of it take a veritable synthetic *a priori* judgment, the following, for instance, of which we have seen the preponderant rôle in the first chapter:

If a theorem is true for the number 1, and if it has been proved that it is true of n + 1 provided it is true of n, it will be true of all the positive whole numbers.

68 Then try to escape from that and, denying this proposition, try to found a false arithmetic analogous to non-Euclidean geometry—it can not be done; one would even be tempted at first blush to regard these judgments as analytic.

69 Moreover, resuming our fiction of animals without thickness, we can hardly admit that these beings, if their minds are like ours, would adopt the Euclidean geometry which would be contradicted by all their experience.

之和等于两直角，但这是因为我们处理的三角形太小了。根据罗巴切夫斯基的观点，这种差值与三角形的面积成正比，当我们对大得多的三角形进行计算时，或当我们的测量变得更精确时，难道这种差值就不会被感觉到吗？因此，欧几里得几何学只不过是一个暂定的几何学而已。

64 为了讨论这一观点，首先我们必须问自己，几何学公理的本性是什么？

65 这些公理是否如康德(Kant)所说的那样，是先验综合判断吗？

66 然后，它们以一种如此强大的力量强加于我们，以至于我们既不能设想相反的命题，也无法在其上建立理论体系。所谓非欧几里得几何学也就不存在了。

67 为确信这一点，让我们举一真正的先验综合判断，例如在第一章①中起重要作用的那一种：

如果定理对 1 为真，若能证明该定理对 n 为真时，对 $n+1$ 亦为真，则该定理对任何正整数都为真。

68 然后，试图否认这个命题从而回避它，试图找到一种类似于非欧几里得几何学的伪算术——那是不可能做到的。人们甚至会在一开始就把这些判断认为是分析性的。

69 此外，再谈谈人们虚构的那些没有厚度的动物吧，人们简直不能认可，如果这些动物的心智同我们一样，它们会采用与它们的一切经验相矛盾的欧几里得几何学。

① 本处“第一章”是《科学与假设》中的第一章，而非本导引第一章。——译者注

70 Should we therefore conclude that the axioms of geometry are experimental verities? But we do not experiment on ideal straights or circles; it can only be done on material objects. On what then could be based experiments which should serve as foundation for geometry? The answer is easy.

71 We have seen above that we constantly reason as if the geometric figures behaved like solids. What geometry would borrow from experience would therefore be the properties of these bodies. The properties of light and its rectilinear propagation have also given rise to some of the propositions of geometry, and in particular those of projective geometry, so that from this point of view one would be tempted to say that metric geometry is the study of solids, and projective, that of light.

72 But a difficulty remains, and it is insurmountable. If geometry were an experimental science, it would not be an exact science, it would be subject to a continual revision. Nay, it would from this very day be convicted of error, since we know that there is no rigorously rigid solid.

73 *The axioms of geometry therefore are neither synthetic* a priori *judgments nor experimental facts.*

74 They are *conventions*; our choice among all possible conventions is *guided* by experimental facts; but it remains *free* and is limited only by the necessity of avoiding all contradiction. Thus it is that the postulates can remain *rigorously* true even though the experimental laws which have determined their adoption are only approximative.

75 In other words, *the axioms of geometry* (I do not speak of those of arithmetic) *are merely disguised definitions.*

70 难道我们会因此得出几何学公理是经验的真理这样的结论吗？可是，人们不会在理想的直线或理想的圆上做实验，只能在实物上做实验。因此，作为几何学基础的实验将基于什么呢？答案很简单。

71 从上面已经看到，我们不断地推理，几何图形就好像固体一样。几何学可以从经验中借鉴的就是这些固体的性质。光的性质及其沿直线传播也产生了一些几何学命题，特别是射影几何学的性质。从这个角度来说，度量几何学是固体的研究，而射影几何学则是光的研究。

72 然而，这里却存在着一个不可克服的困难。假如几何学是一门实验科学，那它就不是一门精确的科学，需要被不断地修正。不仅如此，从现在起它就会被证明是错误的，因为人们知道不存在严格的刚体。

73 因此，几何学公理既不是先验综合判断，也不是实验事实。

74 几何学公理是约定。我们在所有可能的约定中所作的选择是以实验事实为指导的，但选择仍然是自由的，只受到避免一切矛盾的必要性的限制。因此，即使决定采用这些公设的实验定律只是近似的，但那些公设依然严格为真。

75 换句话说，几何学公理(我不是指算术的公理)只不过是伪装的定义而已。

76 Then what are we to think of that question: Is the Euclidean geometry true?

77 It has no meaning.

78 As well ask whether the metric system is true and the old measures false; whether Cartesian coordinates are true and polar coordinates false. One geometry can not be more true than another; it can only be *more convenient*.

79 Now, Euclidean geometry is, and will remain, the most convenient:

1° Because it is the simplest; and it is so not only in consequence of our mental habits, or of I know not what direct intuition that we may have of Euclidean space; it is the simplest in itself, just as a polynomial of the first degree is simpler than one of the second; the formulas of spherical trigonometry are more complicated than those of plane trigonometry, and they would still appear so to an analyst ignorant of their geometric signification.

2° Because it accords sufficiently well with the properties of natural solids, those bodies which our hands and our eyes compare and with which we make our instruments of measure.

Chapter 9 Hypotheses in Physics

80 THE RÔLE OF EXPERIMENT AND GENERALIZATION.—Experiment is the sole source of truth. It alone can teach us anything new; it alone can give us certainty. These are two points that can not be questioned.

81 But then, if experiment is everything, what place will remain for mathematical physics? What has experimental physics to do with such an aid, one which seems useless and perhaps even dangerous?

76 那么，我们应该如何看待这个问题：欧几里得几何学为真吗？

77 这个问题毫无意义。

78 这好比问公制是否正确，旧的度量衡是否错误；笛卡儿坐标是否为真，极坐标是否为假。一种几何学不可能比另一种几何学更真，只能是更方便而已。

79 欧几里得几何学现在是、将来仍然是最为方便的：

1° 因为它是最简单的，而不仅仅是因为我们的思维习惯，或者因为我不知道我们有欧几里得空间的直接直觉；它本身就是最简单的，就像一阶多项式比二阶多项式简单一样；球面三角公式比平面三角公式要复杂得多，对于一个不懂这些公式的几何意义的分析家来说，也会有如此之感觉。

2° 因为它与天然固体的性质完全符合，而这些固体是我们的手和眼睛所能比较的，利用它们人们可以制造测量仪器。

第九章　物理学中的假设

80 **实验与概括的作用**　实验是真理的唯一源泉。只有实验才能给我们一些新的启示，只有实验才能给我们一种确定性，这两点是毋庸置疑的。

81 但是，如果实验就是一切，那么数学物理学还有什么地位呢？实验物理学要这样一个似乎无用并且甚至可能是危险的助手又有何用呢？

82 And yet mathematical physics exists, and has done unquestionable service. We have here a fact that must be explained.

83 The explanation is that merely to observe is not enough. We must use our observations, and to do that we must generalize. This is what men always have done; only as the memory of past errors has made them more and more careful, they have observed more and more, and generalized less and less.

84 Every age has ridiculed the one before it, and accused it of having generalized too quickly and too naïvely. Descartes pitied the Ionians; Descartes, in his turn, makes us smile. No doubt our children will some day laugh at us.

85 But can we not then pass over immediately to the goal? Is not this the means of escaping the ridicule that we foresee? Can we not be content with just the bare experiment?

86 No, that is impossible; it would be to mistake utterly the true nature of science. The scientist must set in order. Science is built up with facts, as a house is with stones. But a collection of facts is no more a science than a heap of stones is a house.

87 And above all the scientist must foresee. Carlyle has somewhere said something like this: "Nothing but facts are of importance. John Lackland passed by here. Here is something that is admirable. Here is a reality for which I would give all the theories in the world." Carlyle was a fellow countryman of Bacon; but Bacon would not have said that. That is the language of the historian. The physicist would say rather: "John Lackland passed by here; that makes no difference to me, for he never will pass this way again."

82 然而，数学物理学还是存在着的，其功劳是不可抹杀的。这是一个有必要说明的事实。

83 事实上，仅有观察是不够的，必须利用这些观察，为此目的必须进行概括。人们一直都是这么做的，只是当人们对过去所犯错误的回忆使人变得越来越谨慎的时候，才会越来越多地从事观察，却越来越少地进行概括。

84 每一个时代的人都曾嘲笑过自己的前辈，指责他们概括得太快和过于天真。笛卡儿(Descartes)曾经怜悯爱奥尼亚人(Ionians)，笛卡儿也让我们嘲笑。毫无疑问，总有一天我们的子孙也会嘲笑我们。

85 既然如此，难道我们就没有办法直接到达终点吗？从而避免可以预见的嘲笑吗？难道我们就不能仅仅只满足于赤裸裸的实验吗？

86 不，那是不可能的，那就完全误解了科学的本质。科学家必须做整理的工作。科学是建立在事实之上的，正如房子是由石头砌成的。但是，事实的积累不能算是科学，正如石头的堆积并不是房子一样。

87 最重要的是科学家必须预见。卡莱尔(Carlyle)在某地曾经说过："只有事实才是重要的。约翰·拉克兰(Jean Sans Terre)①曾经过此地，这是一件值得赞美的事，为此事实我愿献出天下所有的理论。"卡莱尔是培根(Bacon)的同胞，但培根却不会这样说。那是历史学家的口吻。物理学家宁愿说："约翰·拉克兰曾经过此地，此事与我无关，因为他再也不会经过此地了。"

① Jean Sans Terre 是《科学与假设》法文版中的人名，英文版中译为 John Lack Kland。——译者注

88 We all know that there are good experiments and poor ones. The latter will accumulate in vain; though one may have made a hundred or a thousand, a single piece of work by a true master, by a Pasteur, for example, will suffice to tumble them into oblivion. Bacon would have well understood this; it is he who invented the phrase *Experimentum crucis*. But Carlyle would not have understood it. A fact is a fact. A pupil has read a certain number on his thermometer; he has taken no precaution; no matter, he has read it, and if it is only the fact that counts, here is a reality of the same rank as the peregrinations of King John Lackland. Why is the fact that this pupil has made this reading of no interest, while the fact that a skilled physicist had made another reading might be on the contrary very important? It is because from the first reading we could not infer anything. What then is a good experiment? It is that which informs us of something besides an isolated fact; it is that which enables us to foresee, that is, that which enables us to generalize.

89 For without generalization foreknowledge is impossible. The circumstances under which one has worked will never reproduce themselves all at once. The observed action then will never recur; the only thing that can be affirmed is that under analogous circumstances an analogous action will be produced. In order to foresee, then, it is necessary to invoke at least analogy, that is to say, already then to generalize.

90 No matter how timid one may be, still it is necessary to interpolate. Experiment gives us only a certain number of isolated points. We must unite these by a continuous line. This is a veritable generalization. But we do more; the curve that we shall trace will pass between the observed points and near these points; it will not pass through these points themselves. Thus one does not restrict himself to generalizing the experiments, but corrects them;

88 众所周知，有好的实验，也有不好的实验。不好的实验的积累是徒劳的，无论是一百个还是一千个实验，一位真正大师的某项工作——例如巴斯德的工作——就足以让它们全被遗忘。培根应该完全理解这一点，因为是他发明了“判决性实验(experimentum cruces)”这个名词。但卡莱尔理解不了这一点。事实就是事实。一个小学生在毫不在意的情形下读过温度计上的某个数字，不管怎样，他读了它。如果只有事实才是最重要的话，那么这个事实和约翰·拉克兰国王的周游一样可以被称为实在。为什么小学生读数这一事实就没有什么意思？相反，一位熟练的物理学家所做出的另一个读数的事实就非常重要呢？因为从小学生的读数中我们不能推断出任何东西。那么，什么是好的实验呢？好的实验就是除了一件孤立的事实外，还能告诉我们别的东西。好的实验使我们能够预见，使我们能够概括。

89 若无概括，便不可能预见。人们所处的环境将永远不会完全重现。观察到的事实永远不会重复。唯一可以肯定的是，在类似的情况下，会有类似的事实出现。因此，想要预见，必须借助于类比——也就是说，已经进行了概括。

90 无论人们多么胆怯，总要使用插值法。实验只给出了一定数量的孤立的点，必须用一条连续曲线将这些孤立的点关联起来，这才是一个真正的概括。但还有更多的工作要做，所得到的曲线，应在观察点之间及其附近经过而不会通过观察点本身。因此，人们不局限于概括实验，还要矫正实验。若物理学家放弃做这些矫正，而只满足于赤裸裸的实验，那

and the physicist who should try to abstain from these corrections and really be content with the bare experiment, would be forced to enunciate some very strange laws.

91 The bare facts, then, would not be enough for us; and that is why we must have science ordered, or rather organized.

92 It is often said experiments must be made without a preconceived idea. That is impossible. Not only would it make all experiment barren, but that would be attempted which could not be done. Every one carries in his mind his own conception of the world, of which he can not so easily rid himself. We must, for instance, use language; and our language is made up only of preconceived ideas and can not be otherwise. Only these are unconscious preconceived ideas, a thousand times more dangerous than the others.

93 Shall we say that if we introduce others, of which we are fully conscious, we shall only aggravate the evil? I think not. I believe rather that they will serve as counterbalances to each other —I was going to say as antidotes; they will in general accord ill with one another—they will come into conflict with one another, and thereby force us to regard things under different aspects. This is enough to emancipate us. He is no longer a slave who can choose his master.

94 Thus, thanks to generalization, each fact observed enables us to foresee a great many others; only we must not forget that the first alone is certain, that all others are merely probable. No matter how solidly founded a prediction may appear to us, we are never *absolutely* sure that experiment will not contradict it, if we undertake to verify it. The probability, however, is often so great that practically we may be content with it. It is far better to foresee even without certainty than not to foresee at all.

么，他将不得不阐明一些非常离奇的定律。

91 因此，赤裸裸的事实不能让人满意，这就是为什么我们必须要有整理过的科学，或者宁可说要有组织过的科学。

92 人们常说，做实验应该没有成见。但这是不可能的，它不仅会使每一个实验都毫无结果，而且即使我们希望这样做，也难以做到。每个人都有自己的世界观，且不易改变。例如，我们必须使用语言，而我们的语言必然充满了成见而不可能还有别的东西。事实上，那些无意识的成见要比其他成见危险千倍。

93 难道我们可以说，如果加入了我们完全意识到的其他成见，只会使事情变得更糟吗？我不这么认为。我宁可相信这些成见将起到相互平衡的作用——甚至想说是解毒剂。这些成见之间难以一致，且相互冲突，因此会迫使人们从不同角度来思考问题。这足以使人们不再受约束。当人们可以选择自己的主人时，就不再是奴隶了。

94 因此，通过概括，每一个观察到的事实都使我们能够预见大量的其他事实。不过，我们不应该忘记，只有第一个事实是确定的，其他的都只是可能的。在我们看来，无论一个预见有多么可靠的根据，当着手去验证它时，都不能绝对肯定它不会被实验推翻。但其可能性往往是如此之大，以至于实际上我们会对此感到满意。即使是做不确定的预见，也要比完全不去预见好得多。

95 One must, then, never disdain to make a verification when opportunity offers. But all experiment is long and difficult; the workers are few; and the number of facts that we need to foresee is immense. Compared with this mass the number of direct verifications that we can make will never be anything but a negligible quantity.

96 Of this few that we can directly attain, we must make the best use; it is very necessary to get from every experiment the greatest possible number of predictions, and with the highest possible degree of probability. The problem is, so to speak, to increase the yield of the scientific machine.

97 Let us compare science to a library that ought to grow continually. The librarian has at his disposal for his purchases only insufficient funds. He ought to make an effort not to waste them.

98 It is experimental physics that is entrusted with the purchases. It alone, then, can enrich the library.

99 As for mathematical physics, its task will be to make out the catalogue. If the catalogue is well made, the library will not be any richer, but the reader will be helped to use its riches.

100 And even by showing the librarian the gaps in his collections, it will enable him to make a judicious use of his funds; which is all the more important because these funds are entirely inadequate.

101 Such, then, is the rôle of mathematical physics. It must direct generalization in such a manner as to increase what I just now called the yield of science. By what means it can arrive at this, and how it can do it without danger, is what remains for us to investigate.

95 所以，当机会出现时，绝不应轻视验证工作。但凡实验都是漫长而困难的，且从事的人员很少，而我们所需要预见的事实，其数量又是巨大的。相比之下，我们所能做的直接验证的数量永远也只不过是沧海一粟。

96 从这些能直接得到的极少事实，应尽量获取最大的收益。每一实验都必须使我们能够做出尽可能多的预见，且具有最大的可能性。这个问题可以说是在增加科学机器的效益。

97 可以把科学与图书馆作一比较，图书馆的藏书量是不断增加的。由于图书管理员采购图书的资金有限，因此他必须竭尽全力不浪费。

98 就像实验物理学负买书之责，这样实验物理学本身就可以丰富图书馆。

99 至于数学物理学，其职责是编制书目。即使书目编得再好，图书馆也不会因此而更加丰富，但读者却能更有效地利用图书资源。

100 同时，通过向图书管理员展示图书收藏中的空白，这将有助于他明智使用资金，这才是最重要的，因为购书资金是完全不足的。

101 这就是数学物理学的作用，它必须直接概括以增加我刚才所说的科学机器的效益。它是如何做到这一点的，以及如何安全地去做，这就是我们必须研究的问题。

102 THE UNITY OF NATURE.—Let us notice, first of all, that every generalization implies in some measure the belief in the unity and simplicity of nature. As to the unity there can be no difficulty. If the different parts of the universe were not like the members of one body, they would not act on one another, they would know nothing of one another; and we in particular would know only one of these parts. We do not ask, then, if nature is one, but how it is one.

103 As for the second point, that is not such an easy matter. It is not certain that nature is simple. Can we without danger act as if it were?

104 There was a time when the simplicity of Mariotte's law was an argument invoked in favor of its accuracy; when Fresnel himself, after having said in a conversation with Laplace that nature was not concerned about analytical difficulties, felt himself obliged to make explanations, in order not to strike too hard at prevailing opinion.

105 To-day ideas have greatly changed; and yet, those who do not believe that natural laws have to be simple, are still often obliged to act as if they did. They could not entirely avoid this necessity without making impossible all generalization, and consequently all science.

106 It is clear that any fact can be generalized in an infinity of ways, and it is a question of choice. The choice can be guided only by considerations of simplicity. Let us take the most commonplace case, that of interpolation. We pass a continuous line, as regular as possible, between the points given by observation. Why do we avoid points making angles and too abrupt turns? Why do we not make our curve describe the most capricious zig-zags? It is because we know beforehand, or believe we know, that the law to be expressed can not be so complicated as all that.

102 **自然的统一性** 首先我们要注意，一切概括在某种程度上都蕴含了自然具有统一性和简单性的信念。就统一性而言，没有什么困难。如果宇宙的不同部分不像身体的各个部分的话，那么它们之间就既不会相互作用也不会彼此了解，尤其是我们只知道其中的一部分。所以，我们不必思考自然是否具有统一性，而是探究自然是如何统一的。

103 至于简单性这个问题，就不那么容易了。自然是否是简单的还不确定。难道我们能够把自然当作简单的而毫无风险地行动吗？

104 曾有一段时间，马略特(Mariotte)定律的简单性成为支持其准确性的论据。在与拉普拉斯(Laplace)的交谈中，菲涅尔(Fresnel)曾说过自然并不关心解析上的困难，后来为了不触犯当时的流行观点，他不得不对自己所说的话加以解释。

105 如今，观念有了很大的变化。可是那些不相信自然法则一定是简单的人，仍然常常被迫表现得好像他们确实相信一样。如果不是把所有的概括甚至是整个科学都变得不可能的话，那么就不能完全避免这种必要性。

106 很明显，任何事实都可以有无数种方式概括，这只是一个选择的问题。选择只能以简单性为指导。在此举一个最常见的例子，如插值法。我们在观察点之间尽可能有规则地画一条连续的线。为什么要避免造成棱角的点和太尖锐的曲折呢？为什么我们不让曲线出现最反复无常的“之”字形呢？正因为我们事先知道，或者相信自己知道，要表达的定律不可能如此复杂。

107 We may calculate the mass of Jupiter from either the movements of its satellites, or the perturbations of the major planets, or those of the minor planets. If we take the averages of the determinations obtained by these three methods, we find three numbers very close together, but different. We might interpret this result by supposing that the coefficient of gravitation is not the same in the three cases. The observations would certainly be much better represented. Why do we reject this interpretation? Not because it is absurd, but because it is needlessly complicated. We shall only accept it when we are forced to, and that is not yet.

108 To sum up, ordinarily every law is held to be simple till the contrary is proved.

109 This custom is imposed upon physicists by the causes that I have just explained. But how shall we justify it in the presence of discoveries that show us every day new details that are richer and more complex? How shall we even reconcile it with the belief in the unity of nature? For if everything depends on everything, relationships where so many diverse factors enter can no longer be simple.

110 If we study the history of science, we see happen two inverse phenomena, so to speak. Sometimes simplicity hides under complex appearances; sometimes it is the simplicity which is apparent, and which disguises extremely complicated realities.

111 What is more complicated than the confused movements of the planets? What simpler than Newton's law? Here nature, making sport, as Fresnel said, of analytical difficulties, employs only simple means, and by combining them produces I know not what inextricable tangle. Here it is the hidden simplicity which must be discovered.

107 关于木星的质量，可通过其卫星的运动，或大行星的摄动，或小行星的摄动计算出来。如果我们将这三种方法所得到的值分别取平均，就会发现这三个数字非常接近，但并不完全相同。对该结果，可以通过假设引力常数在三种情况下有所不同加以解释，这样观察结果肯定会得到更好的表述。为什么我们要拒绝这种解释呢？不是因为它荒谬，而是因为它过于复杂。我们只是在迫不得已的时候才会这样做，而现在还不必这样。

108 总之，通常每一定律都被认为是简单的，除非出现相反的证明。

109 正如上面所说，这种简单性的思维习惯是强加给物理学家的。但是，如果每天都有新的细节、更丰富和更复杂的发现展示在我们面前时，如何证明这种思维习惯是正确的呢？我们怎么能把它与自然的统一性协调一致呢？因为倘若所有的事物都是相互依赖的，则如此之多的不同因素参与其中的关系就不再是简单的了。

110 如果我们研究科学史，就会发现两种可以说是相反的现象。有时是复杂的外表之下隐藏着简单性；有时相反，简单性的外表掩盖了极其复杂的真实。

111 还有什么比行星错综复杂的运动更加复杂的呢？还有什么比牛顿定律更简单的呢？正如菲涅尔（Fresnel）所说，自然界在那里玩弄解析困难而只使用简单手段，通过这些手段的组合就产生了我所不知道的难以解开的死结。简单性就隐藏在那里，必须去发现它。

112 Examples of the opposite abound. In the kinetic theory of gases, one deals with molecules moving with great velocities, whose paths, altered by incessant collisions, have the most capricious forms and traverse space in every direction. The observable result is Mariotte's simple law. Every individual fact was complicated. The law of great numbers has reestablished simplicity in the average. Here the simplicity is merely apparent, and only the coarseness of our senses prevents our perceiving the complexity.

113 Many phenomena obey a law of proportionality. But why? Because in these phenomena there is something very small. The simple law observed, then, is only a result of the general analytical rule that the infinitely small increment of a function is proportional to the increment of the variable. As in reality our increments are not infinitely small, but very small, the law of proportionality is only approximate, and the simplicity is only apparent. What I have just said applies to the rule of the superposition of small motions, the use of which is so fruitful, and which is the basis of optics.

114 And Newton's law itself? Its simplicity, so long undetected, is perhaps only apparent. Who knows whether it is not due to some complicated mechanism, to the impact of some subtile matter animated by irregular movements, and whether it has not become simple only through the action of averages and of great numbers? In any case, it is difficult not to suppose that the true law contains complementary terms, which would become sensible at small distances. If in astronomy they are negligible as modifying Newton's law, and if the law thus regains its simplicity, it would be only because of the immensity of celestial distances.

115 No doubt, if our means of investigation should become more and more penetrating, we should discover the simple under the complex, then the

112 相反的例子举不胜举。在气体动理论中，人们处理高速运动的分子，这些分子的轨迹由于不断的碰撞而变化，其形状变幻莫测，并在空间中向各个方向运动。而观察到的结果却是马略特的简单定律。每一个事实都很复杂。大数定律在平均中重建了简单性。在这里，简单性只是表观的，是因为我们感官的粗糙妨碍了我们感知复杂性。

113 许多现象遵循比例定律，这是为什么呢？因为在这些现象中，存在某种非常小的东西。因此，所观察到的简单定律只是普遍解析法则的结果，即函数的无限小增量与变量的增量成正比。在现实中，增量并不是无限小而是非常小，比例定律只是近似的，简单性只是表观的。现在所陈述的这些也适用于小运动的叠加法则，它在应用上很有成效，它是光学的基础。

114 至于牛顿定律本身呢？它的隐藏如此之久的简单性也许只是表观的。谁知道它是否由于某种复杂机制的作用，或由于某种由不规则运动引起的难以捉摸的物质的影响，或由于仅仅靠着平均数和大数的作用而变得简单呢？无论如何，若不假设真正的定律包含补充项是困难的，这些补充项在距离较小时可能变得合理。假如在天文学中修正的牛顿定律的补充项是可以忽略的，假如定律因此恢复了它的简单性，那完全是由于天体之间的距离巨大的缘故。

115 毫无疑问，如果我们的研究方法变得日渐深入，就会发现复杂下的简单性，然后是简单下的复杂性，再次出现复杂下的简单性，如此循环不已，而无法预测最后的结果究竟如何。

complex under the simple, then again the simple under the complex, and so on, without our being able to foresee what will be the last term.

116 We must stop somewhere, and that science may be possible we must stop when we have found simplicity. This is the only ground on which we can rear the edifice of our generalizations. But this simplicity being only apparent, will the ground be firm enough? This is what must be investigated.

117 For that purpose, let us see what part is played in our generalizations by the belief in simplicity. We have verified a simple law in a good many particular cases; we refuse to admit that this agreement, so often repeated, is simply the result of chance, and conclude that the law must be true in the general case.

118 Kepler notices that a planet's positions, as observed by Tycho, are all on one ellipse. Never for a moment does he have the thought that by a strange play of chance Tycho never observed the heavens except at a moment when the real orbit of the planet happened to cut this ellipse.

119 What does it matter then whether the simplicity be real, or whether it covers a complex reality? Whether it is due to the influence of great numbers, which levels down individual differences, or to the greatness or smallness of certain quantities, which allows us to neglect certain terms, in no case is it due to chance. This simplicity, real or apparent, always has a cause. We can always follow, then, the same course of reasoning, and if a simple law has been observed in several particular cases, we can legitimately suppose that it will still be true in analogous cases. To refuse to do this would be to attribute to chance an inadmissible rôle.

116 我们必须在某处停下来，为了使科学成为可能，在遇到简单的地方就必须停下来。这是我们能够建立概括大厦的唯一基础。然而这种简单性只是表观的，该基础是否足够坚实？这是必须要研究的问题。

117 为此，看看简单性的信念在概括中起何作用？我们已在许多特例中验证了一条简单的定律，我们不认为这种可经常重复的一致性仅仅是偶然的结果，我们的结论是，该定律在普遍情况下为真。

118 开普勒(Kpler)注意到第谷(Tycho)所观察到的行星的位置都在同一个椭圆上。他从未想到，由于某种奇怪的机缘巧合，第谷每次观察天象时，都是在行星的真实轨道恰好与那椭圆相交的那一刻。

119 至于简单性是真实的还是隐藏了复杂的实在，这又有何关系呢？无论是由于降低了个体差异的大数的影响，还是由于某些可被忽略的若干项的或大或小的作用，它都绝不是由于机遇。这种简单性，无论是真实的还是表观的，总是有原因的。因此，我们将始终能够以同样的方式进行推理。如果一条简单的定律在几个特例中被观察到，我们就可以合理地假设，在相似的情形中，这条定律仍然成立。否认这一点，那就是赋予机遇一种不可承认的作用了。

120 There is, however, a difference. If the simplicity were real and essential, it would resist the increasing precision of our means of measure. If then we believe nature to be essentially simple, we must, from a simplicity that is approximate, infer a simplicity that is rigorous. This is what was done formerly; and this is what we no longer have a right to do.

121 The simplicity of Kepler's laws, for example, is only apparent. That does not prevent their being applicable, very nearly, to all systems analogous to the solar system; but it does prevent their being rigorously exact.

122 THE RôLE OF HYPOTHESIS.—All generalization is a hypothesis. Hypothesis, then, has a necessary rôle that no one has ever contested. Only, it ought always, as soon as possible and as often as possible, to be subjected to verification. And, of course, if it does not stand this test, it ought to be abandoned without reserve. This is what we generally do, but sometimes with rather an ill humor.

123 Well, even this ill humor is not justified. The physicist who has just renounced one of his hypotheses ought, on the contrary, to be full of joy; for he has found an unexpected opportunity for discovery. His hypothesis, I imagine, had not been adopted without consideration; it took account of all the known factors that it seemed could enter into the phenomenon. If the test does not support it, it is because there is something unexpected and extraordinary; and because there is going to be something found that is unknown and new.

124 Has the discarded hypothesis, then, been barren? Far from that, it may be said it has rendered more service than a true hypothesis. Not only has it been the occasion of the decisive experiment, but, without having made the

120 然而，其中仍有区别。如果这种简单性是真实而深刻的，那么它将经受住我们的测量方法越来越精确的考验。倘若我们相信自然在本质上是简单的，那么就必然能从近似的简单性推断出严格的简单性的结论。因这样的事从前就做过，所以我们无须再做了。

121 例如，开普勒定律的简单性仅仅是表观的，但这并不妨碍它们被应用于几乎所有类似于太阳系的系统，尽管它们不是严格精确的。

122 **假设的作用** 所有的概括都是假设。因此，假设发挥了必要的作用，这是谁也不会反驳的。不过它总是应该尽快地、经常地受到证实。当然，如果经不起这种检验，就应该毫不犹豫地抛弃它。这的确是通常的做法，但有时会带有一些不愉快的情绪。

123 其实就连这种不愉快的情绪也是没有道理的。相反，物理学家由于放弃了他的一个假设而感到高兴，因为他得到了一个意想不到的发现机会。我想，他的假设并不是轻易地被采纳的，而是考虑了可能导致这一现象的所有已知因素。如果假设没有得到证实，那是因为存在一些意想不到的和异乎寻常的东西，那是因为将会发现一些未知的和新的事物。

124 这样一来，被抛弃的假设就毫无益处了吗？远非如此，甚至可以说它比真实的假设贡献更大。它不仅是一个决定性实验（decisive experiment）的诱因，而且如果人们不做假设，只是在偶然之中做了这

hypothesis, the experiment would have been made by chance, so that nothing would have been derived from it. One would have seen nothing extraordinary; only one fact the more would have been catalogued without deducing from it the least consequence.

125 Now on what condition is the use of hypothesis without danger?

126 The firm determination to submit to experiment is not enough; there are still dangerous hypotheses; first, and above all, those which are tacit and unconscious. Since we make them without knowing it, we are powerless to abandon them. Here again, then, is a service that mathematical physics can render us. By the precision that is characteristic of it, it compels us to formulate all the hypotheses that we should make without it, but unconsciously.

127 Let us notice besides that it is important not to multiply hypotheses beyond measure, and to make them only one after the other. If we construct a theory based on a number of hypotheses, and if experiment condemns it, which of our premises is it necessary to change? It will be impossible to know. And inversely, if the experiment succeeds, shall we believe that we have demonstrated all the hypotheses at once? Shall we believe that with one single equation we have determined several unknowns?

128 We must equally take care to distinguish between the different kinds of hypotheses. There are first those which are perfectly natural and from which one can scarcely escape. It is difficult not to suppose that the influence of bodies very remote is quite negligible, that small movements follow a linear law, that the effect is a continuous function of its cause. I will say as much of the conditions imposed by symmetry. All these hypotheses form, as it

个实验，则人们将一无所得。人们不会看到异常的东西，只不过又有一个事实被编入实验目录，而不会从中得出任何结论。

125 现在要问，在什么条件下使用假设才没有危险呢？

126 仅仅有坚定从事实验的决心是不够的，此外还有一些危险的假设，最主要的是那些默认的和无意识的假设。由于我们是在不知不觉中就做了假设，因此我们就无法摆脱这些假设。可是在这里，数学物理学可以协助我们。由于数学物理学特有的精确性，即使不借助于它，它也会迫使我们制定所有的假设，但却是在无意识中作出的。

127 我们应当注意，重要的是不要过度地增加假设，只能一个一个地增加。如果我们基于多个假设建立理论，一旦理论被实验推翻，那么哪个前提必须改变呢？这是不得而知的。相反，如果实验成功了，难道我们可以认为实验同时验证了所有的假设吗？难道我们会相信只用一个方程就能确定几个未知数吗？

128 同样还必须注意区分不同种类的假设。第一类假设是非常自然而又难于避免的。我们很难不做这样的假设：非常遥远的物体的影响可以完全忽略，微小运动服从线性规律，结果是其原因的连续函数。同样，有关对称性所强制的条件也是如此。可以说，所有这些假设构成了数学物理学理论的共同基础，而这些假设最终都应该被抛弃。

were, the common basis of all the theories of mathematical physics. They are the last that ought to be abandoned.

129 There is a second class of hypotheses, that I shall term neutral. In most questions the analyst assumes at the beginning of his calculations either that matter is continuous or, on the contrary, that it is formed of atoms. He might have made the opposite assumption without changing his results. He would only have had more trouble to obtain them; that is all. If, then, experiment confirms his conclusions, will he think that he has demonstrated, for instance, the real existence of atoms?

130 In optical theories two vectors are introduced, of which one is regarded as a velocity, the other as a vortex. Here again is a neutral hypothesis, since the same conclusions would have been reached by taking precisely the opposite. The success of the experiment, then, can not prove that the first vector is indeed a velocity; it can only prove one thing, that it is a vector. This is the only hypothesis that has really been introduced in the premises. In order to give it that concrete appearance which the weakness of our minds requires, it has been necessary to consider it either as a velocity or as a vortex, in the same way that it has been necessary to represent it by a letter, either x or y. The result, however, whatever it may be, will not prove that it was right or wrong to regard it as a velocity any more than it will prove that it was right or wrong to call it x and not y.

131 These neutral hypotheses are never dangerous, if only their character is not misunderstood. They may be useful, either as devices for computation, or to aid our understanding by concrete images, to fix our ideas as the saying is. There is, then, no occasion to exclude them.

129 第二类假设可称为中性假设。在大多数问题中，解析学者在计算之初就假定，要么物质是连续的，要么相反认为物质是由原子构成的。无论哪种情况，所得结果都是一样的。只是假设物质是由原子组成时，得到结果要困难一些，仅此而已。如果实验证实了他的结论，难道他就可以说他已经证明了原子的真实存在吗？

130 在光学理论中引入了两种矢量，一个是速度矢量，另一个是涡旋矢量。这是另一个中性假设，因为若假定前者为涡旋矢量，后者为速度矢量，也会得出同样的结论。因此，实验的成功并不能证明第一个矢量就是速度。实验只证明了一件事，即它是一个矢量。这是前提中引入的唯一假设。为了给它一个由于我们心智的薄弱性所要求的这种具体外观，有必要把它看作是一种速度或者是一种涡旋。同样地，用 x 或 y 来表示它也是必要的。但结果既不能证明我们把它当作速度时是对了还是错了，也不能证明称它为 x 而不是 y 时是对的还是错的。

131 只要这些中性假设没有被误解，就永远不会有危险。这些中性假设可能是有用的，要么作为计算的技巧，要么通过具体的图像来帮助我们理解，正如人们所说的那样坚定我们的观念。因此，无需禁止这种假设。

132 The hypotheses of the third class are the real generalizations. They are the ones that experiment must confirm or invalidate. Whether verified or condemned, they will always be fruitful. But for the reasons that I have set forth, they will only be fruitful if they are not too numerous.

133 ORIGIN OF MATHEMATICAL PHYSICS.—Let us penetrate further, and study more closely the conditions that have permitted the development of mathematical physics. We observe at once that the efforts of scientists have always aimed to resolve the complex phenomenon directly given by experiment into a very large number of elementary phenomena.

134 This is done in three different ways: first, in time. Instead of embracing in its entirety the progressive development of a phenomenon, the aim is simply to connect each instant with the instant immediately preceding it. It is admitted that the actual state of the world depends only on the immediate past, without being directly influenced, so to speak, by the memory of a distant past. Thanks to this postulate, instead of studying directly the whole succession of phenomena, it is possible to confine ourselves to writing its "differential equation." For Kepler's laws we substitute Newton's law.

135 Next we try to analyze the phenomenon in space. What experiment gives us is a confused mass of facts presented on a stage of considerable extent. We must try to discover the elementary phenomenon, which will be, on the contrary, localized in a very small region of space.

136 Some examples will perhaps make my thought better understood. If we wished to study in all its complexity the distribution of temperature in a cooling solid, we should never succeed. Everything becomes simple if we reflect that one point of the solid can not give up its heat directly to a distant

132 第三类假设是真正的概括。这些假设必须通过实验加以证实或否证。无论被证实还是被否证，这些假设都会有丰硕的成果。然而基于我所给出的理由，这类假设只有在数量不多的情况下才会如此。

133 **数学物理学的起源** 我们现在要更深入地详细研究促进数学物理学发展的条件。我们立即认识到，科学家总是努力把实验直接给出的复杂现象分解成大量的基本现象。

134 可以有三种不同的方式：第一种方式是在时间里分解。与其把某一现象的渐进发展过程全部一起考虑，人们宁愿把某一时刻与前一时刻联系起来。人们认为世界的现状只取决于紧挨着的过去，而不受遥远过去的记忆的直接影响。根据这一公设，与其直接研究整个现象的连续性，人们宁可只写出它的微分方程式，用牛顿定律代替开普勒定律。

135 第二种方式是在空间中分解。实验所给我们的是一个在相当大的范围内呈现的一堆混乱的事实。我们必须设法找出基本现象，这些现象反而局限于空间的一个极小的区域。

136 举几个例子也许能更清晰地表达我的想法。对于一正在冷却的固体，要想去研究其温度分布的所有复杂情况是永远都不可能做到的。如果我们想到固体上的某一点不能将其热传给遥远的点，那么一切就变得

point; it will give up its heat only to the points in the immediate neighborhood, and it is by degrees that the flow of heat can reach other parts of the solid. The elementary phenomenon is the exchange of heat between two contiguous points. It is strictly localized, and is relatively simple, if we admit, as is natural, that it is not influenced by the temperature of molecules whose distance is sensible.

137 I bend a rod. It is going to take a very complicated form, the direct study of which would be impossible. But I shall be able, however, to attack it, if I observe that its flexure is a result only of the deformation of the very small elements of the rod, and that the deformation of each of these elements depends only on the forces that are directly applied to it, and not at all on those which may act on the other elements.

138 In all these examples, which I might easily multiply, we admit that there is no action at a distance, or at least at a great distance. This is a hypothesis. It is not always true, as the law of gravitation shows us. It must, then, be submitted to verification. If it is confirmed, even approximately, it is precious, for it will enable us to make mathematical physics, at least by successive approximations.

139 If it does not stand the test, we must look for something else analogous; for there are still other means of arriving at the elementary phenomenon. If several bodies act simultaneously, it may happen that their actions are independent and are simply added to one another, either as vectors or as scalars. The elementary phenomenon is then the action of an isolated body. Or again, we have to deal with small movements, or more generally with small variations, which obey the well-known law of superposition. The observed movement will then be decomposed into simple movements, for example, sound into its harmonics, white light into its monochromatic components.

简单了。事实上，该点只会把热传给它最邻近的点，渐渐地热流将达到固体的其它部分。基本现象是两个相邻点之间的热交换。如果认为(这是很自然的)热交换不至于受到可感知距离内的分子温度的影响，那么，基本现象就是严格局域化的，而且相对简单。

137 弯曲一根杆子，其形状变得非常复杂，不可能直接研究。然而可以这样处理这个问题，只须注意它的弯曲只是棒内极小元素形变的结果，并且这些元素的形变仅仅依赖于直接作用在它上面的力，而与作用在其他元素上的力完全无关。

138 我们可以很容易增加更多例子，在上面所举的这些例子中，我们认为不存在超距作用，至少不存在远距离的作用。这只是一种假设。正如万有引力定律所表明的，它并不总是为真。因此，必须加以检验。如果它被证实，即便是近似的，它也是有价值的，因为至少允许我们依据逐次逼近法构建数学物理学。

139 第三种方式是当上述假设经不起考验时，必须寻求其他类似的假设，因为还有其他方法可以找出基本现象。倘若几个物体同时作用，它们的作用可以是相互独立的，并且可以相互叠加，至于叠加方式可以是矢量，也可以是标量。因而基本现象是孤立物体的作用。或者再假设，这是一个微小运动的问题，或者更普遍地说是微小变化的问题，它们遵循众所周知的叠加定律。于是，所观察到的运动将被分解成简单的运动，例如声音被分解为谐音，白光被分解为单色光。

140 When we have discovered in what direction it is advisable to look for the elementary phenomenon, by what means can we reach it?

141 First of all, it will often happen that in order to detect it, or rather to detect the part of it useful to us, it will not be necessary to penetrate the mechanism; the law of great numbers will suffice.

142 Let us take again the instance of the propagation of heat. Every molecule emits rays toward every neighboring molecule. According to what law, we do not need to know. If we should make any supposition in regard to this, it would be a neutral hypothesis and consequently useless and incapable of verification. And, in fact, by the action of averages and thanks to the symmetry of the medium, all the differences are leveled down, and whatever hypothesis may be made, the result is always the same.

143 The same circumstance is presented in the theory of electricity and in that of capillarity. The neighboring molecules attract and repel one another. We do not need to know according to what law; it is enough for us that this attraction is sensible only at small distances, and that the molecules are very numerous, that the medium is symmetrical, and we shall only have to let the law of great numbers act.

144 Here again the simplicity of the elementary phenomenon was hidden under the complexity of the resultant observable phenomenon; but, in its turn, this simplicity was only apparent, and concealed a very complex mechanism.

145 The best means of arriving at the elementary phenomenon would evidently be experiment. We ought by experimental contrivance to dissociate the complex sheaf that nature offers to our researches, and to study with care the elements

140 当我们知道从哪个方面去寻找基本现象时，究竟应该用什么方法才能达到目的呢？

141 首先，常常会发生这样的情况：为了发现它，或者更确切地说为了发现对我们有用的东西，没有必要知道它的机理，大数定律就足够了。

142 再举一个热传播的例子，每个分子都向它邻近的分子辐射，而并不需要知道它是按照什么定律进行辐射的。假如我们在这方面作任何假设，它将是一个中性假设，因此既无用也无法证实。事实上，由于平均的作用和媒质的对称性，所有的差别都被抹平了，而且不管做了哪种假设，结果总是一样的。

143 同样的情况也出现在弹性①理论和毛细管现象理论中。相邻的分子相互吸引或相互排斥，我们不需要知道它们服从什么定律。对我们来说，只要这种吸引力在很小的距离上变得敏感，只要分子非常多，只要媒质是对称的，我们所要做的就是让大数定律发挥作用。

144 在这里，基本现象的简单性再次隐藏在可观察现象的复杂性之下。不过这种简单性只是表观的，它隐藏着一种非常复杂的机制。

145 寻找基本现象的最好方法当然是实验了。通过实验的方法，我们应该分解自然提供给我们研究的复杂情形，并仔细研究那些尽可能纯粹的

① 在《科学与假设》的法文版中，此处是“élasticité”，在 1905 年的英文版中，是“elasticity”，即弹性。——译者注

as much isolated as possible. For example, natural white light would be decomposed into monochromatic lights by the aid of the prism, and into polarized light by the aid of the polarizer.

146 Unfortunately that is neither always possible nor always sufficient, and sometimes the mind must outstrip experiment. I shall cite only one example, which has always struck me forcibly.

147 If I decompose white light, I shall be able to isolate a small part of the spectrum, but however small it may be, it will retain a certain breadth. Likewise the natural lights, called *monochromatic*, give us a very narrow line, but not, however, infinitely narrow. It might be supposed that by studying experimentally the properties of these natural lights, by working with finer and finer lines of the spectrum, and by passing at last to the limit, so to speak, we should succeed in learning the properties of a light strictly monochromatic.

148 That would not be accurate. Suppose that two rays emanate from the same source, that we polarize them first in two perpendicular planes, then bring them back to the same plane of polarization, and try to make them interfere. If the light were *strictly* monochromatic, they would interfere. With our lights, which are nearly monochromatic, there will be no interference, and that no matter how narrow the line. In order to be otherwise it would have to be several million times as narrow as the finest known lines.

149 Here, then, the passage to the limit would have deceived us. The mind must outstrip the experiment, and if it has done so with success, it is because it has allowed itself to be guided by the instinct of simplicity.

元素。例如，自然光通过棱镜分解为单色光，通过偏振器分解为偏振光。

146 不幸的是，这既不总是可能的，也不总是充分的，有时心智就得超越实验。我只举一个一直给我留下深刻印象的例子。

147 如果分解白光，就能隔离出光谱的一小部分。但无论它有多小，它总是有一定的宽度。同样，所谓的单色自然光呈现出一条很细的谱线，但并非无限的细。在对这些自然光的性质进行实验研究时，我们可以设想，通过对越来越细的光线进行操作，最终达到极限，也就是说，我们最终应该获悉严格单色光的性质。

148 然而，这不可能是准确的。假设两束光线来自于同一光源，首先使它们成为两个垂直平面上的偏振光，然后再回到同一个偏振平面中，再试图使它们发生干涉。如果光线是严格单色的，就会发生干涉。但若使用的光线只是接近单色，将不会有干涉发生，而且无论谱线有多窄都不行。为了发生干涉，谱线必须比已知最细的谱线还要窄几百万倍才行。

149 因此，在这里我们被极限过程所欺骗。心智应该超越实验，如果能成功地做到这一点，那是由于心智被简单性的本能所引导的缘故。

150 The knowledge of the elementary fact enables us to put the problem in an equation. Nothing remains but to deduce from this by combination the complex fact that can be observed and verified. This is what is called *integration*, and is the business of the mathematician.

151 It may be asked why, in physical sciences, generalization so readily takes the mathematical form. The reason is now easy to see. It is not only because we have numerical laws to express; it is because the observable phenomenon is due to the superposition of a great number of elementary phenomena *all alike*. Thus quite naturally are introduced differential equations.

152 It is not enough that each elementary phenomenon obeys simple laws; all those to be combined must obey the same law. Then only can the intervention of mathematics be of use; mathematics teaches us in fact to combine like with like. Its aim is to learn the result of a combination without needing to go over the combination piece by piece. If we have to repeat several times the same operation, it enables us to avoid this repetition by telling us in advance the result of it by a sort of induction. I have explained this above, in the chapter on mathematical reasoning.

153 But for this, all the operations must be alike. In the opposite case, it would evidently be necessary to resign ourselves to doing them in reality one after another, and mathematics would become useless.

154 It is then thanks to the approximate homogeneity of the matter studied by physicists that mathematical physics could be born.

155 In the natural sciences, we no longer find these conditions: homogeneity, relative independence of remote parts, simplicity of the elementary fact; and this is why naturalists are obliged to resort to other methods of generalization.

150 知道了基本事实，我们就能把问题用方程的形式表述出来。于是经过组合，就可由方程演绎出可观察到的和可证实的复杂事实。这就是所谓的积分，它属于数学家的工作。

151 人们可能会问，为什么在物理科学中概括能如此容易地采用数学形式？其原因现在已显而易见了。这不仅因为我们要表达的是数值定律，也因为可观测的现象是由大量相似的基本现象叠加而成，正因如此，引入微分方程就十分自然了。

152 仅仅让每一基本现象遵循简单的定律是不够的，所有那些需要组合的现象都必须遵循同样的定律。唯有这样，数学的介入才有用处。事实上，数学教我们如何把同类的事物组合起来。数学的目的是获得组合的结果，而不需要一部分一部分地重新组合。如果我们必须多次重复相同的运算，那么可以通过一种归纳法预先告诉我们结果，从而避免这种重复。关于这一点，我已在前面的数学推理一章中解释过了。

153 为此，一切运算都必须相似。否则，就必须在实际中一个一个地计算它们，这样一来，数学就显得无用了。

154 因此，正是由于物理学家所研究的物质具有近似的均匀性，才诞生了数学物理学。

155 在博物学中，就再也找不到这样的条件：均匀性，远离部分的相对独立性，基本事实的简单性。正因如此，博物学家就不得不求助于别的概括方式。